LOW-DIMENSIONAL NANOELECTRONIC DEVICES

Theoretical Analysis and Cutting-Edge Research

LOW-DIMENSIONAL NANOELECTRONIC DEVICES

Theoretical Analysis and Cutting-Edge Research

Edited by

Angsuman Sarkar, PhD

Arpan Deyasi

AAP | APPLE ACADEMIC PRESS

First edition published 2023

Apple Academic Press Inc.
1265 Goldenrod Circle, NE,
Palm Bay, FL 32905 USA
760 Laurentian Drive, Unit 19,
Burlington, ON L7N 0A4, CANADA

CRC Press
6000 Broken Sound Parkway NW,
Suite 300, Boca Raton, FL 33487-2742 USA
4 Park Square, Milton Park,
Abingdon, Oxon, OX14 4RN UK

Apple Academic Press exclusively co-publishes with CRC Press, an imprint of Taylor & Francis Group, LLC

Library and Archives Canada Cataloguing in Publication

Title: Low-dimensional nanoelectronic devices : theoretical analysis and cutting-edge research / edited by Angsuman Sarkar, PhD, Arpan Deyasi.
Names: Sarkar, Angsuman, editor. | Deyasi, Arpan, 1978- editor.
Description: First edition. | Includes bibliographical references and index.
Identifiers: Canadiana (print) 20220203415 | Canadiana (ebook) 20220203547 | ISBN 9781774638668 (hardcover) | ISBN 9781774638675 (softcover) | ISBN 9781003277378 (ebook)
Subjects: LCSH: Nanoelectronics. | LCSH: Integrated circuits. | LCSH: Photonics. | LCSH: Low-dimensional semiconductors.
Classification: LCC TK7874.84 .L69 2022 | DDC 621.381—dc23

Library of Congress Cataloging-in-Publication Data

CIP data on file with US Library of Congress

ISBN: 978-1-77463-866-8 (hbk)
ISBN: 978-1-77463-867-5 (pbk)
ISBN: 978-1-00327-737-8 (ebk)

About the Editors

Angsuman Sarkar, PhD

Angsuman Sarkar, PhD, is a Professor of Electronics and Communication Engineering at Kalyani Government Engineering College, West Bengal, India. He had earlier served at Jalpaiguri Government Engineering College & Kalyani Government Engineering College as Lecturer, Assistant Professor, and Associate Professor of the Electrical and Computer Engineering Department for more than 20 years. His current research interest includes the study of short channel effects of sub 100nm MOSFETs and nanodevice modeling. He is a senior member of IEEE, a life member of the Indian Society for Technical Education, and associate life member of the Institution of Engineers India, and he currently serves as the Chairman of IEEE Electron Device Society, Kolkata Chapter, India. He has authored five books and has contributed many book chapters, over 70 journal papers in international refereed journals, and over 57 research papers in national and international conferences. He is a member of the boards of editors of several journals as well as a reviewer for several international journals. He is currently supervising six PhD scholars and has already guided five students successfully as principal supervisor. He has delivered invited talks, tutorial speeches, and expert talks at various international conference and technical programs. He has organized IEEE international conferences and several workshops and seminars. He received his MTech degree in VLSI & Microelectronics from Jadavpur University and his PhD from Jadavpur University.

Arpan Deyasi

Arpan Deyasi, MTech, is an Assistant Professor in the Department of Electronics and Communication Engineering at the RCC Institute of Information Technology, Kolkata, India. He has 14 years of professional experience in academics and industry. His work is in the field of semiconductor nanostructure and semiconductor photonics. He has published more than 150 peer-reviewed research papers in journals and conferences, as well as book chapters, and has edited or co-edited five books. He is associated with several international and national conferences in various aspects and was general chair of Frontline Research in Computer, Communication and Device (FRCCD 2015) and the International Conference on Computational Science and Engineering (ICCSE 2016). He is a reviewer of several journals of repute and prestigious conferences in India and abroad. He is a senior member of IEEE, secretary of IEEE Electron Device Society (Kolkata Chapter), member of IE(I), Optical Society of India, IETE, ISTE, etc. He is working as SPOC for several online certification courses and activities at his institute (NPTEL, IIRS, Coursera, EDx, e-outreach, Internshala). He is also working as a faculty adviser of the student chapter of the Institution of Engineers (India) in ECE Department at RCCIIT.

Contents

Contributors

Gufran Ahmad
Department of Electrical Engineering, Dayalbagh Educational Institute, Agra, India

J. Ajayan
Department of Electronics and Communication Engineering, SNS College of Technology, Coimbatore, India

Navjeet Bagga
Department of Electronics & Communication Engineering, PDPM-IIITDM-Jabalpur, India

Somak Bhattacharyya
Department of Electronics Engineering, Indian Institute of Technology, India

Kalyan Biswas
ECE Department, MCKV Institute of Engineering, Howrah, West Bengal, India

Santanu Das
Department of Ceramic Engineering, Indian Institute of Technology, India

Ajit Dash
Department of Electronics and Communication Engineering, National Institute of Science and Technology, Berhampur, Odisha, India

Milan Dhara
Department of Applied Electronics & Instrumentation Engineering, Future Institute of Engineering and Management, Kolkata, India

Rajesh Dutta
Research Student, Electronics and Communication Engineering, Maulana Abul Kalam Azad University of Technology, West Bengal, India

Sambit Kumar Ghosh
Department of Electronics Engineering, Indian Institute of Technology, India

Shrabani Guhathakurata
Department of Electronics and Communication Engineering, National Institute of Science and Technology, Berhampur, Odisha, India

Dipun Jena
Department of Electronics and Communication Engineering, National Institute of Science and Technology, Berhampur, Odisha, India

Arun Kumar
Department of Electrical Engineering, Indian Institute of Technology Patna, Bihar

Sudakshina Kundu
Department of Computer Science and Engineering, Maulana Abul Kalam Azad University of Technology, West Bengal, India

Rohini Mahato
Department of Life Science, Central University of South Bihar, Gaya, Bihar, India

Sandipan Mallik
Department of Electronics Communication Engineering, National Institute of Science and Technology, Berhampur, Odisha, India

Nabin Baran Manik
Department of Physics, Condensed Matter Physics Research Centre, Jadavpur University, Kolkata 700032, India

Sheetikanta Mohanty
Department of Electronics and Communication Engineering, National Institute of Science and Technology, Berhampur, Odisha, India

D. Nirmal
Department of Electronics and Communication Engineering, Karunya Institute of Technology and Science, Coimbatore, India

Abhijit Kumar Pal
Department of Applied Electronics & Instrumentation Engineering, Future Institute of Engineering and Management, Kolkata, India

Debasish Panda
Department of Electronics and Communication Engineering, National Institute of Science and Technology, Berhampur, Odisha, India

Sunipa Roy
Department of Electronics &Communication Engineering, Guru Nanak Institute of Technology, Kolkata, India

Angsuman Sarkar
ECE Department, Kalyani Govt. Engineering College, Kalyani, Nadia, West Bengal, India

Saheli Sarkhel
Department of Electronics and Communication Engineering, Netaji Subhash Engineering College, Kolkata, India

Subhasis Satapathy
Department of Electronics and Communication Engineering, National Institute of Science and Technology, Berhampur, Odisha, 761008, India

Kulbhushan Sharma
VLSI Centre of Excellence, Chitkara University Institute of Engineering and Technology, Chitkara University, Punjab, India

Rajnish Sharma
Chitkara University Institute of Engineering and Technology, Chitkara University, Punjab, India

Prashant Kumar Singh
Department of Electronics and Communication Engineering, University College of Engineering and Technology, V.B.U., Hazaribag, Jharkhand, India

Pramod Kumar Tiwari
Department of Electronics and Communication Engineering, SRM University, Andhra Pradesh

Abbreviations

2D	two-dimensional
3D	three-dimensional
ALD	atomic layer deposition
CB	carbon black
CCG	chemically converted graphene
CdS	cadmium sulfide
CdTe	cadmium telluride
CG	control gate
CIGS	copper indium gallium selenide
CMG	chemically modified graphene
CNT	carbon nanotube
CPC	cross-polarization converter
CPs	conductive polymers
CSC	carrier selective contacts
CTAB	cetyltrimethylammonium bromide
CTL	charge transporting layers
CZTS	copper zinc tin selenide
DC	direct current
DG	double gate
DIBL	drain-induced barrier lowering
DIBT	drain-induced barrier thinning
DLC	diamond-like carbon
DMD	dielectric-metal-dielectric
DOS	density of state
EDLC	electric double-layer capacitor
EM	electromagnetic
EMF	electromotive force
EPBT	energy payback times
ETL	electron transport layer
ETM	electron transporting material
FAB	field fast atom beam
FE	ferroelectric
FETs	field effect transistors

FGH	functionalized graphene hydrogel
FLG	few-layer graphene
GAA	gate all around
GFFs	graphene fiber fabrics
GHs	graphene hydrogels
GO	graphene oxide
GP	graphene paper
GQDs	graphene quantum dots
GSPs	graphene surface plasmons
HIT	heterojunction intrinsic thin layer
HK	knoop hardness
HTL	hole transporting layers
HTM	hole transporting material
IBC	interdigitated back contact
IC	integrated circuit
IoT	internet of things
IHP	inner Helmholtz plane
LGG	linearly graded gate
MDM	metal-dielectric-metal
MIR	mid-infrared
MLG	multilayer graphene
MWCNTs	multiwalled CNTs
NC	negative capacitance
NC-FETs	negative capacitance field effect transistors
NG	nitrogen-enrich graphene
OHP	outer Helmholtz plane
PANi	polyaniline
PCR	polarization conversion ratio
PECVD	plasma-enhanced chemical vapor deposition
PG	polarity gate
PPy	polypyrrole
PTAA	poly-triarylamine
PTh	polythiophene
PV	photovoltaic
QCEs	quantum confinement effects
RCS	radar cross-section
RDF	random dopant fluctuation
RF	radio frequency

rGO	reduced graphene oxide
SCEs	short channel effects
SDBS	sodium dodecylbenzene sulfonate
SEM	scanning electron microscopy
SHJ	Si heterojunction
Si-NT	silicon nanotube
Si-SCs	Si solar cells
SMG	single metal gate
SPP	surface plasmon polariton
SS	subthreshold slope
SS	subthreshold swing
SWCNTs	single-walled CNTs
TBAOH	tetrabutylammonium hydroxide
TE	transverse electric
TFET	tunnel field effect transistor
TF-SCs	thin-film solar cells
TM	transverse magnetic
TMO	transition metal oxide
UTB	ultra thin body
VAGN	vertically aligned graphene nanosheets
WEDs	wearable electronic devices
XAES	X-ray Auger electron spectroscopy
XPS	X-ray photoelectron spectroscopy

Foreword

With the advent of miniaturization at subatomic level, nanoelectronic device-based integrated circuits bring the revolution in civilization by creating enormous potentials for semiconductor industries. The opening of the door for playing with the electronic and photonic properties of these low-dimensional devices becomes realizable owing to the modernization of microelectronic fabrication processes and the ability of tailoring in minute scale. For the perspective of research workers, scientists, technologists, students, and obviously for the people of teaching communities, an efficient yet comprehensive representation of cutting-edge developments becomes extremely crucial at this juncture of the 21st century. From that point of view, the present book exactly fits in the requirement content-wise.

Any adroit review of technological developments in the field of nanoelectronics begins with the fundamental concept of quantum transport process, and mathematical foundations gain the maximum interest for beginners. Though experimental results and their processes are available so far in the market, compact theoretical coverage from various device perspectives are not so far covered within other books. Therefore, a perfect blend of both the sides is the need of the hour, where extra thrust on the theoretical analysis may reveal new insights, precisely for academicians and the research scholars who want to make their footsteps in this field from nascent levels. This uniqueness gives the edge to this present volume compared with the existing counterparts already available.

As a passionate worker and devotee of nano-engineering over the last decade, I believe that the present book will surely make a mark in this competitive market, precisely due to the organization of the contents, technological findings, and compactness along with ease of language suitable for all classes of researchers and practitioners.

—**Dr. Ajit Kumar Panda**
IEEE Distinguished Lecturer of ED Society

Introduction

The information revolution has been started with the miniaturization of semiconductor device dimension, and thereafter continuous progress of microelectronics technology brings us to the threshold of the nanoregime where electron transport can be restricted along different directions as per the requirements of application. The concept of quantum confinement leads researchers and engineers to the edge of the quantum revolution and gives birth to several novel branches of engineering, from quantum electronics to quantum information processing. The ever-increasing demand for data rate and faster computation with less chip area and less power dissipation paves the way for manifestation of quantum integrated circuits, which is probably the heart of future physical devices for telecommunications.

At this brink of the industrial revolution embedded with artificial intelligence and machine learning, it is customary for all academicians and industry people to look at the future roadmap of the information highway as crucial decisions will fuel the next-generation industrial revolution.

To introspect the fact and current progress of nanoelectronics, the crucial role will be played by theoretical engineers, as the fabrication cost of the nanodevice is so high that experimental stage will only be considered after near-perfect design of the device. Though Moore's law at this moment can't be followed, and growth and progress of chip design is getting away from the 1965 famous prediction, it is crucial to follow the defined path as much as possible. This leads to the generation of several new architectures, logic, algorithms, and structures, a perfect amalgamation of all these to furnish the formation of novel quantum devices. Quantum well, wire, dot, dash are a few examples of this, which may be incorporated inside novel MOS devices with much improved gate control mechanisms. Analytical modeling is the key for making these devices into reality, along with the perfect choice of material systems and, therefore, needs much attention. This is the major motivation of making this present bound volume.

In the last few years, several books have been written on the domain of nanoscience and nanotechnology, but priorities in most cases are given to the experimental works. For the beginners like graduate students, or those who are interested in making novel structures for specific applications, unfortunately the published books can hardly solve the purpose. Revolutionary, disruptive technology requires sophisticated infrastructures, but the total process will be in vain if the computation for the design is not perfect. Therefore, stress may be given to the modeling or theoretical investigation in parallel and that requires an alternative insight while making a book.

In the present initiative, focus is intentionally made to that angle, where several devices are analyzed with the underlying physics, material aspect, and architectural orientation; and corresponding electronic and optoelectronic properties entirely depend on that. Cutting-edge novel devices are explored inside these two covers, where the major thrust is given to the theoretical approaches and simulated findings.

All the chapters have been distinctly focused on specific applications, and therefore the edited volume becomes perfect blend for future directions, precisely for learners, academicians, and engineers. Technological revolution can always be initiated with novel ideas, and therefore the open-ended problems discussed in each chapter may lead to the beginning of future novel findings. Herein lies the success story of this initiative.

—Angsuman Sarkar
Arpan Deyasi

Preface

With the shrinking dimensions of devices, electronic and optoelectronic properties of semiconductor devices are modified to a great extent compared with that obtained in their bulk counterparts. The world is now in the regime of quantum electronics, and novel properties are explored from various low-dimensional structures that pave the way for quantum information processing. Theoretical research in nanostructure already showed the immense possibility for advancement in the area of electronics and photonics due to the possibility of making novel devices. Researchers and engineers solve the barriers by overcoming problems through experimental modes or by analytical/numerical approaches.

Device fabrication is a very time-consuming and complex process, and precautions at the highest level are taken prior to fabrication when the dimension of the device is at a nano-level. This is due to the fact that at a small variation of any layer thickness or minute change in the material composition may lead to different electrical and optoelectronic properties of the end product. In order to avoid the error at fabrication of low-dimensional structures, accurate computation is therefore required that considers the complex band structure, composition of the heterostructure system, and more precisely the method of computation.

Several numerical methods are already adopted for solving the quantum well systems, and the finite element method (FEM) or finite domain time difference (FDTD) method provide the best possible solution so far.

In the present book, different detailed theoretical/analytical models of solving the problems of various nanodeivces are tackled in a lucid manner, which will be extremely helpful for graduate students, researchers, and academicians. Moreover, the role of material composition and choice of materials now plays the most critical role in determining the outcome of various low-dimensional devices, and that aspect is covered from different angles in this state-of-the-art book on nanoelectronics, where the applications of those findings will be extremely beneficial for the computing and telecommunication industries.

Beginning with a solid theoretical background for every chapter, this book covers the hottest areas of present-day electronic engineering. Continuous increment of miniaturization of devices, components, and systems requires corresponding cutting-edge theoretical analysis, supported by simulated findings before actual fabrication. That sole purpose is given maximum thrust within these two covers in order to produce an interdisciplinary comprehensive technological product that deals with underlying physics, materials aspect, structures in nanoregimes, and corresponding end-products in the form of devices.

Truly, the nanoworld is amazing! One can get the pleasure of joy while learning and be curious about investigating the real-world problems and challenges the various devoted workers today. Throughout this fascinating journey, as covered within this edited book, one can virtually enter into the realm of quantum electronics. The rapidly evolving facets of nanoelc-tronics are truly amazing and simultaneously provide barriers and make challenges to overcome. Future researchers and engineers will surely get much of value from this book, and there in lies the success of making this product, solely for the benefit of civilization of next decade.

CHAPTER 1

Design Topologies for Low-Frequency and Low-Noise Neural Signal Processing

KULBHUSHAN SHARMA*

VLSI Centre of Excellence, Chitkara University Institute of Engineering and Technology, Chitkara University, Punjab, India

**E-mail: kulbhushan.sharma@chitkara.edu.in*

ABSTRACT

With the advent of advancement in semiconductor electronics neural signal processing research has expanded in recent years to encompass detection of brain related diseases. In this chapter, various blocks of multi-channel Neural Signal Processing Unit (NSPU) have been discussed. High focus has been laid on electrodes, different amplifier architectures, Operational Transconductance Amplifier topologies and pseudo-resistor topologies. Particularly, amplifier architectures namely Capacitive Coupled Capacitive Feedback, Miller Integrator Feedback Network and multi-channel two-stage Current-reuse Complimentary Input have been investigated in detail. In addition to this, chapter also focuses on 2nd order filter architecture for understanding the basics of neural signal filtration. Results investigated for different amplifier and filter architectures have been discussed in detail towards the end of chapter. The main objective of this chapter is to evaluate the merits as well as demerits of various architectural blocks used in NSPU. The detailed information presented in this chapter is anticipated to provide the reader with a direction to understand the intricacies in this challenging field of neural signal processing.

1.1 INTRODUCTION

Neuroscientists have been developing effective and new tools to understand the functioning and behavior of brain. This will lead to early detection of brain-related diseases, advances in prosthetic devices and dependable brain–computer interfaces (BCI).[1,2] The next generation of neural systems demand for the suitable conditioning of neural signals coming in from diagnosing apparatus. Chronic diseases, such as Parkinson's disease, epilepsy, paralysis, and spinal cord injuries are best diagnosed using biomedical instrumentation techniques, such as electroencephalography (EEG), in vivo multielectrode recordings, electrocorticography (ECoG), and optogenetic stimulation (OS) etc.[3,4] Integrated systems based on multiple-channel multiple-stage neural signal conditioning circuit blocks play a vital role in setting the power, noise, and supply voltage level of the diagnosing equipments used for the treatment of aforesaid chronic diseases.[5,6]

An incessant evolution of integrated neuroscience research made the invention of data acquisition system for neural recording applications possible to suitably amplify, filter, and digitize the neural signals which inherently have very low amplitudes of the order of 1 μV–5 mV within the frequency band of 0.5 Hz–7 kHz.[7] The neural signals originate due electrochemical activities taking place in the brain. Based on the method of measurement and characteristics features of neural signals, these can be classified into four basic categories, viz., Electroencephalography (EEG), Electrocorticography (ECoG), Extracellular action potentials (EAPs), and Local field potentials (LPFs). Typically, EEGs have the amplitude ranging from 10 to 100 μV with bandwidth in the range of 0.5 Hz–100 Hz. ECoGs have amplitude less than 100 μV with bandwidth ranging from 0.5–200 Hz. LPFs have the amplitude less than 5 mV with bandwidth ranging from 1 to 200 Hz. Besides these, EAPs have the amplitude less than 500 μV with bandwidth ranging from 0.1 Hz to 7 kHz.[8] The EEG is noninvasively measured, ECoG is semi-invasively measured whereas LPFs and EAPs are invasively measured. The evolution of neuroscience began with the experiment of intracellular recordings performed by Hodgkin and Huxley.[9] Till late 80s, many researchers[10–13] had demonstrated various types of experimental architecture to study the Central Nervous System (CNS). Mountcastle[11] and Hubel et al.[10] reported single-channel recordings that helped to know the significance of the biopotentials for the transmission of information in the CNS. Several neuroscientists worked upon the link between the limb movement and the neural signals in 1980s.[14–16] This had been giving direction to the research efforts in years to come and created a kind of revolution for the multiunit recordings. These systems were capable of simultaneous recording of the signals coming in from several neurons for neural recording-based applications.[17, 18] With the progression in research, numerous noninvasive as well as implantable multichannel neural recording systems have been invented to control the neuronal activities within the brain for better diagnosis of any brain-related chronic diseases. Recent examples of such neural recording systems are wireless neural interface System on Chip (SoC),[19] closed-loop systems,[20] implantable wireless SoC,[21] and multichannel neural recording probes.[22] However, most of the reported neural signal recording systems face challenges to obtain low values of power consumption and input referred noise. It is necessary that each blocks of neural recording system should consume low power and simultaneously achieve low values of noise. The total biological and thermal noise accompanying the array of electrodes from neuronal activities and electrode–electrolyte interface, respectively is generally less than 10 μV_{rms}.[23] The noise in neural recording system is mainly dominated by an array of amplifiers and filters. Each electrode has a neural amplifier attached to it. The input referred noise voltage for the array of amplifiers and filters must be in the range of 1–4 μV_{rms}.[7,8,2324] Besides this, the power dissipation by each amplifier and filter unit must be $< 10\ \mu W$ to ensure safety of the patients. Increase in power levels of complete neural recording system may lead to rise in temperature and can cause necrosis. The maximum acceptable temperature rise limit for all commercial invasive systems is $\leq 2°C$[25]. Owing to the circuit design trade-off between noise and power, it becomes quite challenging for the design engineer to reduce the values of both noise and power. Not only noise-power trade-off but also noise-stability trade-off is challenging in designing of neural system. A design engineer should pay enough effort to optimize noise-power and noise-stability tradeoffs in a multichannel neural recording system. Integrated systems based on neural amplifiers consisting of multiple stage and multichannel neural recording amplifier arrays must have ultra-low power low-noise operation for effective acquisition, amplification, and filtration of neural signal obtained from different parts of the brain. Moreover, the power supply interference from 50 Hz/60 Hz frequency components effects the information residing in the neural signals. A notch filter can be used along with gm-C filter stage to attenuate the power supply interference from 50 Hz/60 Hz frequency components.[26] High offset rejection, power supply rejection ratio (PSRR) and common-mode rejection ratio (CMRR) are also very important in a multichannel neural recording system comprising multiple stages for better conditioning of neural signals. For achieving higher values of aforesaid parameters, impedance matching of multichannel amplifiers with multiple stages is of pertinent interest. This large impedance mismatch majorly vitiates the total CMRR regardless of the amplifier's intrinsic CMRR.[27] The midband gain, PSRR, and CMRR of the first-stage amplifier must be large for better conditioning of neural signals. As far as filtration of neural signal is concerned, a filter circuit must have an accurate frequency response, large time constant, high linearity, high dynamic range (DR), and low-power and low-harmonic distortion. On the other hand, while designing an analog-to-digital convertor, researchers[28,29] face major challenge to design low-power comparator circuits with lower values of offset voltage. Various architectures like preamplifier-based comparator, multistage open-loop CMOS comparator, and dynamic latch comparators, etc. have been proposed to resolve this issue. In this chapter, various blocks of multichannel neural signal processing unit (NSPU) have been discussed. High focus has been laid on electrodes, different amplifier architectures, operational transconductance amplifier (OTA) topologies, pseudoresistor topologies which have been discussed in detail. In addition to this, chapter also focuses on the second-order filter architecture for understanding the basics of neural signal filtration. This chapter is divided as follows: Section 1.2 describes the multichannel neural signal processing unit. The discussion related to electrodes, amplifiers, and filters can be found in Sections 1.3–1.5 respectively. All the results related to various blocks of multichannel neural signal processing unit are discussed in Section 1.6. At the end of this chapter, important conclusions have been drawn and those have been discussed in Section 1.7.

1.2 MULTICHANNEL NEURAL SIGNAL PROCESSING UNIT

An integrated multichannel neural recording system helps in conditioning the information obtained from the recording sites by means of noninvasive or invasive sensing elements. Figure 1.1 shows the block diagram of multichannel neural signal processing unit which comprises different blocks, viz., electrode array, multichannel neural amplifiers, tunable filters, analog-to-digital converters (ADCs), multiplexer, and digital signal processing (DSP). The whole system developed in this way is used to drive a neuroprosthetic device or a display device for diagnosing various chronic diseases. The neuroprosthetic device can be a robotic arm, artificial limbs, and wheel chair, whereas display device can be a computer, cathode ray oscilloscope (CRO) and liquid crystal device (LCD) screen. In some applications, a wireless transmitter is also used as the last stage of complete neural processing unit for transmitting the recording data wirelessly to the receiver end. It can be envisaged from Figure 1.1 that the electrical signals in EEG, LFP, ECoGs, and EAPs which inherently possess amplitude values in the range of 10 μV–1 mV are being acquired with the help of multichannel microelectrode array.[7] Electrodes/array of electrodes couple the neural signals from the brain. Owing to small values of amplitude of neural signals and noise picked up by the electrodes, there is always a requirement to amplify and filter these signals while retaining the information residing inside the interspike intervals. The details of electrodes, integrated circuit blocks, such as amplifiers and filters used in multichannel NSPU have been described in the Sections 1.3–1.5 of this chapter respectively.

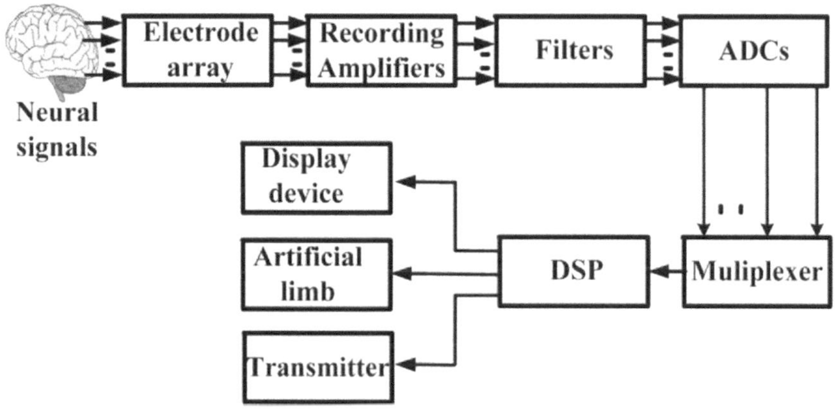

FIGURE 1.1 Block diagram of multichannel neural recording system for neuro-prosthetic application.

1.3 ELECTRODES

Electrodes serve as a sensing element in NSPU for coupling neural signals both resistively as well as capacitively.Various types of electrodes have been reported in literature which are categorized on the basis of the method of measurement, that is, noninvasive, semi-invasive, or invasive. The hydrogel-based Ag/Agcl electrodes,[30] flexible silicon-based dry electrodes[31] and noncontact electrodes[32] are examples of noninvasive electrodes. Flat electrodes[33] and 252-channel thin polyimide foil substrate electrode array[34] are examples of semi-invasive electrodes. Utah electrode array[26] and platinum-tipped extracellular electrodes[35] are examples of invasive electrodes. For understanding the equivalent circuit model of electrodes, let us consider the hydrogel-based Ag/AgCl wet electrodes which are widely used in noninvasive EEG applications. Figure 1.2 shows the skin–electrode interface circuit model for hydrogel-based Ag/AgCl wet electrodes. It can be observed from Figure 1.2 that the hydrogel-based Ag/Agcl electrodes are characterized by a series tissue resistance (R_{Tissue}), series combination of parallel resistive and capacitive elements, half-cell potential (E_{HC}), and resistance of lead (R_{Lead}).[32, 36] The stratum corneum is the resistive layer of the skin whose resistance is reduced by hydrating it with gel (hydrogel Ag/Agcl). The value of resistance (R) and capacitance (C) of stratum corneum is around 100 Ω and 10 nF, respectively while the resistance of gel (R_{Gel}) is approximately around KΩ. The other categories of electrodes, such as dry electrodes and noncontact electrodes can also be represented in the form of equivalent circuit model in a similar way as illustrated in Figure 1.2. However, the layered resistive and capacitive structures will differ in each case. Details of different equivalent circuit model for electrodes can be found in the review of Chi et al.[32]. After successful sensing of the neural signal from the brain, the same need to be amplified by the amplification stage. Section 1.4 describes the various neural amplifier architectures which can be used for raising the strength of neural signals acquired by the electrodes.

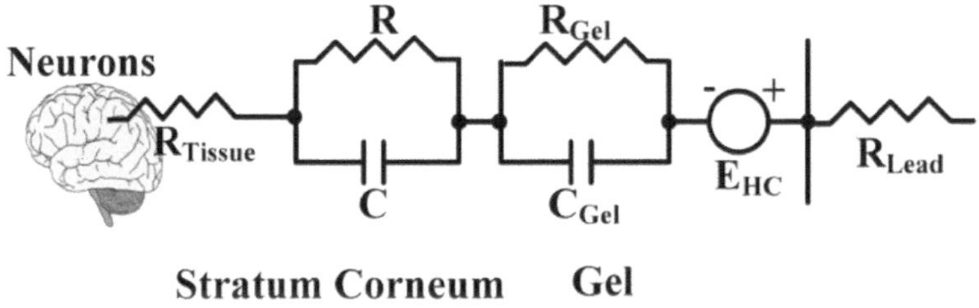

FIGURE 1.2 Skin-electrode interface circuit model for hydrogel-based Ag/AgCl wet electrodes.[32]

1.4 AMPLIFIERS

Neural amplifier arrays are power hungry circuit blocks in the multichannel neural processing unit. Besides this, the neural amplifier architectures dominate in noise contribution than any other circuit blocks. The optimization of noise-power trade-off, frequently referred as noise efficiency factor (NEF), is one of the challenging parameter in neural amplifier architecture. Optimization of noise-stability trade-off and implementation of accurate frequency response is also a challenging agenda in the neural amplifier architecture. The other parameters of neural amplifier, such as CMRR, PSRR, and DR should be high. To circumvent these issues, researchers have proposed various amplifier architectures, namely, capacitively coupled capacitive feedback (CC-CF) architecture,[37] Miller integrator feedback network (MIFN) architecture[38] and multichannel two-stage current-reuse complimentary input (CRCI)-based architecture.[27] The merits of using CC-CF architecture are high values of gain and CMRR, the MIFN architecture establishes its benefits in terms of DC offset and unwanted low-frequency input voltages rejection. The multichannel CRCI-based two-stage architecture reduces the input referred noise (IRN) voltage and helps in better offset rejection. However, the CC-CF architecture suffers from large area consumption owing to the use of capacitors for determining gain, MIFN architecture consumes high power owing to increased power requirements due to the use of feedback amplifier and multichannel CRCI-based two-stage architecture suffers from low common mode rejection ratio owing to mismatch at the input of CRCI amplifier. Olsson et al.[39] demonstrated the CC-CF-based amplifier architecture for the first time in 2002. As shown in Figure 1.3a, the CC-CF neural recording amplifier architecture typically highlights the presence of an operational amplifier, two series connected MOS pseudoresistors (R_F), and capacitances (C_{in} and C_F).The input capacitor (C_{in}) and feedback capacitance (C_F) help in setting the stable gain ($A_1 = C_{in}/C_F$). The lower cut-off frequency (f_L) of the CC-CF neural recording amplifier architecture depends upon the C_F and R_F connected in the feedback path which forms a high-pass filter and helps in rejecting DC offsets. The f_L can be given by[37] the equation as follows:

$$f_L = \frac{1}{2\pi R_F C_F} \tag{1.1}$$

Figure 1.3b shows the MIFN neural amplifier architecture. This architecture consists of a feedback loop which includes a low pass filter (LPF) circuit use to sense the DC signal observed at the output node of the amplifier (A_1). Further, the feedback loop is responsible for deducting the sensed DC signal from the input signal to accomplish filtering action. The 3 dB cut-off frequency (f_H) of the MIFN amplifier architecture depends upon the time constant ($\tau = R_F C_I$). The mid-band gain of A_1 and its f_L depends upon the w_{p1} of A_1, that is, $w_{p1} = g_m/A_{01}C_L$, where w_{p1} is the dominant pole of A_1, g_m denotes the transconductance of OTA of A_1, and A_{01} is the open-loop gain of A_1, and C_L is the load capacitance. Figure 1.3c shows the multichannel two-stage neural amplifier architecture. This architecture comprises current-reuse complimentary input (CRCI) amplifier with sharing architecture as the first stage and fully differential CC-CF amplifier based on current mirror OTA with common mode feedback (CMFB) as the second stage. This architecture reduces IRN voltage and helps in better offset rejection. However, it suffers from low CMRR. A detailed literature survey[23, 38, 40, 41, 42] reveals that OTA topology significantly contributes to power and noise in the overall architecture of any neural amplifier. The important low-voltage and low-noise OTA topologies used in various neural amplifier architectures have been described in the subsequent subsections.

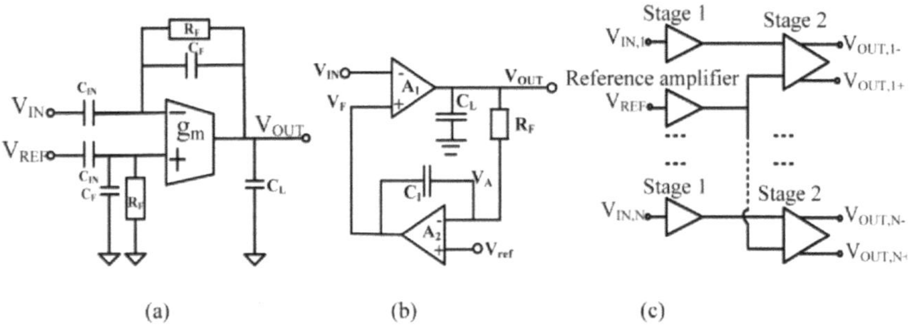

FIGURE 1.3 Neural amplifier architectures. (a) CC-CF,[35] (b) MIFN,[38] and (c) multichannel CRCI-based two-stage architecture.[27]

1.4.1 LOW-VOLTAGE OTA TOPOLOGIES

OTA is an integral and important part of neural amplifier architecture. To understand the functionality of low-voltage OTA structures, it is important to first understand the working of conventional gate-driven current mirror OTA structure.

1.4.1.1 GATE-DRIVEN CURRENT MIRROR OTA STRUCTURE

Figure 1.4a and b shows the schematic and small-signal model of conventional gate-driven current mirror OTA structure[37]. The current mirror topology shown in Figure 1.4a describes the property of equal current distribution between the two branches of a circuit to which the external current source along with the current mirror is applied. For the current mirror operation, it is required that the transistor devices M_1 should be matched, that is, the transistors should possess the same W/L ratio. If both the transistors possess the same W/L ratio, we can conclude that the currents I divide equally into I/2 through two branches of differential pair. PMOS devices M_1 with large gate area have been used in differential input pair to reduce flicker noise.[37] Flicker noise can be reduced by increasing W/L ratio and increasing the product $W. L$ too which ultimately increases the gate area. Fortunately, this approach drives the differential pair into subthreshold region where current efficiency (g_m/I_d) is maximum.[24] Thermal noise can be reduced by operating MOS transistors with highest current efficiency (g_m/I_d).[24] The current mirror pairs M_{2a} and M_{2b}, M_{3a} and M_{3b} are responsible for mirroring the current to output. In order to minimize the noise contribution of transistors pairs M_{2a} and M_{2b}, M_{3a} and M_{3b}, W/L ratios of these transistors have been chosen smaller than W/L ratios of the differential pair, which ultimately minimizes their transconductance values (g_{m2} and g_{m3}), current efficiency (g_m/I_d) ratio, and hence the noise. Reducing noise in this way shall be done while taking into account the important trade-off between noise and stability. At the output, the transistors M_4 and M_5 are placed in a cascode structure (one above the other) in order to increase the output impedance and thus the output gain of the OTA is obtained. The difference in signal currents is obtained at the output due to applied input voltage at the differential pair.

In Figure 1.4a, if we denote transconductance of M_1 devices as g_{m1}, M_{2a} devices as g_{m2a}, M_{2b} devices as g_{m2b}, M_{3a} device as g_{m3a}, and M_{3b} device as g_{m3b}, then the small-signal equivalent model of conventional current mirror OTA shown in Figure 1.4a can be illustrated as in Figure 1.4b.

From Figure 1.4b, the transfer function of current mirror OTA shown in Figure 1.4a can be given by the following equation:

$$\frac{V_O(s)}{V_I(s)} = \frac{g_{m1}.R_{out}}{\left(1+sR_{AB}.C_{AB}\right)\left(1+sR_{out}.C_{out}\right)} \tag{1.2}$$

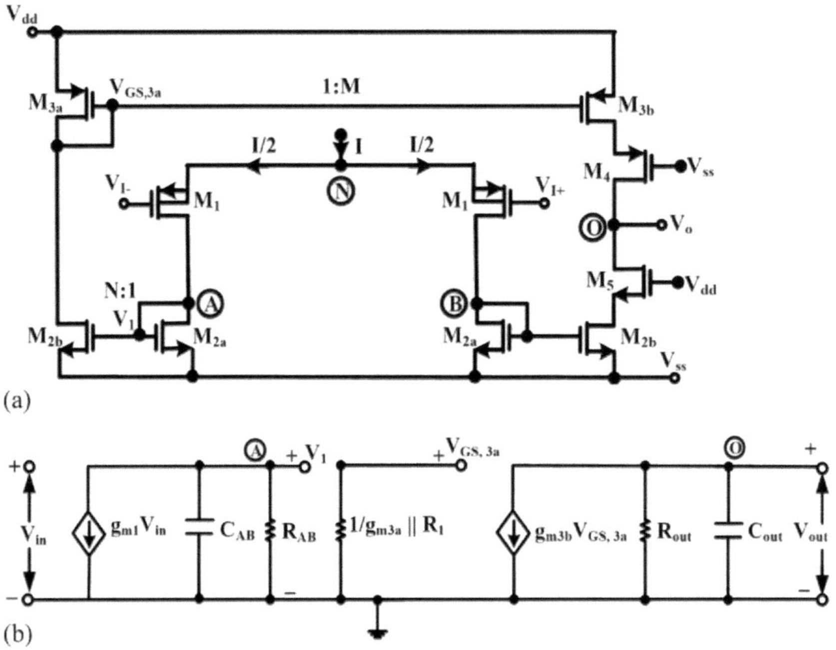

FIGURE 1.4 Conventional current mirror OTA: (a) schematic,[35] (b) small-signal model.

The expression for dc gain of current mirror OTA can be represented as:

$$A_O = \left| \frac{V_O}{V_I} \right|_{s=0} = g_{m1}.R_{out} \tag{1.3}$$

The nondominant pole (W_{p1}) of current mirror OTA can be given by:

$$W_{p1} \approx \frac{g_{m2a}}{C_{AB}} \tag{1.4}$$

The dominant -3 dB pole ($W_{3\,\mathrm{dB}}$) of current mirror OTA can be expressed as:

$$W_{3dB} \approx \frac{1}{R_{out} C_L} \tag{1.5}$$

Using eqs 1.3 and 1.5, the gain bandwidth product (GBW) can be expressed as:

$$\mathrm{GBW} \approx \frac{g_{m1}}{C_L} \tag{1.6}$$

The thermal noise voltage of the current mirror OTA topology can be expressed as:

$$\overline{V^2_{ni,(th)}} = \left[\frac{1}{g^2_{m,1}} \times \frac{16kT}{3} \left(g_{m,1} + 2g_{m2} + g_{m,3} \right) \right] \Delta f \quad (1.7)$$

The flicker noise voltage of the current mirror OTA topology can be expressed as:

$$\overline{V^2_{ni,(1/f)}} = \left[\frac{1}{g^2_{m,1}} \times \frac{2K_f}{c_{ox}} \times \frac{1}{f} \left(\frac{g^2_{m,1}}{W_1 L_1} + 2\left(\frac{g^2_{m,2}}{W_2 L_2} \right) + \frac{g^2_{m,3}}{W_3 L_3} \right) \right] \Delta f \quad (1.8)$$

where k is the Boltzmann constant (1.38×10^{-23} J K^{-1}), T is the temperature in Kelvin, and K_f is process-dependent constant in V^2F. For minimal thermal and flicker noise influence, the input differential devices of current mirror OTA are biased in subthreshold region and sized with larger gate area, respectively.

1.4.1.2 BULK-DRIVEN (BD) CURRENT MIRROR OTA WITH PARTIAL POSITIVE FEEDBACK

BD current mirror OTA employing partial positive feedback shown in Figure 1.5[43] is similar to the current mirror OTA shown in Figure 1.4, the difference lies in a way input signal is fed through bulk rather than gate and partial positive feedback being offered by the devices M_{2b}. The bulk-driven technique facilitates low-voltage operation and partial positive feedback provides flexibility to the circuit designer to narrow down the trade-off between noise and stability by controlling the transconductance of current mirror devices M_{2a}-M_{2c}.

As the transconductance of M_{2b} device is increased, the stability/PM decreases, but noise voltage reduces. On the other hand, common-mode response will improve if the tail devices M_{2a}, M_{2b}, and M_{2c} offer high impedance and are better matched. However, very high value of CMMR cannot be achieved owing to the fact that the transconductance of M_{2b} devices has to be less than M_{2a} devices. Moreover, the low impedance offered by the partial positive feedback network may lead to offset generation.

It is worth to underline here that the transconductance of M_{2b} devices should be less than that of the M_{2a} devices to prevent the OTA circuit from instability. In case if transconductance of M_{2b} is equal/greater than that of the M_{2a} devices, the amplifier becomes a positive feedback system. For the same reason, the $(W/L)_{2a} > (W/L)_{2b}$ at most in a ratio of 5:4.

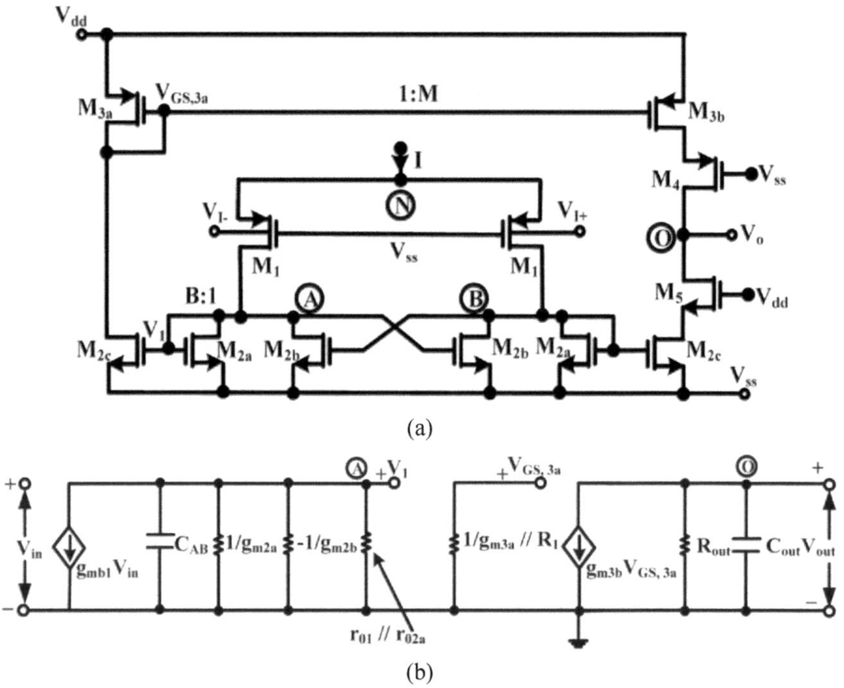

FIGURE 1.5 Current mirror OTA structure with partial positive feedback: (a) schematic, (b) small-signal model.[43]

In Figure 1.5a, if we denote the transconductance of M_1 devices as g_{mb1}, M_{2a} devices as g_{m2a}, M_{2b} devices as g_{m2b}, M_{2c} devices as g_{m2c}, M_{3a} device as g_{m3a}, and M_{3b} device as g_{m3b}, then the small-signal equivalent model of BD partial positive feedback current mirror OTA shown in Figure 1.5a can be depicted as in Figure 1.5b. The effective transconductance (g_{meff}) of this OTA topology can be given by:

$$g_{meff} = \frac{g_{mb}}{g_{m2a} - g_{m2b}} \times g_{m2c} = \frac{B}{1-K} \times \eta g_m \tag{1.9}$$

where $g_{mb} = \eta g_m$, $K = \frac{g_{m2b}}{g_{m2a}}$ which signifies $(W/L)_{2b}/(W/L)_{2a}$, where B represents the gain of $M_{2c}.M_{2a}\eta$ is the transconductance coefficient whose value varies between 0.2 and 0.4.[44] The BD technique offers low-voltage power saving operation. However, owing to low transconductance of bulk-driven technique as compared with gate-driven technique, the noise level of input differential (M_1 devices) increases. The partial positive feedback technique reported by Carillo et al.[43] tends to provide a solution to the low transconductance of input differential pair used in the OTA. Partial positive feedback enhances the effective trananconductance of M_1 devices by a factor B/1-K. From Figure 1.5b, the transfer function of BD current mirror OTA employing partial positive feedback technique (shown in Figure 1.5a) can be given by:

$$\frac{V_O(s)}{V_I(s)} = \frac{B}{1-K} \times \frac{\eta g_{m1}.R_{out}}{\left(1+sR_{AB}.C_{AB}\right)\left(1+sR_{out}.C_{out}\right)} \tag{1.10}$$

The expression for the DC gain of this OTA topology can be represented as:

$$A_O = \left|\frac{V_O}{V_I}\right|_{s=0} = \frac{B}{1-K} \times \eta g_{m1}.R_{out} \tag{1.11}$$

The nondominant pole (W_{p1}) of this OTA topology can be given by:

$$W_{p1} \approx \frac{(1-K)g_{m2a}}{C_{AB}} \tag{1.12}$$

The dominant -3 dB pole ($W_{3\text{ dB}}$) of BD current mirror OTA with partial positive feedback can be expressed as follows:

$$W_{3dB} \approx \frac{1}{R_{out}C_L} \tag{1.13}$$

Using eqs 1.11 and 1.13, the GBW can be expressed as follows:

$$\text{GBW} \approx \frac{B}{1-K} \times \frac{\eta g_{m1}}{C_L} \tag{1.14}$$

The thermal and flicker noise of the circuit shown in Figure 1.5a can be given by eqs 1.15 and 1.16, respectively as follows:

$$\overline{V^2_{ni,(th)}} = \left[\frac{1}{g^2_{m,eff}}\left(1+\left(\frac{g_{m,2a}}{g_{m,2b}}\right)^2\right) \times \frac{16kT}{3}\left(g_{m,1}+\left(g_{m,2a}+g_{m,2b}+g_{m,2c}\right)+g_{m,3}\right)\right]\Delta f \tag{1.15}$$

$$\overline{V_{ni,(1/f)}^2} = \left[\frac{1}{g_{m,eff}^2} \left(1 + \left(\frac{g_{m,2a}}{g_{m,2b}} \right)^2 \right) \times \frac{2K_f}{c_{ox}} \times \frac{1}{f} \left(\frac{g^2_{m,1}}{W_1 L_1} + \left(\frac{g^2_{m,2a}}{W_{2a} L_{2a}} + \frac{g^2_{m,2b}}{W_{2b} L_{2b}} + \frac{g^2_{m,2c}}{W_{2c} L_{2c}} \right) + \frac{g^2_{m,3}}{W_3 L_3} \right) \right] \Delta f \tag{1.16}$$

1.4.1.3 BULK-DRIVEN QUASI-FLOATING GATE (BDQFG) MILLER OTA

Figure 1.6a and b shows the schematic and small-signal model of BDQFG Miller OTA, respectively. The BDQFG strategy-based PM_3 and PM_4MOS devices form the input differential pair of OTA as shown in Figure 1.6a.[45] In addition to low-voltage operation, this strategy enhances the transconductance, reduces the noise when compared with the BD strategy.[46] The quasi-floating gate and capacitively connected bulk terminal (via capacitor, C) of PM_3 and PM_4 are subjected to low frequency input signals. The biasing voltage has been applied to the BDQFG devices (PM_3 and PM_4) through large-valued complimentary pseudoresistors.[47, 48] The enhanced transconductance values using BDQFG strategy can be expressed as[46]:

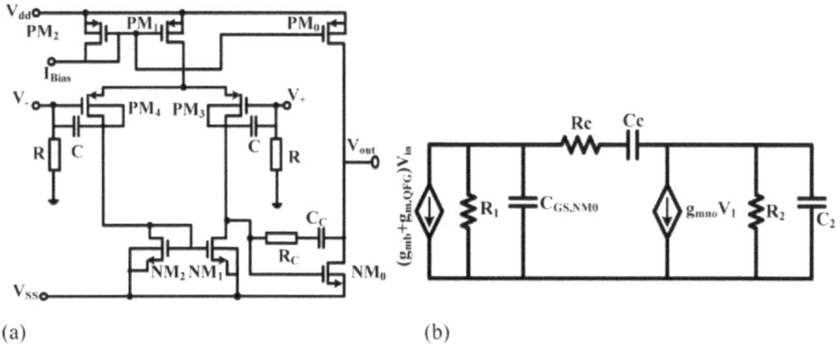

FIGURE 1.6 BDQFG Miller OTA: (a) schematic, (b) small-signal model.[52]

$$g_{m,\text{BDQFG}} = g_{mb} + g_{m,\text{QFG}} \tag{1.17}$$

where $g_{m,\text{QFG}}$ and g_{mb} are the transconductances of quasi-floating gate and bulk-driven MOS devices, respectively. The overall transconductance of the BDQFG Miller OTA ($g_{m,\text{OTA}}$) can be expressed as follows:

$$g_{m,\mathrm{OTA}} = \frac{(g_{mb} + g_{m,\mathrm{QFG}})(1 - g_{m,\mathrm{NM0}}\mathrm{R_C})\mathrm{R_1C_C}\left(s - \dfrac{g_{m,\mathrm{NM0}}}{C_C(1-g_{m,\mathrm{NM0}})}\right)}{s^2 + \left[\dfrac{1}{R_1C_{\mathrm{GS,NM0}}} + \dfrac{C_{\mathrm{GS,NM0}}+C_C}{R_CC_CC_{\mathrm{GS,NM0}}}\right]\mathrm{s} + \left[\dfrac{1}{R_1R_CC_CC_{\mathrm{GS,NM0}}}\right]} \quad (1.18)$$

where $R_1 = r_{o,\mathrm{PM3}} \| r_{o,\mathrm{NM1}}, r_{o,\mathrm{PM3}}$ is the output resistance of device $\mathrm{PM_3}$, $r_{o,\mathrm{NM1}}$ is the output resistance of device $\mathrm{NM_1}$, $g_{m,\mathrm{NM0}}$ is the transconductance of device $\mathrm{NM_0}$, $\mathrm{C_{GS,NM0}}$ is the gate-source capacitance of device $\mathrm{NM_0}$, R_C is the MOS device-based tunable two series connected compensation pseudoresistor comprising.[49] The open-loop transfer function of BDQFG Miller OTA from small-signal model can be calculated as follows*:

$$\frac{V_O(s)}{V_I(s)} = \frac{\left(g_{mb} + g_{m,QFG}\right)g_{mno}\left(1 - \dfrac{sC_C}{g_{m,NM0}}\right)}{As^2 + Bs + 1} \quad (1.19)$$

The stability of OTA depends upon the poles and zeroes of the transfer function which can be calculated with the aid of the following expression:

$$A_V = \frac{V_O(s)}{V_I(s)} = \frac{A_0}{\left(1 + \dfrac{s}{W_{p1}}\right)} \times \frac{\left(1 - \dfrac{s}{W_Z}\right)}{\left(1 + \dfrac{s}{W_{p2}}\right)} \quad (1.20)$$

Where A_0 is the DC gain, W_{p1}, W_{p2}, and W_Z are the nondominant pole, dominant pole, and zeroes respectively.

The nondominant pole (W_{p1}) of this OTA structure depends upon $g_{m,\mathrm{NM0}}$, Miller capacitor (C_C), $C_1(C_{\mathrm{DB,PM2}} + C_{\mathrm{DB,NM1}} + C_{\mathrm{GS,NM0}} \approx C_{\mathrm{GS,NM0}})$, and $C_2(=C_{\mathrm{BD,PM0}} + C_{\mathrm{BD,NM0}})$ which can be given by:

$$W_{p1} \approx -\frac{C_C g_{m,NM0}}{C_1 + C_2} \quad (1.21)$$

The dominant -3 dB pole ($W_{3\,\mathrm{dB}}$) of BDQFGMiller OTA depends upon $g_{m,NM0}$, C_C, R_1 and R_2 which can be expressed as:

---

* $A = C_2C_CR_1R_2 + C_{GS,NMO}C_CR_1R_2 + C_{GS,NMO}C_2R_1R_2 + C_C^2R_1R_2$
$B = C_CR_2 + C_2R_2 + C_{GS,NM0}R_1 + C_CR_1 - C_C^2R_1R_2 + C_Cg_{m,NMO}R_1R_2$

$$W_{3dB} \approx -\frac{1}{g_{m,NM0} C_c R_1 R_2} \tag{1.22}$$

where $R_2 = r_{o,\mathrm{PM0}} \| r_{o,\mathrm{NM0}}$, $r_{o,\mathrm{PM0}}$ is the output resistance of device PM_0, $r_{o,\mathrm{NM0}}$ is the output resistance of device NM_0.

The zero of the transfer function is associated with capacitor C_C as follows:

$$W_Z \approx -\frac{g_{m,NM0}}{C_c} \tag{1.23}$$

It can be observed from eq 1.23 that W_Z is located at the right half-plane, so the stability of OTA is affected due to this additional zero. Therefore, R_C [two series connected tunable MOS device[49] is connected in series to C_C for achieving higher values of phase margin. The open-loop transfer function of OTA with the addition of R_C in series with C_C gets modified to eq 1.24 as follows**:

$$\frac{V_O(s)}{V_I(s)} = \frac{\left(g_{mb} + g_{m,QFG}\right) g_{m,NM0} \left(1 + \frac{sC_C}{g_{m,NM0}} - sR_C C_C\right)}{Ps^4 + Qs^3 + Rs^2 + Ss} \tag{1.24}$$

It can be analyzed from (1.24) that the $R_c (>1/g_{m,NM0})$ mitigates the effect of zero introduced by C_C. Same has been demonstrated in.[50].

The DC gain (A_0) can be determined using eq 1.25 as follows:

$$A_O = \left|\frac{V_O}{V_I}\right|_{s=0} = \left(g_{mb} + g_{m,QFG}\right) g_{m,NM0} R_1 R_2 \tag{1.25}$$

Eq 1.25 shows the increment in DC gain as compared with bulk-driven technique. Using eqs 1.22 and 1.25, the gain bandwidth product (GBW) can be expressed as:

$$GBW \approx \frac{g_{mb} + g_{m,QFG}}{C_C} \tag{1.26}$$

** $P = 2R_C^3 R_1 R_2 C_2 C_C^4 + R_C^4 C_C^2 R_1 R_2 + 2R_C^3 C_C^3 R_1 R_2 C_{GS,NMO} + R_C^3 C_C^4 R_1 R_2 + R_C^3 C_C^3 R_1 C_{GS,NMO}$

$Q = 2R_C^2 C_C^3 R_1 R_2 C_2 + 2R_C^3 C_C^3 R_1 R_2 + 2R_C^3 C_C^3 R_2 C_2 + R_C^3 C_C^3 R_1 C_2 + R_C^3 C_C^3 R_2 + R_C^2 C_C^3 R_1 R_2 C_{GS,NMO} + R_C^2 C_C^2 R_1 R_2 C_2 C_{GS,NMO}$

$R = R_C^2 C_C^2 (R_1 R_2 + R_1 C_2 + R_2 C_2 + R_2) + 2R_1 R_2 R_C C_2 C_C^2 + R_1 R_2 C_C R_C^2 + R_1 R_2 R_C C_C C_{GS,NMO} + R_1 R_2 R_C C_C^2 + R_C C_C R_1 R_2 C_2 C_{GS,NMO} + R_C C_C^2 g_{m,NMO}$

$S = 2R_1 R_2 (3C_2 C_C + R_C C_C + C_{GS,NMO}) + 2R_1 C_2 R_C C_C + 2R_C C_C - C_C g_{m,NMO}$

The enhancement in the transconductance ($g_{mb} + g_{m,QFG}$) increases the GBW of the OTA.

The IRN ($\overline{V^2_{ni,BDQFG}}$) of BDQFG MOS device can be expressed as:

$$\overline{V^2_{ni,\text{BDQFG}}} = \left(\left(\frac{g_m}{g_{mb}}\right)^2 + \left(\frac{C_{Total}}{C}\right)^2\right)\overline{V^2_{ni,GD}} \tag{1.27}$$

where $\overline{V^2_{ni,GD}}$ is the noise of the OTA using gate-driven technique and C_{Total} is the total capacitance of BDQFG MOS device used in OTA. Though power and supply voltage requirements of OTA is reduced using BDQDG technique at differential pair, yet, the noise values are greater than that produced by the gate-driven techniques. The noise in OTA can be reduced using the design techniques discussed in subsequent subsection.

1.4.2 LOW-NOISE OTA TOPOLOGIES

1.4.2.1 COMPLIMENTARY INPUT (CRCI) AMPLIFIER

Figure 1.7a shows the CRCI amplifier. The gain (A_1) of the single-ended CRCI amplifier can be set by the capacitor ratio, C_1/C_2.[27] By driving the gates of both PMOS and NMOS input devices M_0 and M_1 respectively, the CRCI technique reuses the current flowing through these devices. Further, this technique doubles the effective transconductance, and reduces the IRN. The devices M_0 and M_1 are sized with larger aspect ratio which facilitates in reducing the flicker noise. However, for achieving high value of CMRR, large value of capacitors C_1 and C_2 is required for better matching.

A low-pass RC filter is formed by the capacitor (C_2) and pseudoresistor (R_2) at the gate of M_0 which filters out the noise from the M_2 and R_1. Considering $g_{m0} = g_{m1} = g_m$ and the input referred noise voltage can be given by[27]:

$$\overline{V^2_{ni}} = \left[\left(\frac{C_1 + C_2}{C_1}\right)^2\left(\frac{kT}{Kg_m}\right)\right]\Delta f \tag{1.28}$$

Thus, owing to enhanced g_m, the input referred noise voltage ($\overline{V^2_{ni}}$) decreases which proves to be useful for resolving interspike intervals of amplified neural signal of interest.

1.4.2.2 FLOATING GATE (FG) OTA

Figure 1.7b shows the floating-Gate OTA[51] which comprises complementary differential pair devices (M_A, M_D) and ($M_B M_E$), in right and left branches of the node. This strategy enhances transconductance and power efficiency. The programmable current source is implemented using floating-gate PMOS device (M_P). The (M_{CMFB}) device is floating-gate NMOS device which provides common-mode feedback and is responsible for improving the CMRR. The floating-gate MOS devices used in OTA provides improved noise performance. However, charge deposited at the gate of the MOS devices due to capacitors may cause DC offsets.

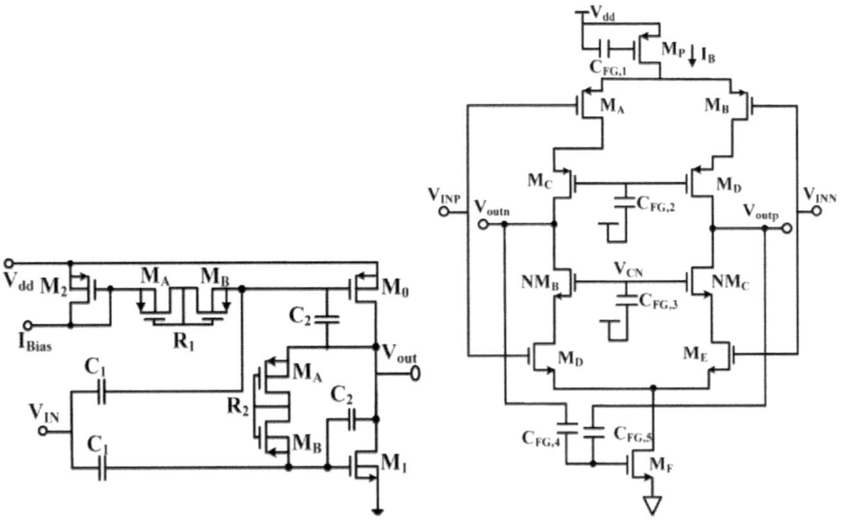

FIGURE 1.7 Schematic of low-noise OTA topologies: (a) current-reuse complimentary input amplifier,[27] (b) floating gate.[51]

1.4.3 PSEUDORESISTOR TOPOLOGIES

Pseudoresistors are MOS-based circuits which are capable of emulating high value of incremental resistance (IR).[37] Tunable pseudoresistor is required for controlling the low cut-off frequency of the neural amplifier architecture. It is desired to have IR of pseudoresistor in Tera-Ohm range for achieving lower cut-off frequency well below than 1 Hz. One of the simplest topologies for gate-controlled MOS pseudoresistor is shown in

Figure 1.8a.[47] This pseudoresistor is frequently referred to as tunable one MOS pseudoresistor structure. Different values of IR for this pseudoresistor can be obtained by suitably tuning it to various values of V_{Tune}. The IR range for tunable one MOS pseudoresistor structure lies between 2.989 Ω and 56.1 MΩ over $-1\ V \leq V_{BA} \leq 1\ V$.[52] However, the value of IR in the case of OM pseudoresistor decreases for source-drain (V_{SD}) which is greater than 0 V. This issue is known as common-mode voltage shifting. Figure 1.8b[47] shows the gate-voltage-controlled two series connected MOS devices. This pseudoresistor topology is capable of emulating higher value of IR resistance as compared with OM pseudoresistor topology. Similar to OM pseudoresistor, the value of emulated IR can be controlled by V_{Tune}. The approximate IR for two series connected MOS pseudoresistor ranges from 18.7 KΩ to 3.25 GΩ over $-1\ V \leq V_{BA} \leq 1\ V$.[52].

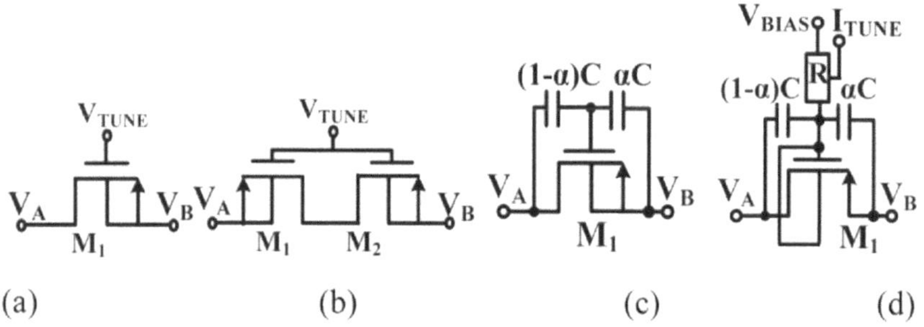

FIGURE 1.8 Pseudoresistor structure: (a) gate-voltage-controlled One MOS, (b) gate-voltage-controlled two series connected MOS,[47] (c) balanced[51] and (d) BDQFG MOS.[53]

Figure 1.8c shows the schematic representation of balanced pseudoresistor structure.[51] This topology uses scaled capacitive potential divider comprises capacitor $(1-\alpha)C$ and αC. The improvement in linearity is observed when the gate voltage of balanced pseudoresistor structure varies in accordance with the value of the source and drain voltages at $\alpha = 0.5$. The emulated IR of balanced pseudoresistor is of the order of very few Gigaohms for ±277 mV voltage swing with 1% variation.[51] Figure 1.8d shows the schematic representation of BDQFG pseudoresistor. This topology not only relies on scaled capacitive potential divider technique but also uses the biasing strategy for tuning gate and bulk terminals of BDQFG MOS device through a large value current-controlled resistance.[53] This sort of biasing strategy improves the linearity of BDQFG pseudoresistor structure and removes the problem of common-mode voltage shifting around 0 V. High linearity is observed for MOA-based BDQDG pseudoresistor when its both gate and bulk voltages are varied equally in accordance with the source and drain voltages at $\alpha = 0.8$.[45] BDQFG pseudoresistor is capable of emulating constant value of 1 T Ω over -1 V≤ V_{BA}≤ 1 V.[45] and is most suitable to be used in feedback path of CC-CF neural amplifier architecture.

1.5 FILTERS

Signal amplified by the adoption of any of the above-mentioned amplification techniques constitutes low as well as high frequency components. Practically, for neural signal applications to pass desired low frequency band of interest, a filter circuit is required. The switched capacitor (SC) filter,[54, 55], gm-C filter,[56, 57] and inductor–capacitor (LC) filter[58] are examples of various filter circuit architectures. The gm-C filter is widely used in neural recording applications owing to its merits like high linearity, low-harmonic distortion, large time constant, and high accuracy, etc.[59, 60] The filter circuits also rely on the OTA topologies. The Miller OTA,[61] telescopic OTA[62] and folded-cascode OTA[63] topologies are useful for implementing filter circuits. Even the OTA topologies discussed in Sections 1.4.1 and 1.4.2 can also be used in the design of filter architectures. For understanding the nuances related to filter circuits, let us consider the schematic of second-order low-pass gm-C filter illustrated in Figure 1.9 which relies on biquad topology.[64] The output signal obtained at the output of amplifier is fed to the input terminal (V_{IN}) of the filter architecture as shown in Figure 1.9, while the V_{REF} terminal is grounded.

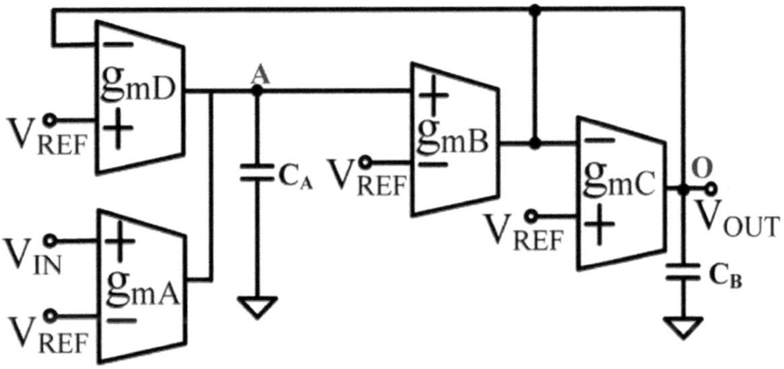

FIGURE 1.9 Second-order LPF architecture.[64]

The second-order LPF architecture comprises transconductors g_{mA}, g_{mB}, g_{mC}, and g_{mD} and capacitors C_A and C_B. The output of transconductors g_{mA} and g_{mD} is connected to C_A, while that of g_{mC} and g_{mD} is connected to C_B. The transfer function of filter circuit construed in Figure 1.9 can be given by[64]:

$$\frac{V_O(s)}{V_I(s)} = \frac{\dfrac{g_{mA} \cdot g_{mB}}{C_A C_B}}{s^2 + s\dfrac{g_{mC}}{C_B} + \dfrac{g_{m2} g_{mD}}{C_A C_B}} \tag{1.29}$$

If the allover transconductance of filter circuit shown in Figure 1.9 is denoted by g_m, then the cut-off frequency (f_c) of the second-order low-pass gm-C filter can be given by

$$f_c \approx \frac{g_m}{2\pi\sqrt{C_A C_B}} \tag{1.30}$$

The quality factor (Q) of gm-C filter can be expressed as:

$$Q \approx \sqrt{\frac{C_B}{C_A}} \tag{1.31}$$

The outputs of g_{m1} and g_{m4} is connected to node A, while the outputs of g_{m2} and g_{m3} are connected to node O. Owing to same output nodes each pair can share output stage and common-mode feedback circuitry which reduces area and power consumption. The Figure of Merit (FOM) of the filter architecture can be expressed as[64]:

$$\text{FOM(filter)} \approx \frac{P_{diss}}{pQ_{\max} f_c \text{DR}^2} \tag{1.32}$$

where DR, P_{diss}, and p are designated as DR, power dissipation, and the number of poles of filter. For a power-efficient filter, low value of FOM is desired. Upon successful filtration of neural signals, the same are being fed to next stage of digitization. Many researchers[65–67] have proposed various ADC architectures, such as flash ADC, pipeline ADC, sigma-delta ADC, and successive approximation register (SAR) ADC, etc. Among all these researched configurations, SAR ADC is considered to be the preferred option for medium-to-high resolution, low-power and low-cost applications.[68, 69]

1.6 RESULTS AND DISCUSSIONS

This section presents results obtained for amplifier and filter architectures discussed in Section 1.3 and 1.4, respectively. The measured results for different neural amplifier architectures have been discussed in subsection 1.6.1, while those for filter architectures of different orders used in NSPU have been discussed in subsection 1.6.2.

1.6.1 RESULTS OF AMPLIFIER ARCHITECTURES

Figure 1.10 depicts the gain versus frequency plots of CC-CF,[37] MIFN,[38] and two-stage (1st stage: CRCI and 2nd stage: CC-CF)[27] closed-loop neural amplifier architectures. It can be clearly observed from Figure 1.10 that CC-CF architectures achieve gain of 39.5 over bandwidth of 0.025 Hz—7.2 kHz, MIFN architectures achieve gain of 70 dB over bandwidth of 100 Hz—9.2 kHz and the two-stage architecture achieves gain of 58.7 dB over bandwidth of 0.49 Hz–10.5 kHz. The CC-CF and two-stage architectures are capable of achieving very low values of lower cut-off frequency as compared with MIFN architecture.

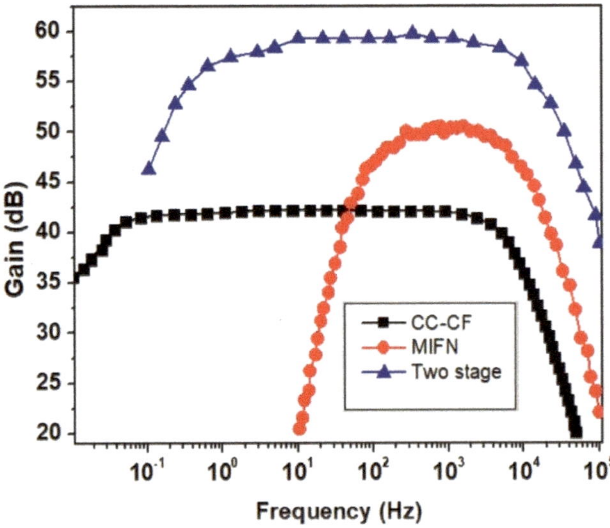

FIGURE 1.10 Gain versus frequency plots of closed-loop neural amplifier architectures: (a) CC-CF,[37] (b) MIFN,[38] and (c) multichannel CRCI-based two-stage architecture.[27]

Figure 1.11 illustrates input referred noise (IRN) voltage versus frequency plot of CC-CF,[37] MIFN[38], and two-stage (1st stage: CRCI and 2nd stage: CC-CF)[27] closed-loop neural amplifier architectures. Integration under the curve shown in Figure 1.11 for CC-CF, MIFN, and two-stage architectures yields root mean square noise voltage of 2.2 μV_{rms}, 5.4 μV_{rms}, and 3.04 μV_{rms}, respectively. The CC-CF architecture achieves lowest noise values in comparison to MIFN and two-stage architectures.

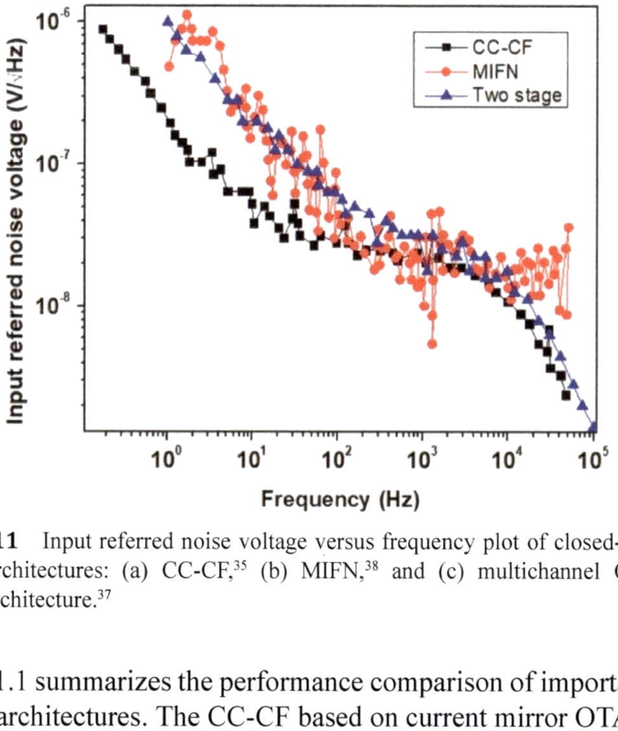

FIGURE 1.11 Input referred noise voltage versus frequency plot of closed-loop neural amplifier architectures: (a) CC-CF,[35] (b) MIFN,[38] and (c) multichannel CRCI-based two-stage architecture.[37]

Table 1.1 summarizes the performance comparison of important neural amplifier architectures. The CC-CF based on current mirror OTA and two series connected MOS pseudoresistor reported by Harrison et al.[37] has mid-band gain and IRN of 39.5 dB and 2.2 μV_{rms} over 0.025 Hz–7.2 kHz bandwidth, respectively. This amplifier architecture consumed 80 μW of power from ±2.5 V voltage supply with NEF of 4. In comparison to CC-CF architecture, the MIFN architecture reported by Gosselin et al.[38] is capable of achieving high mid-band gain of 70 dB over bandwidth of 100 Hz–9.2 kHz. However, MIFN architecture has high values IFN and NEF of 5.4 μV_{rms} and 4.9, respectively. This architecture consumes 8.6 μW of power from a 1.8 V supply. A close look at Table 1.1 clearly infers

TABLE 1.1 Performance Comparisons of Neural Amplifier Architectures.

Parameters	CC-CF architecture[37]	MIFN architecture[38]	CRCI-based two-stage architecture[27]	FCCFN architecture[70]	LSF-based architecture[71]
Technology (μm)	1.5	0.18	0.09	0.13	0.065
Gain (dB)	39.5	70	58.7	46	52
IRN (μV_{rms})	2.2	5.4	3.04	3.8	5.18
Bandwidth (Hz)	0.025–7.2 k	100–9.2 k	0.49–10.5 k	192–7.4 k	10 k
NEF/PEF	4	4.9	1.93/3.72	2.16/5.59	2.91/5.19
THD with Input voltage	1% 16.7 mVpp	–	1.6% 1 mVpp	0.08% 3 mVpp	–
DR (dB)	69	–	–	–	–
CMMR (dB)	≥83	45	≥45	85	>75
PSRR(dB)	≥85	–	≥50	75	>60
Area (mm^2)	0.16	0.05	0.137	0.053	0.032
Supply voltage (V)	±2.5	1.8	1	1.2	0.6
Supply current (μA)	16	4.8	2.85	–	1.78
Power (μW)	80	8.6	2.85	1.92	1.07

the CRCI-based two-stage architecture improvement in NEF. This architecture has mid-band gain, IRN, NEF, and power efficiency factor (PEF) of 58.7 dB, 3.04 μV_{rms}, 1.93, and 3.72 over bandwidth of 0.49–10.5 kHz. This architecture has the advantage of low power consumption (2.85 µW) from 1 V power supply. The feed forward compensated capacitive feedback network (FCCFN) reported by Ruiz-Amaya et al.[70] yields the gain of 46 dB over bandwidth of 192 Hz–7.4 k Hz with IRN, NEF, PEF, and power consumption of 3.8 μV_{rms}, 2.16, 5.59, and 1.92 µW, respectively. The amplifier architecture based on the level shifted feedback (LSF) technique reported by Lyu et al.[71] consumes least power (1.07 µW) from 0.6 V supply voltage among other architectures depicted in Table 1.1. However, this architecture has high values of IRN of 5.18 μV_{rms} with mid-band gain of 52 dB over bandwidth of 10 kHz. In all the architectures depicted in Table 1.1, the FCCFN architecture shows the highest value of CMRR and CC-CF architecture shows the highest value of PSRR.

The in vivo results obtained using penetrating electrode (BlackRock Microsystems LLC) and FCCFN amplifier architecture from visual cortex of adult male long Evans rat have been shown in Figure 1.12.[70] Flat ribbon cables were used to connect electrodes with amplifier. For better insight of in vivo results obtained from the visual cortex of adult male long Evans rat, one of the spike is zoomed.

1.6.2 RESULTS OF FILTER ARCHITECTURES

Table 1.2 depicts the performance comparison of various filter architectures. It can be observed from Table 1.2 that the second-order low-pass filter based on Gm-C circuit topology reported by Lu et al.[64] has the DR of 84.03 dB over bandwidth of 2.5 kHz with IRN of 91.8 μV_{rms}. This architecture works on 3.3 V supply voltage and consumes high power (47.2 µW) and has the advantage of lowest value of FOM (33.73 aJ) when compared with other architectures depicted in Table 1.2. The second-order low-pass filter architecture based on OTA-C topology reported by Hanumantha Rao and Rekha[72] has the DR of 55.1 dB over bandwidth of 2.5 kHz with IRN of 360 μV_{rms}. This filter architecture works at 0.5 V supply voltage and offers the advantage of low power consumption (47 nW) with FOM of 0.412 pJ. Table 1.2 clearly illustrates that the fifth-order LPF architecture reported by Pinto et al.[73] has the lowest IRN of 4.7 μV_{rms} over bandwidth of 100 Hz with DR of 57 dB and power consumption of 603 nW from a 0.25 V supply voltage. The value of FOM achieved by this filter architecture is

1.70 pJ. The latest fourth-order LPF architecture reported by Rajan et al.[74] has the IRN of 39.92 μV_{rms} over bandwidth of 250 Hz with DR of 62.1 dB and power consumption of 208 nW from a 1 V supply voltage. This architecture is capable of achieving FOM of 0.163 pJ.

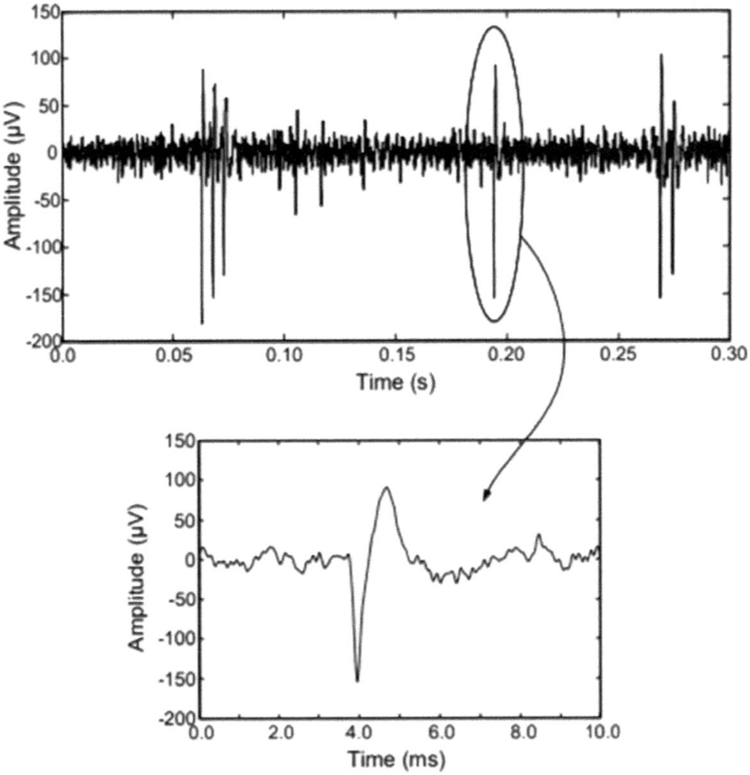

FIGURE 1.12 In vivo recording from visual cortex of adult male long Evans rat using penetrating electrode and FCCFN amplifier architecture with zoomed single spike

Source: Reproduced with permission from Ruiz-Amaya et al.,[70] 2015, Open access. (https://creativecommons.org/licenses/by/4.0/)

1.7 CONCLUSION

The important blocks of NSPU, viz., electrodes, neural amplifier, and filter were discussed in detail in this chapter. The design equation associated with various neural amplifier architectures were discussed along with the advantages and disadvantages of each architecture. The nitty-gritty of various low-voltage and low-noise OTA topologies as well as techniques to implement linear pseudoresistor structures were addressed in detail. For understanding the basics of filter architecture, the second-order low-pass filter architecture has been taken as design example. Results obtained for different amplifier and filter architectures have been discussed in detail toward the end of the chapter. Merits as well as demerits of various architectural blocks used in NSPU have been discussed. The detailed information presented in this chapter is anticipated to make the reader understand the intricacies in this challenging field of neural signal processing.

TABLE 1.2 Performance Comparison of Filter Architectures.

Parameters	Gm-C[64]	OTA-C[72]	OTA-C[73]	OTA-C[74]
Technology (μm)	0.6	0.065	0.13	0.18
Order	2	2	5	4
DC gain (dB)	–	–0.14	≈ 0	–0.090
IRN (μV_{rms})	91.8	360	4.7	39.92
Bandwidth (Hz)	2.5 k	100	100	250
DR (dB)	84.03	55.1	57	62.1
Area (mm^2)	0.17	0.065	0.67	0.46
Supply voltage (V)	3.3	0.5	0.25	1
Power (W)	47.2 μ	47n	603 n	208n
FOM (J)	33.73 a	0.412 p	1.70 p	0.163 p

KEYWORDS

- **ADC**
- **low-pass filter**
- **neural amplifier**
- **neural signal processing**
- **OTA**

REFERENCES

1. Serruya, M. D.; Hatsopoulos, N. G.; Paninski, L.; Fellows, M. R.; Donoghue, J. P. Brain-machine Interface: Instant Neural Control of a Movement Signal. *Nature* **2002**, *416* (6877), 141.
2. Wessberg, J. et al. Real-time Prediction of Hand Trajectory by Ensembles of Cortical Neurons in Primates. *Nature* **2000**, *408* (6810), 361.
3. Hochberg, L. R. et al. Neuronal Ensemble Control of Prosthetic Devices by a Human with Tetraplegia. *Nature* **2006**, *442* (7099), 164.
4. Kennedy, P. R.; Bakay, R. A.; Moore, M. M.; Adams, K.; Goldwaithe, J. Direct Control of a Computer from the Human Central Nervous System. *IEEE Trans. Rehab. Eng.* **2000**, *8* (2), 198–202.
5. Bagheri, A.; Salam, M. T.; Velazquez, J. L. P.; Genov, R. Low-frequency Noise and Offset Rejection in DC-coupled Neural Amplifiers: A Review and Digitally-assisted Design Tutorial. *IEEE Trans. Biomed. Circuits Syst.* **2016**, *11* (1), 161–176.
6. Dong, H.; Yuanjin, Z.; Rajkumar, R.; Dawe, G.; Minkyu, J. 0.45 V 100-Channel Neural-Recording IC with Sub-mW/Channel Consumption in 0.18 mm CMOS. Paper presented at the Proceedings of the IEEE International Solid-State Circuits Conference Digest of Technical Papers (ISSCC'13), San Francisco, CA, USA, 2013.
7. Sharma, K.; Sharma, R. Design Considerations for Effective Neural Signal Sensing and Amplification: A Review. *Biomed. Phys. Eng. Express* **2019**, *5* (4), 042001.
8. Holleman, J. Design considerations for neural amplifiers. Paper presented at the 2016 38th Annual International Conference of the IEEE Engineering in Medicine and Biology Society (EMBC), 2016.
9. Schwiening, C. J. A Brief Historical Perspective: Hodgkin and Huxley. *J. Physiol.* **2012**, *590* (Pt 11), 2571.
10. Hubel, D. H.; Wiesel, T. N. Receptive Fields of Single Neurones in the Cat's Striate Cortex. *J. Physiol.* **1959**, *148*(3), 574.
11. Mountcastle, V. B. Modality and Topographic Properties of Single Neurons of Cat's Somatic Sensory Cortex. *J. Neurophysiol.* **1957**, *20* (4), 408–434.
12. Palay, S. L. Synapses in the Central Nervous System. *J. Biophys. Biochem. Cytol.* **1956**, *2* (4), 193.
13. Sidman, R. L.; Dickie, M. M.; Appel, S. H. Mutant Mice (quaking and jimpy) with Deficient Myelination in the Central Nervous System. *Science* **1964**, *144* (3616), 309–311.
14. Bioulac, B.; Lamarre, Y. Activity of Postcentral Cortical Neurons of the Monkey during Conditioned Movements of a Deafferented Limb. *Brain Res.* **1979**, *172* (3), 427–437.
15. Sattelle, D.; Lane, N. Architecture of Gastropod Central Nervous Tissues in Relation to Ionic Movements. *Tissue Cell* **1972**, *4* (2), 253–270.
16. Tanji, J.; Taniguchi, K.; Saga, T. Supplementary Motor Area: Neuronal Response to Motor Instructions. *J. Neurophysiol.* **1980**, *43* (1), 60–68.
17. DeMott, D. W. An Inexpensive, Multi-channel, Electrophysiological Recording System. *Electroencephalogr. Clin. Neurophysiol.* **1961**, *13* (3), 467–470.
18. Polikov, V. S.; Tresco, P. A.; Reichert, W. M. Response of Brain Tissue to Chronically Implanted Neural Electrodes. *J. Neurosci. Methods* **2005**, *148* (1), 1–18.

19. Jia, Y.; Guler, U.; Lai, Y.-P.; Gong, Y.; Weber, A.; Li, W.; Ghovanloo, M. 26.8 A Trimodal Wireless Implantable Neural Interface System-on-Chip. Paper presented at the 2020 IEEE International Solid-State Circuits Conference-(ISSCC), 2020.
20. Zhou, A. et al. A Wireless and Artefact-free 128-channel Neuromodulation Device for Closed-loop Stimulation and Recording in Non-human Primates. *Nat. Biomed. Eng.* **2019**, *3* (1), 15–26.
21. Yeon, P.; Bakir, M. S.; Ghovanloo, M. Towards a 1.1 mm 2 free-floating wireless implantable neural recording SoC. Paper presented at the 2018 IEEE Custom Integrated Circuits Conference (CICC), 2018.
22. Lopez, C. M. et al. A Neural Probe With Up to 966 Electrodes and Up to 384 Configurable Channels in 0.13 μm SOI CMOS. *IEEE Trans. Biomed. Circuits Syst.* **2017**, *11* (3), 510–522.
23. Chandrakumar, H.; Marković, D. A High Dynamic-range Neural Recording Chopper Amplifier for Simultaneous Neural Recording and Stimulation. *IEEE J. Solid-state Circuits* **2017**, *52* (3), 645–656.
24. Chaturvedi, V.; Amrutur, B. An Area-efficient Noise-adaptive Neural Amplifier in 130 nm CMOS Technology. *IEEE J. Emerging Sel. Top. Circuits Syst.* **2011**, *1* (4), 536–545.
25. Bharucha, E.; Sepehrian, H.; Gosselin, B. A Survey of Neural Front End Amplifiers and Their Requirements toward Practical Neural Interfaces. *J. Low Power Electron. Appl.* **2014**, *4* (4), 268–291.
26. Harrison, R. R. The Design of Integrated Circuits to Observe Brain Activity. *Proc. IEEE* **2008**, *96* (7), 1203–1216.
27. Yang, T.; Holleman, J. An Ultralow-power Low-noise CMOS Biopotential Amplifier for Neural Recording. *IEEE Trans. Circuits Syst. II: Express Briefs* **2015**, *62* (10), 927–931.
28. Babayan-Mashhadi, S.; Lotfi, R. Analysis and Design of a Low-voltage Low-power Double-tail Comparator. *IEEE Trans. Very Large Scale Integration (VLSI) Syst.* **2013**, *22* (2), 343–352.
29. Miyahara, M.; Asada, Y.; Paik, D.; Matsuzawa, A. A low-noise self-calibrating dynamic comparator for high-speed ADCs. Paper presented at the 2008 IEEE Asian Solid-State Circuits Conference, 2008.
30. Posada-Quintero, H. F.; Rood, R. T.; Burnham, K.; Pennace, J.; Chon, K. H. Assessment of Carbon/Salt/Adhesive Electrodes for Surface Electromyography Measurements. *IEEE J. Transl. Eng. Health Med.* **2016**, *4*, 1–9.
31. Yu, Y.-H.; Lu, S.-W.; Liao, L.-D.; Lin, C.-T. Design, Fabrication, and Experimental Validation of Novel Flexible Silicon-based Dry Sensors for Electroencephalography Signal Measurements. *IEEE J. Transl. Eng. Health Med.* **2014**, *2*, 1–7.
32. Chi, Y. M.; Jung, T.-P.; Cauwenberghs, G. Dry-contact and Noncontact Biopotential Electrodes: Methodological Review. *IEEE Rev. Biomed. Eng.* **2010**, *3*, 106–119.
33. Scholvin, J.; Kinney, J. P.; Bernstein, J. G.; Moore-Kochlacs, C.; Kopell, N.; Fonstad, C. G.; Boyden, E. S. Close-packed Silicon Microelectrodes for Scalable Spatially Oversampled Neural Recording. *IEEE Trans. Biomed. Eng.* **2015**, *63* (1), 120–130.
34. Castagnola, E. et al. PEDOT-CNT-coated Low-impedance, Ultra-flexible, and Brain-conformable Micro-ECoG Arrays. *IEEE Trans. Neural Syst. Rehab. Eng.* **2014**, *23* (3), 342–350.

35. Harrison, R.; Watkins, P.; Kier, R.; Lovejoy, R.; Black, D.; Normann, R.; Solzbacher, F. A low-power integrated circuit for a wireless 100-electrode neural recording system. Paper presented at the 2006 IEEE International Solid State Circuits Conference-Digest of Technical Papers, 2006.
36. Spach, M. S.; Barr, R. C.; Havstad, J. W.; Long, E. C. Skin-electrode Impedance and Its Effect on Recording Cardiac Potentials. *Circulation* **1966**, *34* (4), 649–656.
37. Harrison, R. R.; Charles, C. A Low-power Low-noise CMOS Amplifier for Neural Recording Applications. *IEEE J. Solid-state Circuits* **2003**, *38* (6), 958–965.
38. Gosselin, B.; Sawan, M.; Chapman, C. A. A Low-power Integrated Bioamplifier with Active Low-frequency Suppression. *IEEE Trans. Biomed. Circuits Syst.* **2007**, *1* (3), 184–192.
39. Olsson, R.; Gulari, M.; Wise, K. Silicon neural recording arrays with on-chip electronics for in-vivo data acquisition. Paper presented at the 2nd Annual International IEEE-EMBS Special Topic Conference on Microtechnologies in Medicine and Biology. Proceedings (Cat. No. 02EX578), 2002.
40. Denison, T.; Consoer, K.; Santa, W.; Avestruz, A.-T.; Cooley, J.; Kelly, A. A 2 μW 100 nV/rtHz Chopper-Stabilized Instrumentation Amplifier for Chronic Measurement of Neural Field Potentials. *IEEE J. Solid-state Circuits* **2007**, *42* (12), 2934–2945.
41. Mollazadeh, M.; Murari, K.; Cauwenberghs, G.; Thakor, N. Micropower CMOS Integrated Low-Noise Amplification, Filtering, and Digitization of Multimodal Neuropotentials. *IEEE Trans. Biomed. Circuits Syst.* **2009**, *3* (1), 1–10. doi:10.1109/TBCAS.2008.2005297
42. Ng, K. A.; Xu, Y. P. A compact, low input capacitance neural recording amplifier with C in/gain of 20fF. V/V. Paper presented at the 2012 IEEE Biomedical Circuits and Systems Conference (BioCAS), 2012.
43. Carrillo, J. M.; Torelli, G.; Valverde, R. P.-A.; Duque-Carrillo, J. F. 1-V Rail-to-rail CMOS opamp with Improved Bulk-driven Input Stage. *IEEE J. Solid-state Circuits* **2007**, *42* (3), 508–517.
44. Khateb, F.; Dabbous, S. B. A.; Vlassis, S. A Survey of Non-conventional Techniques for Low-voltage Low-power Analog Circuit Design. *Radioengineering* **2013**, *22* (2), 415–427.
45. Sharma, K.; Pathania, A.; Pandey, R.; Madan, J.; Sharma, R. MOS Based Pseudo-resistors Exhibiting Tera Ohms of Incremental Resistance for Biomedical Applications: Analysis and Proof of Concept. *Integ. VLSI J.* **2020** (accepted).
46. Khateb, F.; Vlassis, S.; Kumngern, M.; Psychalinos, C.; Kulej, T.; Vrba, R.; Fujcik, L. 1 V Rectifier Based on Bulk-driven Quasi-floating-gate Differential Difference Amplifiers. *Circuits Syst. Signal Process.* **2015**, *34* (7), 2077–2089.
47. Rezaee-Dehsorkh, H.; Ravanshad, N.; Lotfi, R.; Mafinezhad, K. A linear tunable amplifier for implantable neural recording applications. Paper presented at the 2011 IEEE 54th International Midwest Symposium on Circuits and Systems (MWSCAS), 2011.
48. Rezaee-Dehsorkh, H.; Ravanshad, N.; Lotfi, R.; Mafinezhad, K.; Sodagar, A. M. Analysis and Design of Tunable Amplifiers for Implantable Neural Recording Applications. *IEEE J. Emerging Sel. Top. Circuits Syst.* **2011**, *1* (4), 546–556.
49. Zou, X.; Xu, X.; Yao, L.; Lian, Y. A 1-V 450-nW Fully Integrated Programmable Biomedical Sensor Interface Chip. *IEEE J. Solid-state Circuits* **2009**, *44* (4), 1067–1077.

50. Cortes, F. P.; Fabris, E.; Bampi, S. Analysis and Design of Amplifiers and Comparators in CMOS 0.35 µm Technology. *Microelectron. Reliab.* **2004**, *44* (4), 657–664.
51. Wang, T.-Y.; Liu, L.-H.; Peng, S.-Y. A Power-efficient Highly Linear Reconfigurable Biopotential Sensing Amplifier Using Gate-balanced Pseudoresistors. *IEEE Trans. Circuits Syst. II: Express Briefs* **2015**, *62* (2), 199–203.
52. Sharma, P.; Sharma, K.; Jatana, H.; Madan, J.; Pandey, R.; Sharma, R. A 1.1 µW Biopotential Amplifier Based on Bulk-driven Quasi-floating Gate Technique with Extremely Low-value of Offset Voltage. *Analog Integrated Circuits Signal Process.* **2020**, *103* (2), 303–313.
53. Sharma, K.; Jamuar, S.; Sharma, R. Design and Simulation of Pseudo-resistor with Extremely High Linearity for an Improved Neural Signal Recording. *Rev. Sci. Instrum.* **2020**, *91* (6), 066102.
54. Payandehnia, P.; Yeknami, A. F.; Meng, X.; Yang, C.; Temes, G. C. A passive CMOS low-pass filter for high speed and high SNDR applications. Paper presented at the 2015 IEEE International Symposium on Circuits and Systems (ISCAS), 2015.
55. Wakaumi, H. A switched-capacitor filter with dynamic switching bias OP amplifiers. Paper presented at the 2017 IEEE 8th Latin American Symposium on Circuits & Systems (LASCAS), 2017.
56. Qian, X.; Xu, Y. P.; Li, X. A CMOS Continuous-time Low-pass Notch Filter for EEG Systems. *Analog Integ. Circuits Signal Process.* **2005**, *44* (3), 231–238.
57. Yehoshuva, C.; Rakhi, R.; Anto, D.; Kaurati, S. 0.5 V, ultra low power multi standard G m-C filter for biomedical applications. Paper presented at the 2016 IEEE International Conference on Recent Trends in Electronics, Information & Communication Technology (RTEICT), 2016.
58. Lo, T.-Y.; Hung, C.-C. F. *1V CMOS Gm-C Filters: Design and Applications*; Springer Science & Business Media, 2009.
59. Acosta, L.; Jiménez, M.; Carvajal, R. G.; Lopez-Martin, A. J.; Ramírez-Angulo, J. Highly Linear Tunable CMOS Gm-C Low-Pass Filter. *IEEE Trans. Circuits Syst. I: Regular Papers* **2009**, *56* (10), 2145–2158.
60. Silva-Martinez, J.; Solis-Bustos, S. Design considerations for high performance very low frequency filters. Paper presented at the 1999 IEEE International Symposium on Circuits and Systems (ISCAS), 1999.
61. Ferreira, L. H. C.; Pimenta, T. C.; Moreno, R. L. An Ultra-Low-Voltage Ultra-Low-Power CMOS Miller OTA With Rail-to-Rail Input/Output Swing. *IEEE Trans. Circuits Syst. II: Express Briefs* **2007**, *54* (10), 843–847. doi:10.1109/TCSII.2007.902216
62. Shahrokhi, F.; Abdelhalim, K.; Serletis, D.; Carlen, P. L.; Genov, R. The 128-channel Fully Differential Digital Integrated Neural Recording and Stimulation Interface. *IEEE Trans. Biomed. Circuits Syst.* **2010**, *4* (3), 149–161.
63. Wattanapanitch, W.; Fee, M.; Sarpeshkar, R. An Energy-efficient Micropower Neural Recording Amplifier. *IEEE Trans. Biomed. Circuits Syst.* **2007**, *1* (2), 136–147.
64. Lu, J.; Yang, T.; Jahan, M. S.; Holleman, J. A low-power 84-dB dynamic-range tunable Gm-C filter for bio-signal acquisition. Paper presented at the 2014 IEEE 57th International Midwest Symposium on Circuits and Systems (MWSCAS), 2014.
65. Boulemnakher, M.; Andre, E.; Roux, J.; Paillardet, F. A 1.2 V 4.5 mW 10b 100MS/s Pipeline ADC in a 65nm CMOS. Paper presented at the ISSCC, 2008.

66. El-Chammas, M.; Murmann, B. A 12-GS/s 81-mW 5-bit Time-interleaved Flash ADC with Background Timing Skew Calibration. *IEEE J. Solid-state Circuits* **2011**, *46* (4), 838–847.
67. Lee, S.-Y.; Cheng, C.-J. A Low-voltage and Low-power Adaptive Switched-current Sigma–delta ADC for Bio-acquisition Microsystems. *IEEE Trans. Circuits Syst. I: Regular Papers* **2006**, *53* (12), 2628–2636.
68. Yang, J.; Naing, T. L.; Brodersen, R. W. A 1 GS/s 6 bit 6.7 mW Successive Approximation ADC Using Asynchronous Processing. *IEEE J. Solid-state Circuits* **2010**, *45* (8), 1469–1478.
69. Zhu, Z.; Liang, Y. A 0.6-V 38-nW 9.4-ENOB 20-kS/s SAR ADC in 0.18-μm CMOS for Medical Implant Devices. *IEEE Trans. Circuits Syst. I: Regular Papers* **2015**, *62* (9), 2167–2176.
70. Ruiz-Amaya, J.; Rodriguez-Perez, A.; Delgado-Restituto, M. A Low Noise Amplifier for Neural Spike Recording Interfaces. *Sensors* **2015**, *15* (10), 25313–25335.
71. Lyu, L.; Wang, Y.; Chen, C.; Shi, C.-J. R. A 0.6 V 1.07 μW/Channel Neural Interface IC Using Level-shifted Feedback. *Integration* **2020**, *70*, 51–59.
72. Hanumantha Rao, G.; Rekha, S. A 0.8-V, 55.1-dB DR, 100 Hz Low-Pass Filter with Low-Power PTAT for Bio-Medical Applications. *IETE J. Res.* **2019**, 1–11.
73. Pinto, P. M.; Ferreira, L. H.; Colletta, G. D.; Braga, R. A. A 0.25-V Fifth-order Butterworth Low-pass Filter Based on Fully Differential Difference Transconductance Amplifier Architecture. *Microelectron. J.* **2019**, *92*, 104606.
74. Rajan, V. S.; Kishore, K. H.; Sanjay, R.; Kumaravel, S.; Venkataramani, B. A Novel Programmable Attenuator Based Low Gm-OTA for Biomedical Applications. *Microelectron. J.* **2020**, *97*, 104721.

CHAPTER 2

Sensing Using the Memristive Approach, Governing Physics, and Instrumentation Challenges

ABHIJIT KUMAR PAL[1] and SUNIPA ROY[2*]

[1]*Department of Applied Electronics & Instrumentation Engineering, Future Institute of Engineering and Management, Kolkata, India*

[2]*Department of Electronics & Communication Engineering, Guru Nanak Institute of Technology, Kolkata, India*

[*]*Corresponding author. E-mail: sunipa_4@yahoo.co.in*

ABSTRACT

The foremost attraction to research across the globe now featured about the resistive memory device (Re-RAM or RRAM) where the memory performance predominantly more instigated from "resistive" change compared with "capacitive" one (CMOS device) reckoning to be latent alternative for CMOS, has paved the system for several proposition for memory storage devices. The forked process of internet–IoT technology requires more high-density memories, The Von Neumann architecture relies in modern computers for in-memory computation, thus eradicating the energy-intensive and time-overwhelming data movement, the vulnerable state-of-the-art technology involving in neuromorphic computing plagued an energy efficient high storage memory design–all can be attributing from resistive switching memory, an ultimate nonvolatile memory device liable to its faster speed, less power consumption, enhanced storage density, high temperature stability, potential applications in numerous fields, and exceptional scalability. In this literature, the long voyage to the RRAM toward the enhancement of data storage capacity is reviewed in a comprehensive manner. Starting with the pivotal aspects resembling the field of incipient memory techniques, switching mechanism and electrical characteristics, innumerable issues on endurance, retention, and uniformity, various material preparation techniques for RRAM development are deliberated in this discussion. This chapter is also emphasized on memory storage competency of RRAM incorporating multilevel cell (MLC) to achieve improved storage density. Physical mechanisms involved to attain reliable operation of MLC operation are elaborated in this review with a conclusion to relative merits and elucidate challenges involved in future aspects of RRAMs.

2.1 INTRODUCTION

At Bell Labs, S.M Sze along with D. Kahng first introduced the uses of floating gate (FG) as a memory in 1967.[1] Metal oxide semiconductor field effect transistor (MOSFET) in combination with an extra floating gate inside a memory cell formed the silicon-based flash memories and representing state-of-the-art memories with nonvolatility, holds the present ancillary memory technology for its low cost and larger density. From this time, a tremendous revolutionary growth is established in electronic technology by the introduction of semiconductor memory devices since Moore et al. used the first 64-bit RAM chip in an IBM cache memory.[2] For the neoteric electronic system, semiconductor memory relays an important role as a requisite component and backbone today to store data temporarily or permanently. Storage system is essential for any kinds of computational platform starting from nanodevices to large supercomputers.[3] Initially, punch card were used to store data of lesser bytes, but now, storage system essentially has extended up to multiterabytes to petabytes of data storage capabilities regardless with narrow space with lesser power consumption, reduced power loss, fast storage capacity regarding application aspects[4,5] for solid-state disks (SSDs) one of the most common component is flash memory as it is more cost-effective due to dipping prices and growth in densities.[6] The memory hierarchy may represent by a pyramidal structure shown in Figure 2.1 that can be recognized as fundamental understanding of diverse memory types. Variation of pyramid width specifies the storage abilities for each particular level. Data storage for the on-and off chip memory is quite lesser than that of the off and on memory storage memory. Movement toward the base of the structure advises reduction in cost per

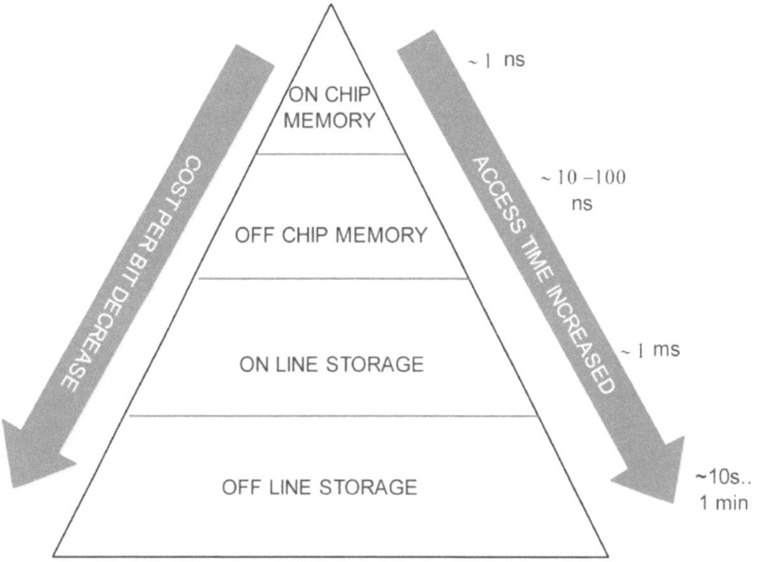

FIGURE 2.1 Memory hierarchy pyramid where width indicates the storage.

bit along with the increment in access time.[5] Conversely, operational speed of lower values (write/erase time: 1 ms/0.1 ms), higher writing voltage (>10 V) and inadequate endurance (writing speed of 10^6/erase cycles) are the main hindrances that a flash memory suffers.[7] Furthermore, in near future, in that way, flash memories will grasp their miniaturization limit not only for technical criterion but also for physical restriction like leakage currents of higher values. For discussion like flash memory as an example of nonvolatile memory sources, memory technologies may achieve higher performance by the combination of advantages and disadvantages, such as in case of dynamic RAM (DRAM) which has high capacity and density, but it is volatile in nature which means it requires refresh operation every few milliseconds to retain data resulting in increasing power consumption of the device which is undesirable. On the other hand, fast memory technology–Static RAM (SRAM) may overcome this dilemma, but its larger size is encumbering the implementation in designing of memory cells in a broader scale. NOR and NAND flash memory are the two types of flash memories. In a CMOS NOR flash, cells are parallelly linked with bit lines followed by individual read and programming. Cells are allied in series with the bit lines for CMOS NAND architecture. In comparison, series connection required lesser space compared with the parallel connection resulting in the reduction in cost.[7] Considering all of them, both architectures suffer from lower operational speed (write/erase time), inadequate endurance of 10^6 write/erase cycles and high writing voltage ($>10^9$).[8] Density, speed, and power consumption are the three important parameters essential for the design of semiconductor memory devices. Another most import aspect of the semiconductor memory device is the economic significance with qualified importance of the various types. Semiconductor memory devices are lingering for a few past decades and attend increased storage densities, minimum power, higher speeds, and large performance with decreasing the cost.[9,10] However, limitations of semiconductor based memory devices are also recognized simultaneously. Lots of incipient technologies are pointing to remove those limitations and possibly substitute all or maximum of the present memory technologies using semiconductor to develop a universal semiconductor memory technology. Apart from this to hold the control of enormous electronic market that can be elaborated from computer applications in computer to several electronic products for consumers, such a type of memory device development is necessary. In the invited paper, Wu et al. reviewed one of the important motivations on emerging nonvolatile memories (NVM) which is the consumption of energy for memory where the requirement of current is in the range of 10 μA during programming. In perspective of issues regarding scaling, designing of flash memory in 30 nm technology node progressively becomes a challenging factor regarding the challenges in consistent detection of charge.[11] Further Zahhor et al.[12] suggested about existing charge-based storage memory devices facing challenges for the design in the 10 nm node range or beyond that range results in the loss of stored charge in a memory cell and thus the performances are degraded. As a result, a modern class of memory technology known as emerging memory technology is currently developed primarily in the industry to revolutionize the existing class of memory hierarchy.[13] To integrate the switching speed and to improve the storage class capacity, like the flash memories emerging technology, are the best alternatives to current NVM. An ideal memory device may be classified depending upon the following characteristics–lower operating voltage (<1 V), higher endurance (10^{17} cycles), enriched data retention time (>10 years), consumption of energy should be low with larger scalability (<10 nm). Irrespective of charge-based resistive memory based on memory cell resistance variability concept of memristor technology is highly incorporated for the development of most encouraging emerging technology for subsequent generation of NVM. Although not a single memory device has the characteristics of ideal one, rather, it has been actively investigated that the emerging memory technology meets a part of ideal one.[14] Depending upon the change in resistance for data storage, emerging memory technology may be categorized as spin-transfer-torque magnetic random access memory (STT-MRAM),[15] phase change memory (PCM),[16] and resistive random access memory (RRAM)[8, 13] which may be considered as most consistent candidates for imminent NVM. some of the most common features of emerging NVM are (1) they are nonvolatile in nature with two terminals and (2) their switching mechanism is transit between off-state or high resistance region (HRS) and on-state or low resistive region (LRS). All those switching transition is accomplished in terms of applied electrical inputs. Instead of that the physics beyond the switching mechanism is quite different for various emerging NVM. In the case of STT-MRAM to obtain a change in resistance, a parallel and an antiparallel configuration of two ferromagnetic layers are designed to create change in resistance in between them where they are separated by a thin tunneling layer. To switch between the phase in crystalline condition (referred as LRS) and the phase in amorphous condition (referred as HRS) for PCRAM, a chalcogenide material is used; in the case of RRAM, a conductive filament is formed (LRS) and rupturing this conductive filament (for HRS) in between two electrodes in the insulator. A comparative study of mainstream memory and emerging memory on different aspects are deliberated in Table 2.1.

RRAM further classified depending upon the composition of conductive filament (CF) as conductive bridge random access memory (CBRAM) and oxygen vacancy-based RRAM. Over a long research, it is concluded that RRAM has a huge potential in memory applications with cell dimensions in the nanometer range,[17–19] operating current in the range of some μA with the operating voltage of 2–3 V,[20–23], consists of switching speed in the nanosecond range[18, 24] with good reliability and endurance and data retention.[25, 26] From 2013, Panasonic starts mass production of RRAM due to the outclass performance of this device, and in the year 2014, Adesto reported a recorded large unit of CBRAM shipment (near about 1 million). A high-density test chip of 16 GB and 32 GB storage size using RRAM for the storage class memory (SCM) applications is also reported.[27] Now a days a lot of effort is given on the development of 3D vertical RRAM which overtakes the 3 D NAND flash in various aspects especially in

TABLE 2.1 Comparison of Emerging Memory Technologies.

	Conventional memories				Emerging memories		
	SRAM	DRAM	FLASH		STT-MRAM	PC RAM	RRAM
			NOR	NAND			
Cell area (F^2)	>100	6	10	<4 (3D)	60-20	4-20	<4 (3D)
Multiple bit	1	1	2	3	1	2	3
Voltage (V)	<1	<1	>10	>10	<2	<3	<3
Read time (ns)	≈1	≈10	≈50	≈10,000	<10	<10	<10
Write time (ns)	≈1	≈10	10,000	100,000	<5	≈50	<10
Retention	N/A	≈64 ms	>10 y	>10 y	>10 y	>10 y	>10 y
Endurance	$>10^{16}$	$>10^{16}$	$>10^{5}$	$>10^{4}$	$>10^{15}$	$>10^{9}$	$>10^{6}$~$>10^{12}$
Write energy (pJ/bit)	≈0.001	≈0.01	≈100	≈0.01	≈0.1	≈10	≈0.1

speed, operation voltage, and endurance–though full commercialized production is a long run wait.

Numerous papers have been published after inspecting the RRAM for the past couple of years to review the different facets of RRAM.[28, 29] The physics behind this pair of fundamental resistive switching behavior is broadly reviewed by Waser et al. in 2009 about redox-based reactions and ionic transport[28]. Later, a comprehensive review was done by Kim et al. on switching mechanism in binary oxide thin film where unipolar switching mechanism is specifically observed.[30] Besides this, Wong et al. engrossed on the performance of the device and enlisted the accomplishment in binary metal oxide RRAM.[31] In their review, Jeong et al. highlighted on various switching mechanism and materials with variety and increased demand on switching behavior.[29] Regarding the potentiality in industrialization, the characteristics of CBRAM is demonstrated by Valov et al.[32] Considering all, we can believe an update on the development of RRAM array, its uses, switching mechanism along with various applications of this device, all have a high research domain in current years. Accompanied with large compensations in various arenas of performance, memristors show tremendous potential in the future generation of mainframe storage class devices. However, for application purpose, it is quite difficult to use this as this resistance voltage distributed randomly.

2.2 RRAM CLASSIFICATION AND MECHANISM FOR PHYSICAL STORAGE

2.2.1 FUNDAMENTAL REQUIREMENTS OF RRAM

A matrix is formed to organize a memory cells in a RAM. Rows of the matrix represent the word lines whereas bitlines are represented by columns that arc connecting to the electronic amplifier within the matrix to bear the operation for read and write. Basically, the word and bitlines are linked with each node to form the passive crossbar matrix. In the state of low resistance, signal circumvents by cells, elements connected serially with a specific nonlinearity, such as diodes or varistor type elements (Fig. 2.2a) must be added with specific degree of nonlinearity for individual node to evade the parasitic path problem.[33] If a memory cell is not organized, A select transistor is used to form an active matrix to organized RAM for decoupling memory cell if it is not addressable (Fig. 2.2b). A capacitive type metal (*M*)-insulator

(I)-metal structure is formed for the resistive switching memory cells in a memristor, where the resistive or insulating material is sandwiched between two different metal electrodes. Oxides of metal or higher chalcogenides having specific ion conductivity are used for insulating material I. To activate the switching property, an initial electroforming cycle in general is used for switching in two distinct resistance states for this structure. The cell will SET low resistance (ON) state from its high resistance by the application of writing voltage pulse or return into the OFF or RESET state. The fascinating fact is that the resistance switching mechanism can be designated as memristor after the prediction of Chua as it suggested the presence of a basic fourth circuit element symmetrical with inductor, resistor, and capacitor after that it is described as memristive device.[34] In this article, it is described that a flux φ can be represented as the time integral of voltage $V(t)$ and it is associated with charge q as a time integral of current $I(t)$. If V/I be the resistance and is depending on the charge $q(t)$ which is passing throughout the device is considered as memresistance $M(q(t))$ and mathematically it may be represented as

$$V(t) = M(q(t))/t.$$

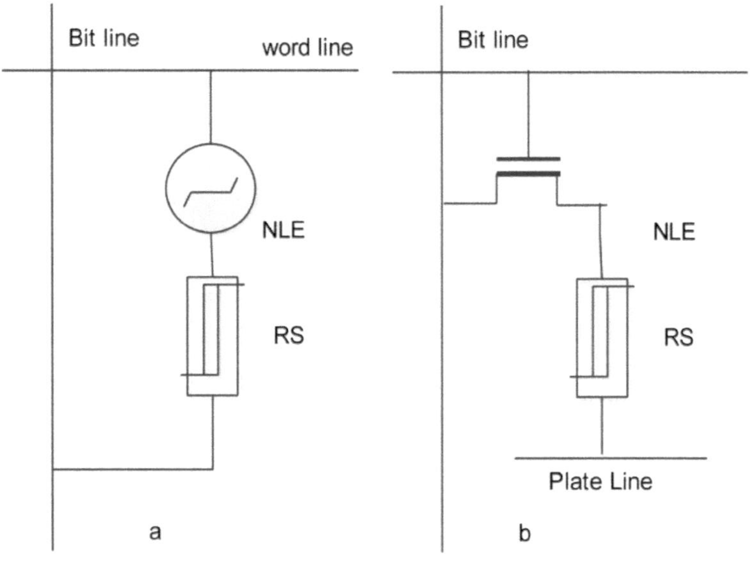

FIGURE 2.2 A storage node matrix for RRAM, where resistive switching cell is denoted by RS. (a) Passive matrix: where specific nonlinearity is represented by a serial element NLE. (b) Active matrix: where T is the select transistor.

First Stanley and his associates observed the relation between RRAM switching mechanism and Chua's perception and the article of the literature discussed.[35] This depiction is tremendous and useful for the resistance having multilevel values or may store analog voltages. For the resistive switching techniques, with respect to the electrical polarity, two different schemes are used for the memristive cells. For writing voltage to be polarity independent, the switching method is termed as unipolar switching (Fig. 2.3b), where the compliance current typically limited the current throughout the SET process. A higher current is required with voltage below SET potential for the RESET operation. For the bipolar switching (Fig. 2.3a), SET process occurs at one potential and reverse potential is applied for RESET operation. To achieve the bipolar switching behavior in the MIM system, some asymmetrical structure is needed such as dissimilar electrode material "M" or during the electroforming process, a dedicated voltage polarity must be supplied. In the literature, further it is referred as active electrode (AE) and an (ohmic) counter electrode (CE) in which the bipolar switching takes place. Operation conditions by which systems can be reformed in between unipolar and bipolar operation have been named as nonpolar.[36]

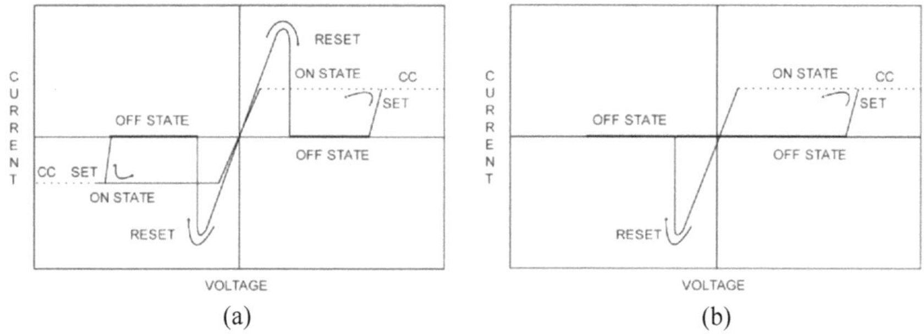

FIGURE 2.3 Resistance switching schemes for RRAM memory cells. I–V curves for (a) bipolar switching and (b) unipolar switching. CC: denoted the compliance current.

2.2.2 MECHANISMS FOR RESISTIVE SWITCHING AND THE FUNCTION OF DEFECTS

In all the cases, the driving force of the electrically induced resistance switching is relatively different. A single molecule may result in the development or modification of the molecular configuration into a resistance memory.[37] Depending upon the ferroelectric direction or a domain polarization of ferromagnet, a tunnel current will be generated.[38] In the case of phase change memories (PCM), temperature-induced change has been observed between an ON-state (crystalline phase) and OFF-state (amorphous) for particular selenides and tellurides.[39] And three classes of chemical effects are included, that is, voltage or temperature or the combination of two are triggered which affects in the MIM cell redox processes. First, the bipolar electrochemical metallization (ECM) mechanism is anticipated electrochemically on an active electrode metal, for example, Ag, the Ag^{+} cations with high mobility drifted toward the "I" layer consisting of ion conductivity, thus leading to the growth of Ag dendrites (Fig. 2.4a–d) toward the counter electrode (inert) by discharging, resulting into a highly conductive filament formation cell which is in ON-state.[40] The filament dissolution occurred electrochemically when the reversed potential is applied by altering the polarity. The transition metal oxides (TMO) are triggered in the case of valence change mechanism (VCM) by the anions migration to create oxygen vacancies. Consequently, a redox reaction takes place for this stoichiometric change which can be represented by the cation sublattice valence change and leads to the change in electrical conductivity A voltage pulse is induced for this bipolar memory switching where the direction of change such as oxidation or reduction is depending upon the applied pulses polarity. Another class considered as memory effect or thermochemical mechanism (TCM), which indicates the stoichiometric changes due to increase of temperature for current induction. The major resemblances in this review will describe about the variation of these three mechanisms, overall progress for the current state of knowledge, and also the future work in this area will be outlined.

Various ways are followed for grouping of these classes. One group may comprise PCM, TCM, and VCM systems with emphasis on the material, in which insulating or resistive materials "I" using oxides and higher chalcogenides have significant role as insulating or resistive materials "I" and a secondary role will be played by the electrode materials. The metal electrodes "M" of metal types has a critical role in ECM systems and also for the memories depending upon the effects of electronic or pure electrostatic along with the tributary influence of material "I" in these systems. Depending upon the switching schemes, a classification had been made where unipolar switching belongs to PCM, TCM, and VCM having the mode of bipolar switching. This review also covers the three redox-based close effects which are important for grouping, which is conceivable for PCM. In all cases, this concerns the major role of defects in lattice which are essential for the microscopic switching mechanism. Crystalline phase vacancies of up to 25% are shown in the PCM accommodation by the chalcogenides. During switching, the kinetics of phase transformation is accelerated by several magnitude orders to the occurrence of these lattice defects. Dearnaley and others comprehensively reviewed that for resistance switching, hysteretic behaviors are shown during the selection of materials with large diversity for an MIM configuration,[41], Oxley,[42] and Pagnia.[43] In most of the papers, thermochemical memory switching (TCM), the switching effects are described as unipolar, although definite task is impossible as in most of the cases, experimental data are insufficient. In the late 1990s, Asamitsu et al., Kozicki et al, and Beck et al introduced a novel era in the research activity started in the late 1990s,[40,44,45] which is recently appraised by Waser[46] and Sawa.[47] In this current review, focus has been made on redox-based ion-migration understandability correlated with the process of resistive switching and the presence of particular defects which have a crucial role in these switching phenomena.

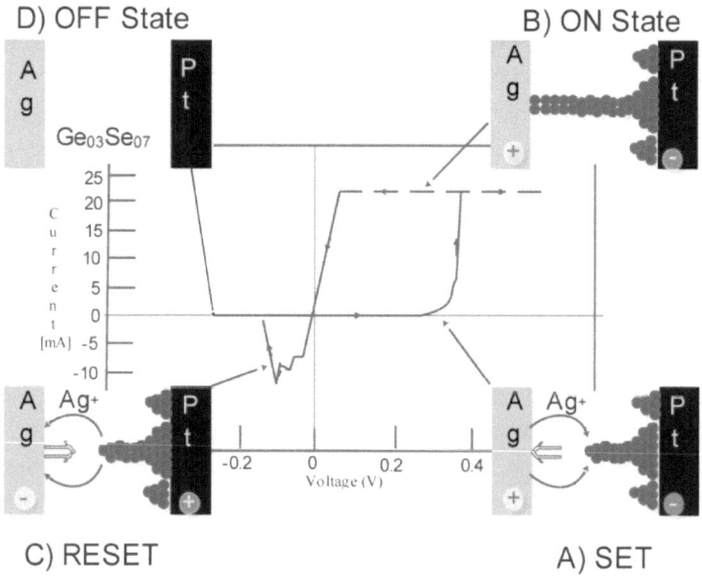

FIGURE 2.4 Electrochemical metallization cell using Ag/Ag-Ge-Se/Pt showing typical current–voltage behavior due to the uses of a triangular voltage sweep. (a–d) Different switching procedure.

2.2.3 ELECTROCHEMICAL METALLIZATION SYSTEMS

For resistive switching performance, metal deposition and dissolution in electrochemical process is exploited for electrochemical metallization (ECM) cells or popularly called conductive bridging (CB) cells or programmable metallization cell (PMC). For resistive switching in metal ion-based RRAM or CBRAM, the physical mechanism has centered the metal ions migration and consequent reduction/oxidation (redox) reactions.[48] For the structural view in the case of CBRAM, a metal oxide layer is sandwiched between an anode or oxidizable top, such as Ag, Ni, Cu, and Ag, and a relatively inert bottom electrode or cathode, such as Pt, Wt. In such types of memory cells the process of filament formation takes place due to dissolution of active metal electrodes (Ag or Cu), transportation of cations (Cu^+ or Ag^+), and toward the inert bottom electrode the reduction or deposition of their consequent.[49] Therefore, formation and dissolution of metal films is the main characterization for such types of RRAM. A typical Ag/a-ZnO/Pt Resistive RAM cells were fabricated to explain the metal ion-based CBRAM switching mechanism.[50] In Figure 2.5, the switching process of CBRAM is illustrated. Figure 2.5a represents the CBRAM memory cell in pristine or pure state. In the process of filament formation, at the top position, electrode of Ag material is used (TE) which will act as an active component while at the bottom side, electrode of Pt material are used which act as an inert component..Whenever the positive potential is applied to the electrode of Ag at the top position, the oxidation process begins ($Ag \rightarrow Ag^+ + e^-$) at this electrode and the deposition of cations of generated Ag^+ are observed within the layer of dielectric (a-ZnO) from the electrode of Ag. Similarly attraction of this positive Ag+ cations by the bottom electrode (BE) of Pt which are kept at negative potential, and in this way, occurrences of the reduction reaction ($Ag^+ + e- \rightarrow Ag$) is followed at the electrode of bottom side. In this process, atoms of Ag are formed due to reduction of the cations (Ag^+), and till the formation of conducting bridge, accumulation is continuing (Fig. 2.5b–d) and low resistive state (LRS) exhibited by the RRAM device that can be referred as "SET" process. During the application of reverse polarity, the device is said to be switched to high resistance state (HRS) as the highly conducting filament dissolves completely, and this process is specified as "RESET" and is illustrated in Figure 2.5e.

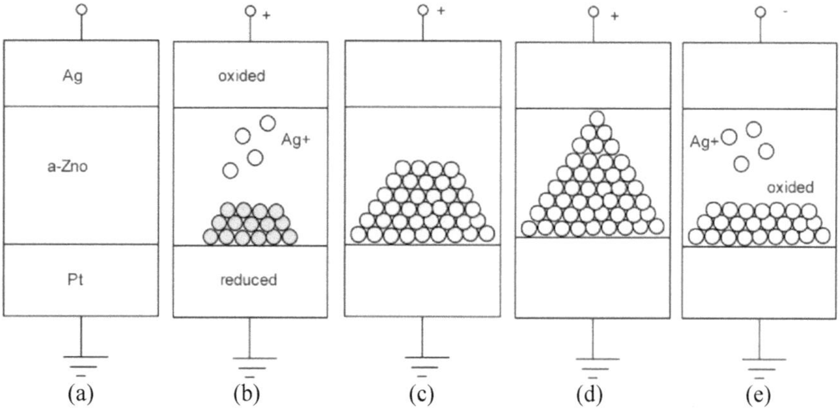

FIGURE 2.5 Conductive bridge RAM switching mechanism: (a) Device in pure condition, (b, c) oxidation of silver (Ag), reduction and migration of Ag^+ ions toward cathode, (d) accumulation growth of highly conductive filament of atoms of Ag leading by Pt electrode, and (e) opposite polarity is applied for dissolution of filament.

2.2.4 VALENCE CHANGE MECHANISM

In OxRRAM or oxygen vacancy-based RRAM, the physical mechanism associated with the resistive switching is the creation of vacancies of oxygen (V_o^{2+}) and subsequent ions of oxygen (O^{2-}) relocation, thus ensures the development of conductive filament within both electrodes of RRAM cell.[31] At the beginning of this fabricated RRAM cell, a soft breakdown is essentially required during the forming process. During this soft dielectric breakdown, atoms of oxygen are bumped out from the lattice, as at the anode interface, a high electric field exists and oxygen ions (O^{2-}) are formed, whereas in the layer of oxide, the oxygen vacancies (V_o^{2+}) are left. For anode, if noble materials are used to develop the oxide interfacial layer, the ions of oxygen (O^{2-}) interact with the anode materials or neutral nonlattice oxygen may be discharged. Thus, the interface at electrode/oxide performs like an "oxygen reservoir."[51] Next, the formation of conductive filament (CF) due to oxygen vacancies (V_o^{2+}) accumulates within bulk oxide, the RRAM cell switches to the state of low resistance (LRS), and inside the device, a considerable amount of current will flow. From the interface of anode, the ions of oxygen (O^{2-}) migrate back to bulk oxide to switch back the device into HRS, the reset operation takes place either by combining with the oxygen vacancies (V_o^{2+}) or the CF precipitates by metallic oxidation, and thus rupturing the filament partially, thereby the RRAM cell switching back into HRS. In unipolar switching, inside the cell, the oxygen ions (O^{2-}) are diffused and activated thermally by the current from Joule heating, and due to the presence of concentration gradient, such oxygen ions diffusion takes place from the region around the CF or from the interface. It is interesting to note that a relatively higher reset current is required for unipolar switching of RRAM to rise the confined temperature around CF. In the case of bipolar switching, a diffusion barrier is presented by the interfacing layer, and also the process of thermal diffusion is not adequate, henceforth, the revised electric field is aided to the oxygen ions (O^{2-}). But in both the cases, rupture of CF takes place partially and the RRAM cell switches into the HRS. For the electrons, a tunneling gap is formed from the formation of primary oxygen vacancies (V_o^{2+}) and poor region. In the gap region, an indulgent breakdown is created for which the reconnection of CF is possible and the electrodes revert back the device to the LRS (SET process).

2.2.5 COMPARISON BETWEEN ECM AND VCM

Most popularly known electrochemical metallization memory (ECM) or CBRAM relies on metal electrodes which are electrochemically active like Ni, Cu, or Ag for the formation of CF which is metal cation-based CF.[52] Similarly, for "OxRRAM" also identified as valence change memory (VCM) due to the migration of anions within the storage material itself,[53] metal atoms are composed of oxygen vacancy defects. Although the switching mechanism for both OxRRAM and CBRAM are deliberated, till now, certain differences may be present in these types of RRAM pairs,[54] for example, for set process where the growth of CF started or where to break in the reset process, respectively, and to form the CF, how oxygen vacancies/metal atoms gathered. The current–voltage (I–V) characteristics for 20 successive switching cycles of Ta/TaOx/Pt[55]-based RRAM has been carried out, and a strong deviation has been observed for both types of switching procedure during successive switching cycles–resulting in the decrease in the entire memory window, thus, the overall performance of the RRAM is degraded. During the SET and RESET process, as CF is formed randomly as well as it is ruptured, variability in resistance is primarily attributed for cycle-to-cycle. Depending upon the different operating parameters, a detailed comparison of both these RAMs is given in Table 2.2. For these RRAM memory types, an interesting difference in endurance characteristics is observed. The reason for this divergence is due to the fact that the drift and diffusion of metal atoms in the conducting filaments of CBRAM are much easier compared with oxygen vacancies, which cause the degradation of the retention time and endurance characteristics of CBRAM in comparison with the OxRRAM. For these two RRAM types, irrespective of the differences in the mechanism of switching, many similar characteristics are shown by both the RRAM types. The only important difference is that an expressively higher endurance has been observed for OxRRAM compared with CBRAM.

TABLE 2.2 Comparison Between CBRRAM and Metal Oxide RRAM.

Parameter	Metal-oxide RRAM	Conductive bridge RRAM
Speed (ns)	5 [56]	1 [60]
Operation voltage (V)	~3 [57]	~7 [61]
Operation current (μA)	5 [57]	10 [62]
Endurance (cycles)	10^{12} [51]	10^{6} [50]
On/off ratio	10^{7} [58]	10^{7} [63]
Retention@85°C (s)	10^{6} [59]	10^{6} [50]
Multilevel capacity	Possible	Possible
Compatibility with CMOS	Possible	Possible
Fabrication	Simple	Simple
Scalability	Excellent	Excellent

2.3 RESISTANCE SWITCHING MATERIALS

A clear observation suggests that in some insulators under the application of electrical potential, their resistance change is observed and this property of change of resistance is recently utilized for the development of future nonvolatile memories.[64] For various oxide materials, the phenomenon for resistance switching has been observed, but for their compatibility with CMOS BEOL processing for the future NVM applications, binary metal oxides are extensively used. Numerous materials based on metal oxide revealing the resistance switching nonvolatility, for example, hafnium oxide (HfO*x*),[65–70] titanium oxide (TiO*x*),[71–78] tantalum oxide (TaO*x*),[79–83] nickel oxide (NiO),[84–87] zinc oxide (ZnO),[50, 59, 88–90] zinc titanate (Zn_2TiO_4),[77, 78], manganese oxide (MnO*x*),[91, 92] magnesium oxide (MgO),[93] aluminum oxide (AlO*x*),[57, 94, 95] and zirconium dioxide (ZrO_2)[96–99] have drawn most of the attention and have been discussed broadly for the last couple of years. Pulse laser deposition (PLD), atomic layer deposition (ALD), and reactive sputtering techniques are used for the deposition of these metal oxides. Although, the uniformity and thickness of the thin film is extensively possible using ALD.[31] The transportation path of carriers for conventional electronic devices nd the selection of the electrode material are essentially important. Also, the switching characteristics of RRAM device is greatly affected depending upon the selection of electrode material. For example, a steady resistive switching behavior was observed in (Cu/P_3HT: PCBM/ITO) structure, however, this switching behavior disappeared after Cu electrode has been replaced with Pt electrode.[100] An extensive variation is observed in the selection of materials used as electrode for RRAM. Primarily based on the composition, the materials for electrode can be clustered into five categories, which also includes the fundamental substance electrodes, electrodes based on silicon, alloy, based upon oxides, and electrodes based on nitrites. Electrodes which are available abundantly and are very common and mostly used are elementary substance electrodes, which include the substances, such as Al,[57] Ti,[92] Cu,[76] graphene,[101] carbon nanotubes,[102] Ag,[50] W,[86] and Pt.[88] Silicon-based electrodes comprise only two types of electrodes, such as n-type Si and p-type Si.[103, 104] To stabilize the resistive switching behavior, electrodes of alloy material usually are used and they mainly include Cu–Ti,[105] Cu–Te,[106] and Pt–Al.[107] The nitride-based electrodes that are extensively preferred are TiN and TaN.[103, 108, 109] The abundant oxide-based electrodes including Al-doped ZnO,[109] Ga-doped ZnO,[110] and ITO[111] are also used. For RRAM devices fabrication, the group of metal oxide materials that are used in recent years along with different material combinations casting for both top and bottom electrode are given in Table 2.3. Thus, it can be predicted that resistance switching with nonvolatility is broadly observed in various metal oxides. For the construction of RRAM, metal oxide metal (MOM) structures can be constructed using these oxide materials that are currently used in the semiconductor technology. As Pt is very hard to etch,it is used as bottom electrode material. Same bottom electrodes are used for the single device structure, whereas separate bottom electrodes are used at bottom side for individual devices in the crossbar architecture. For its deposition and successive lift-off, physical vapor deposition method is utilized. For the top electrode deposition and formation of resistive switching layer, atomic layer deposition (ALD) or physical vapor deposition (PVD) is used. In the thin film of oxides, such as SiO, Al_2O_3, Ta_2O_5, ZrO_2 and TiO_2, negative resistance in a MIM structure is shown and is described by Hickmott in 1962.[112] After the discovery of Hickmott's, materials consisting the switching properties were explored in materials like thin films of NiO, and further researches were followed on the properties of Al_2O_3, Ta_2O_5, and SiO.[41, 113] Later, in the 1990s, oxides of perovskite, Cr-doped SrTi (Zr)O_3, $Pr_{0.7}Ca_{0.3}MnO_3$(PCMO), and $PbZr_{0.52}Ti_{0.48}$ were discovered, though they are unstable for the applications in memory devices.[114] In 2004, binary-transition metal oxide (TMO) is used for the development of OxRRAM.[114] Full compatibility with the CMOS technology and integration with the 0.18 μm CMOS technology is investigated for OxRRAM. Further experiment suggests that up to 300°C the OxRRAM was stable and the operating voltage level is generally lesser than 3 V.[114] This groundbreaking work inspired additional research into RRAM materials and devices–oxides of binary and ternary metal, chalcogenides, polymers, nitrides, and 2D layered materials are the new varieties of materials that have been explored meanwhile from 2004.[115, 116] In comparison for the compatibility with CMOS processes, binary metal oxides are found to be more compatible when compared with nonbinary oxide materials. However, it is worthy to point out that materials such as La_2CuO_4, HoScO*x*, Ag_2S, poly-o-anthranilicacid (PARA), poly 3,4-ethylenedioxythiophene, polystyrene sulfonate (PEDOT:PSS), etc. are also being explored for RRAM applications.[117–120] As in the dielectric MIM stack transition, metal oxides are thoroughly used. Due to various advantages of resistive switching devices based on TMO comprising the properties, such as flexibility, high thermal dissipation, chemical stability and transparency 2D materials, hexagonal boron nitride (h-BN), Graphene oxide (GO), MoS_2, etc. are used.[121–123] It has been reported that h-BN has several advantageous properties, such as bipolar and forming-free, with enhanced reliability along with threshold-type resistive switching, outstanding thermal conductivity (600 W/mK), chemical constancy (>1500°C), high mechanical stress controllability (500 N/m), short variability, greater resistance ratio (up to 10^4), low operational voltages

TABLE 2.3 RRAM Materials and Comparison of Different RRAM Types.

Year	Electrode (Top/bottom)	Oxide material	HRS/LRS ratio	Mode of operation	Retention	Endurance	References
2018	Ag/Pt	MnO/Ta_2O_5	10^6	Bipolar	~10^4 s	100 cycles	[125]
2018	Pd/TiN	HfO_x/Ag NPs	10^3	Bipolar	>10^4 s	>10^8 cycles	[126]
2018	ITO/Pt	Zn_2TiO_4	NS	Bipolar	>10^4 s	>500 cycles	[90]
2018	Ti/TiN	HfO_2	>10	Bipolar	>10	10^4 cycles	[65]
2018	ITO/Pt	a-TiO_2	>10	Bipolar	<10^3 s	NS	[77, 78]
2017	Au/Ti/Au	TiO_{2-x}	>104	Bipolar	NS	NS	[127]
2017	TiN/Ti/Au	TiO_{2-x}	<10^3	Bipolar	>10^5 s	>50 cycles	[74]
2017	Al/Al	HfO_x	~10^4	Unipolar	NS	NS	[58]
2016	Ti/W/Ru	Ti/W TiO*x*/MgO	<32	Bipolar	NS	>10^3 write cycles	[73]
2016	Ag/Pt	a-ZnO	>10^7	Bipolar	10^4	NS	[50]
2016	TiN/Ti/Au	TiO_{2-x}	~10^5	Bipolar	NS	10^5 cycles	[128]
2016	W/TiW/Cu	WO_3/Al_2O_3	10^4	Bipolar	NS	~300 cycles	[62]
2016	ITO/FTO	TiO*x*/Ag NPs	10^3	Bipolar	NS	>300 cycles	[71]
2015	Ti/Au	Ta_2O_5/TiO_2 NPs	2370	Bipolar	NS	<40 cycles	[61]
2015	Ti/TiN	HfO_2	10^5	Bipolar	10^4 s	10^{10} cycles	[129]
2015	W/Pt	Ta/TaO*x*	>10^2	Bipolar	>10^4 s	>10^8 cycles	[83]
2015	Pt/TiN	TaO*x*	NS	Bipolar	NS	NS	[81]
2015	W/Zr/TiN	HfO_2	NS	Bipolar	NS	>10^6 write cycles	[69]
2014	TiN/Pt	HfO_2	10^6	Bipolar	10^4 s	NS	[130]

TABLE 2.3 *(Continued)*

Year	Electrode (Top/bottom)	Oxide material	HRS/LRS ratio	Mode of operation	Retention	Endurance	References
2013	Ta/Ti	$TaOx/TiO_2$	10^5	Bipolar	$>10^4$ s	$>10^{12}$ cycles	[129]
2012	Pt/Pt	ZnO	10^6	Bipolar	$>10^6$ s	$>10^6$ cycles	[59]
2011	TiN/Pt	HfOx/AlOx	NS	Bipolar	NS	10^6 cycles	[131, 132]
2011	TaN/Pt	Al_2O_3/Ru NCs	$>10^5$	Bipolar	10^5 s	NS	[133]
2011	Au/ITO	ZnO	10^4	Bipolar	$>10^4$ s	10^2 cycles	[89]
2010	Au/Ag	ZrO_2	10^4	Bipolar	$\sim 10^4$ s	>500 cycles	[134]
2010	Pt/Pt	ZnO	$\sim 10^2$	Unipolar	NS	NS	[88]
2010	Al/Ti/Pt	Al_2O_3	>10	Bipolar/Unipolar	10^4 s	$>10^3$ cycles	[57]
2009	Ti/Pt	MnO_2	~ 102	Bipolar	$>10^4$ s	$>10^5$	[92]
2009	TiN/Pt	ZrO_2	NS	Bipolar	NS	$>10^2$	[134]
2009	Ti/Pt	ZrO_2	NS	Bipolar/Unipolar	NS	$>10^2$	[99]
2009	Cu/Pt	TiO_2	~30	Bipolar	NS	NS	[76]
2008	Pt/Pt/Ti	NiO	NS	Unipolar	NS	NS	[86]
2008	Pt/Ti/Pt	Al_2O_3	NS	Bipolar	NS	NS	[95]
2008	TiN/TiN	TiOx/HfOx	>103	Bipolar	$\sim 10^5$ s	>106 cycles	[56]
2008	Pt/Pt	ZnO	<105	Unipolar	NS	>102 cycles	[90]
2007	Ti/Pt	ZrO_2	NS	Bipolar	NS	>104 cycles	[97]

(below 0.4 V), and extended retention times (over 10 h)[116, 124] along with the modest fabrication process. For the growth of multilayer h-BN sheets, chemical vapour deposition (CVD) is used inherently containing some defects that are valuable in resistive switching, as they decrease the energy to breakdown, thus permitting the devices to be forming-free.[116] During the growth of CVD deviation, thickness of boundary grains is observed that creates device-to-device variability, however, it is low cycle-to-cycle.[121]

2.4 MULTILEVEL RESISTIVE RANDOM ACCESS MEMORY (RRAM)

2.4.1 MULTILEVEL PER CELL (MLC) STORAGE

Due to smaller physical size and lesser power requirement, RRAM devices can be potentially used in future memory applications. Increase in storage density is one of the most reliable solutions for low-cost and high-density nonvolatile data storage and that may be accommodated in RRAM cells for multiple resistive states. Presently, lots of exploration is being carried out in the field of RRAM to scale down the dimensions in structural density for memory arrays. Earlier, by decreasing the device size the storage density of RRAM has been increased, though for experimental procedure, the involved complexity prohibited its successful implementation. Three-dimensional (3D) crossbar architecture is another type of suggested method. "Vertical" and "crossbar" RRAM are the two types of architectures that have been proposed for multilevel RRAM designing[135, 136]; though advancement in fabrication procedures restricts for both architecture as these are not desirable. Multilevel cell storage technology (MLC) is a very simple alternative to increase storage density in RRAM devices which enables more storage, that is storing more than one bit per cell without any reduction in cell dimension. One of the most promising properties of RRAM is MLC which can favorably rise the storage density of the memory cell.[131, 132, 137–140] Thus, multiple ON- and OFF-state can be obtained instead of a single HRS and LRS. It is also possible to achieve multiple resistances, without affecting the dimensions of the device. For reliable MLC operation, it is important precisely to control the resistance of various resistance material. A precise control over the resistance of the various resistance level is important for reliable MLC operation, else due to variability in resistance and issues related with reliability mainly due to the random behavior of the conductive filament formation during the switching process, the device will be affected.[141]

2.4.2 APPROACHES TO OBTAIN DIFFERENT MULTILEVEL PER CELL (MLC) IN RRAM

The MLC performance of RRAM for high-density applications can be exploited. Three methods can be employed to obtain MLC behavior in RRAM (i) compliance current changing, (ii) reset voltage controlling, and (iii) deviation of pulse width for programing/erasing.

2.4.3 MLC BY COMPLIANCE CURRENT CHANGING

In general, 1-RRAM (1R) cell configuration[50] or 1-Transistor 1-RRAM (1T-1R) cell configuration[65] is controlled along with the RRAM device. In 1R configuration compliance current (I_{cc}) may be changed during "set" operation to obtain MLC, whereas by controlling and by changing the transistor gate voltage, compliance current is controlled for set operation of RRAM for 1T-1R structure of the cell to obtain MLC.The typical I–V curves for obtaining MLC curves of $Ti/Ta_2O_5/Pt$[142] -based RRAM cell are illustrated in Figure 2.6. Whenever the compliance current (I_{cc}) is incremented from 150 μA to 1 mA, different LRS are obtained at six different position like I_{cc} = 150 μA, 200 μA, 300 μA, 500 μA, 700 μA, and at 1 mA, respectively due respective current increment during LRS (I_{LRS}) while maintaining the constant HRS, and for all LRS levels, this HRS current (IHRS) remains same. For RRAM configured by $Ti/Ta_2O_5/$ Pt, increase inof the level of I_{cc} results in the increment of reset current (I_{reset}) maximally, and subsequently, the set voltage is maintained almost constant. Further decrement in the resistance of LRS was observed while the increment in reset current owing to the formation of stronger filament with the increased I_{cc}.The CF formation and its consistent widening with rise in *I*cc is the recognized MLC mechanism in compliance current (I_{cc}) mode which is depicted schematically in Figure 2.5. As the CF size is increasing due to increase in I_{cc}, it causes decrement in the CF resistance, and henceforth, is resulting in multiple LRS levels for variousI_{cc} values. It has been also perceived that for rupturing the CF having larger diameter, increasing I_{cc} increasesthe I_{reset} current as the requirement for power is high (Fig. 2.7).

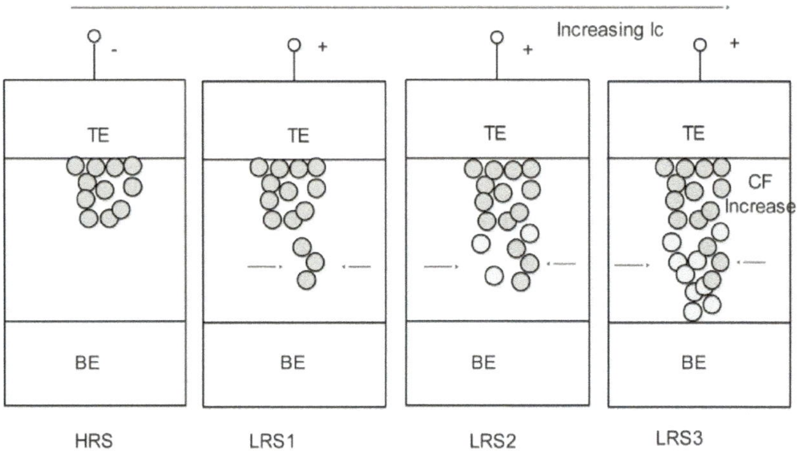

FIGURE 2.6 Illustration of RRAM cells with multiple resistance states in obtained by varying compliance current "*I*cc."

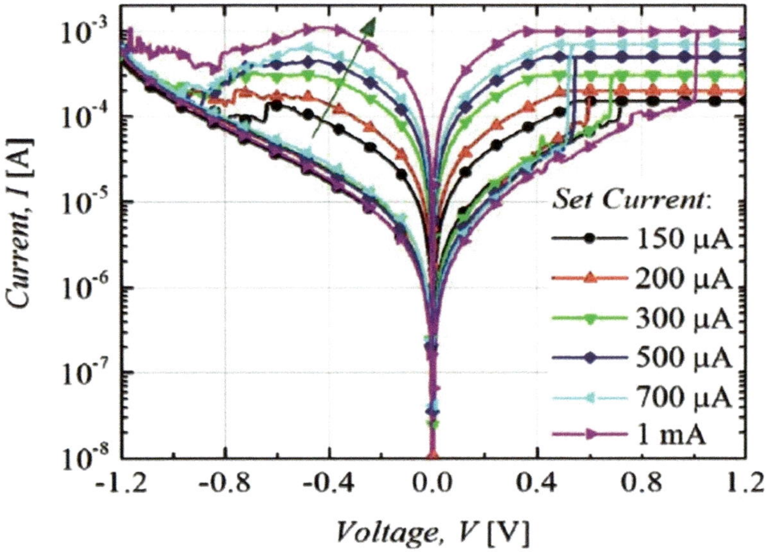

FIGURE 2.7 Ti/Ta_2O_5/Pt RRAM showing multilevel characteristics obtained by controlling the compliance current.

Source: Reprinted from Ref. [12].

2.4.4 MLC BY CONTROLLING RESET VOLTAGE

Like compliance current by governing the reset voltage (V_{reset}) while maintaining I_{cc} at constant value, MLC characteristics are also available in RRAM. For this case, the typical I–V curves are obtained to achieve MLC in TiN/HfO*x*/AlO*x*/Pt-based RRAM cell[131, 132] by applying different reset voltages. It is identified that HRS current will decrease due to the increment in reset voltage. In this way, with the same LRS resistance, multiple HRS levels are obtained. Additionally, as V_{reset} is increased, the set voltage (V_{set}) is also increased while I_{reset} remains almost constant. Whenever the separation in between the electrode and tip of the CF is increased, I_{HRS} is primarily decreased with increase in reset voltage as shown in Figure 2.8. Thus, higher value of resistance will be obtained for increased V_{reset} values by increasing the gap between the bottom electrodes and conducting filament tip. Therefore, increase in separation in between bottom electrode (BE) and CF tip results in multiple levels of resistance of HRS due to the reset voltage. It is perceived that during the reset operation, instead of abrupt change in reset current, a regular change is observed within the devices. In such devices, whenever reset voltage is increased, HRS resistance will be changed due to the decrement in size of the CF. Essentially, in view of cross-point architecture, this approach is more worthwhile for relatively low complex circuitry.

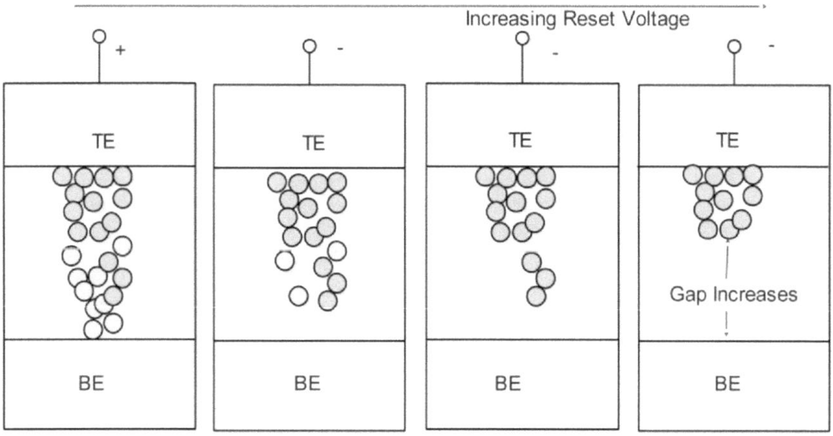

FIGURE 2.8 Illustration of variation of reset voltage causes multiple resistance states in RRAM cell.

2.4.5 MLC BY VARYING PROGRAM/ERASE PULSE WIDTH

By varying the program or erase pulse width variability and maintaining the pulse amplitude constant, MLC characteristics are also obtained.[130] Three different levels of HRS were confirmed in Hafnium oxide-based RRAM by changing the reset pulse span to 5 μs from 50 ns.[131, 132] This process is quite easier to obtain the MLC. However, this is an energy inefficient scheme which also restricts to obtain reliable characteristics in the RRAM cell. In comparison to the transient responses between the amplitude of reset pulse and controlling of width pulse, for that particular case, the RRAM device consumes higher energy. The basic reason behind this phenomenon is that unwanted higher energy is dissipated as thermal energy in the material used for resistive switching. A detailed survey regarding the multiple resistance states shown by RRAM devices has been reported.However, only eight resistive states have been verified in a single RRAM cell till date either by changing I_{cc} or the reset voltage. It is worthwhile that in the RRAM cell, there is a large opportunity for increasing the number of resistance states, to improve its density of storage.

2.5 APPLICATIONS OF RRAM

It is seen that among the emerging memory technologies, RRAM has one of the pioneer candidates that perceive the potentiality for reforming the memory hierarchy in primary stage for its high speed, nonvolatile data storage capability, enhancement of data storage density, and logical computing. In this section, some of these novelty in application perspective of RRAM have been discussed.

2.5.1 NONVOLATILE LOGIC

A computer system comprises von Neumann architecture as the instructive codes and the data being transferred by utilizing buses between the number of units within the computer which consists of separate computing and memory unit. Thus, it causes time delay and increased energy consumption during the data transferring process, which is primarily termed as "von Neumann bottleneck." RRAM crossbar array is one of the better suggestion for reducing the influence of von Neumann bottleneck[143] to acquire the high integration density storage and low cost. The 4*F*2 array architecture,[144] and the compact device structure of two terminal RRAM are greatly beneficial. As an illustration, simple Boolean logic functions like –logical OR, NOT, and AND realization necessitates the use of multiple transistors. Two or three RRAM cells are sufficient for the implementation of these logical functions, of an approximated area of nearly about 10*F*2 only.[145] Till date, for the realization of Boolean logic functions, numerous methods have been suggested.[146] Compared with the existing non-Boolean computing paradigms, this Boolean computation is significantly more established like neuromorphic quantum computation. Consequently, energy enhancement and CPU cost-efficiency will be possible without the necessity for new algorithm development or software implementation, though in a crossbar array, implementation of complex computing process suffers from the lack of technical solution. Hence, implementation and demonstration of basic logical levels are basically focused in most of the literature using RRAM as its complexity will arise for the implementation of an entire computer unit using the arrays of RRAM.

2.5.2 NEUROMORPHIC COMPUTATION USING RRAM

To overwhelm the "von Neumann bottleneck," one of the most promising technology is brain-inspired neuromorphic computing which has exposed extreme potential in a broader assortment of cognitive and complex tasks like audio/visual recognition, self-driving, and big-data analytics in real time in comparison with neuromorphic network based on CMOS. Neuromorphic computation based on RRAM-array gives benefit in terms of on-chip weight storage, online training, and scaling up much larger array size.[147, 148] Additionally, the processing speed of RRAM is improved by three orders of magnitude with reduced power consumption rate by four orders of magnitude.[149] For hardware implementation of neuromorphic computing paradigms, two methods are suggested, among them, one of the strategies parodists the structural and working mechanism of biological neural networks, and the other method works on the accumulation of existing algorithms on artificial neural network (ANN). For neural network, for transferring of spikes between numerous neurons and to store the information in terms of transferring weights, a synapse of a neuron is used. Although in the literature a little bit works reported, demanding to imitate such types of learning rules on RRAM devices, however, this types of bioinspired learning rules realization to a complex task is more complicated as the algorithm exists theoretically is not strong enough, rather an approach by mapping an ANN directly to a neuromorphic computing based on RRAM is practically more feasible. Although certain innovative tasks based on this method have been established on pattern and speech recognition.[150, 151] Though a very favorable technology, RRAM-based synapse is still far behind hand in the application as various problems, such as optimization of materials, designing of control circuit, designing of algorithm and architecture, and computation for analog domain are to be addressed efficiently.

2.5.3 SECURITY APPLICATION

Today, security aspect specially in the arena of information technology is an important and major concern due to the rapid development in this sector. Thus, a hardware-based integrated security system is much necessary. In comparison to CMOS logic-based security circuits which exhibit the process of manufacturing of semiconductor randomly, RRAM-based security circuits are found to be more robust to attacks of several types owing to their switching mechanisms randomly.[152, 153] Physical unclonable function (PUF)-a novel security feature commonly based on RRAM in general suggested,[154] where this proposal is basically for the authentication of device (strong PUF) and application on the generation of the key (weak PUF). For the purpose of strong PUF pairs, the input–output or challenge response pair (CRP) are required in large number, while less number of CRPs having higher reliability are tremendously necessary for weak PUF.[155] Although, Further evaluation and strong practical demonstration is necessary for further research on the maturity of this novel primitives inside the domain of hardware security, although incredible performance have been established by PUFs based on RRAM. SRAM and DRAM consume nearly about half of the static power for modern mobile SoC chips. To overcome this and to have a fast parallel memory operations, device with reduced area and smaller energy consumption is essentially required, and this is the basic resolution for the proposal of RRAM-based nonvolatile SRAM (nvSRAM),[156] where a pair of two RRAM cells are arranged on eight transistors to form an 8T2R structure. Recently, nonvolatile ternary content-addressable memory (TCAM) consisting of a structure of 4T2R cell[157] and flip flops with nonvolatility accompanied by compact stress time and RRAM technology-based write power have been verified.[158]

2.6 CONCLUSION

In this chapter,the memory architecture advancement in recent years is briefly reviewed. Also detailed study has been carried out on the recent developments and the restrictions while providing a valued intuition into the area of emerging memory technologies. A brief discussion, limelighting the prominence of RRAM, its construction, process of working, and classification is simultaneously reflected over here. Also, the major performance parameters and their consequence on the operation of RRAM have also been comprehended inside the present literature. An extravagant revision on the capability of RRAM MLC accompanied by the methodology has been deliberated. Further the enactment of RRAM for numerous imperative applications, for example, neuromorphic computation, nonvolatile logic, security, and SRAM nonvolatility have been emphasized. Although, substantial success has been accomplished in the case of RRAM technology, however, further effort is necessary as several challenges are still limiting the operation of RRAM in terms of higher operational current, lower resistance ratio, and concerns over reliability. Large exertions should be contributed in studies aiming to the development of faster erasing/programing, requirement of low power consumption, accelerating the density of storage by the implication of MLC and further development in the fabrication process for enriched regularity. Apart from this, focuses should be renewed toward the use of RRAM for nonvolatile logical operations and in embedded memory as the breakthroughs available in these fields are more significant and exciting. With continued effort and enhancements, it is clear that RRAM devices will be an incredible technology for future applications of nonvolatile memory.

KEYWORDS

- **resistive memory device**
- **neuromorphic computing**
- **multilevel cell**
- **memristive approach**
- **storage density**

REFERENCES

1. Sze, S. M. A Floating Gate and Its Application to Memory Devices. *Bell. Syst. Tech.* May 16, **1967**, 1288–1295.
2. Ayling, J. K. A High-performance Monolithic Store Tech Papers. In *IEEE Int Solid-State Circuits*; Philadelphia. Piscataway: IEEE, 1969; pp. 36–37.
3. Bez, R. A Introduction to Flash Memory. *Proc. IEEE 2003* **2003**, *91* (4), 489–502.
4. Aritome, S. Advanced Flash Memory Technology and Trends for Files Storage Application. *IEEE IEDM Tech Dig: IEEE 2002*. SanFrancisco. Piscataway; 2000; pp. 763–766.
5. Burr, G. W. Storage Class Memory, June 2010. Retrieved from http://geoffreyburr.org/papers/Burr_IITC2010_slides.pdf.
6. SNIA. Solid State Storage 101: An Introduction to Solid State Storage. San Francisco: SNIA, 2009.
7. Pan, F.; Gao, S. Recent Progress in Resistive Random Access Memories: Materials, Switching Mechanisms, and Performance. *Mater. Sci. Eng. R* July **2014**.
8. Jagan Singh Meena, S. M.-Y. Overview of Emerging Nonvolatile Memory Technologies. *Nanoscale Res. Lett. Springer Open J.* September **2014**.
9. Kim, K.; Koh, G. The Prospect on Semiconductor Memory in Nano Era. *IEEE. 7th Inter Conf Solid-State Integer Circuits Technol*, 2004; pp. 662–667.
10. Lam, C. H. The Quest for the Universal Semiconductor Memory. *IEEE Conf Electron Devices and Solid-State Circuits*, Hong Kong.Piscataway: IEEE, December 19–21, 2005; pp. 327–333.
11. Huaqia ng Wu, X. A. (n.d.). Resistive Random Access Memory for Future Information Processing System. *Invited Paper, IEEE.*
12. Furqan Zahoor, T. Z. Resistive Random Access Memory (RRAM): An Overview of Materials, Switching Mechanism, Performance, Multilevel Cell (mlc) Storage, Modeling, and Applications. *Nano Scale Res. Lett.* **2020.**
13. Philip Wong, H.-S.; Salahuddin, S. Memory Leads the Way to Better Computing. *Nat. Nanotechnol.* March **2015**, *10*, 191–194.
14. Yu, S. *Resistive Random Access Memory (RRAM) From Devices to Array Architectures*; Kris Iniewski, S. E., Ed. Morgan & Claypool Publishers; March 2016.
15. Zhu, J. Magnetoresistive Random Access Memory: The Path to Competitiveness and Scalability. *Proc. IEEE* **2008**, *96* (11), 1786–1798.
16. Wong, H.-S. P.; Raoux, S. Phase Change Memory. *Proc. IEEE* **2010**, *98* (12), 2201–2227.
17. Ho, C.; Hsu, C.-L.; Chen, C.-C.; Liu, J.-T.; Wu, C.-S.; Huang, C.-C.; … Yang, F.-L. 9nm Half-pitch Functional Resistive Memory Cell with <1 μA Programming Current Using Thermally Oxidized Sub-stoichiometric WOx film. 2010 International Electron Devices Meeting, 2010; pp. 19.1.1–19.1.4.
18. Govoreanu, B.; Kar, G.; Chen, Y. 10 × 10 nm^2 Hf/HfOx Crossbar Resistive RAM with Excellent Performance, Reliability and Low-Energy Operation. *IEDM Tech*, 2011; pp. 31.6.1–31.6.4.
19. Redolfi, L. G. A Novel CBRAM Integration Using Subtractive Dry-etching Process of Cu Enabling High-performance Memory Scaling Down to 10 nm Node. Symposium on VLSI Technology Digest of Technical Papers, 2015; pp. T134–T135.

20. Chenga, H.; Tsai, C. Y. High Performance Ultra-Low Energy RRAM with Good Retention and Endurance. *IEDM*, 2010; pp. 19.4.1–19.4.4.
21. Fang-Yuan Yuan et al. Conduction Mechanism and Improved Endurance in HfO_2-Based RRAM with Nitridation Treatment. *Nanoscale Res. Lett.* **2017**.
22. Kim, W.; Park, S. I.; Zhang, Z.; Yang-Liauw, Y.; Sekar, D.; Wong, H.-S. P.; Wong, S. S. Forming-free Nitrogen-doped AlOX RRAM with Sub-μA Programming Current. 2011 Symposium on VLSI Technology-Digest of Technical Papers, Honolulu, HI, USA, 2011; pp. 22–23.
23. Luo, Q.; Xu, X.; Liu, H.; Lv, H.; Gong, T.; Long, S. et al. Demonstration of 3D Vertical RRAM with Ultra Low-leakage, High-selectivity and Self-compliance Memory Cells. 2015 IEEE International Electron Devices Meeting (IEDM), Washington, DC, USA, December 7–9, 2015; pp. 10.2.1–10.2.4.
24. Tsunoda, K. K. Low Power and High Speed Switching of Ti-doped NiO ReRAM under the Unipolar Voltage Source of less than 3 V. *IEDM Tech.*, 2007; pp. 767–770.
25. Goux, L.; Belmonte, A.; Celano, U.; Woo, J.; Folkersma, S.; Chen, C. Y. et al. Retention, Disturb and Variability Improvements Enabled by Local Chemical-potential Tuning and Controlled Hour-Glass Filament Shape in a Novel $W\backslash WO_3\backslash Al_2O_3\backslash Cu$ CBRAM. 2016 IEEE Symposium on VLSI Technology, 2016; pp. T126–T127.
26. Ho, C.; Shen, T. Y.; Hsu, P. Y.; Chang, S. C.; Wen, S. Y.; Lin, M. H. et al. Random Soft Error Suppression by Stoichiometric Engineering: CMOS Compatible and Reliable 1Mb HfO_2-ReRAM with 2 Extra Masks for Embedded IoT Systems. 2016 IEEE Symposium on VLSI Technology, 2016; pp. 1–2.
27. Fackenthal, R.; Kitagawa, M. A 16Gb ReRAM with 200MB/s Write and 1GB/s Read in 27nm Technology. 2014 IEEE International Solid-State Circuits Conference, 2014; pp. 338–340.
28. Waser, R.; Dittmann, R. Redox-based Resistive Switching Memories–nanoionic Mechanisms, Prospects, and Challenges. *Adv. Mater.* **2009**, *21*, 2632–2663.
29. Jeong, D. S.; Thomas T. Emerging Memories: Resistive Switching Mechanisms and Current Status. *Rep. Progress Phys.* **2012**, *75* (9).
30. Kim, M.; Jeong, D. S. Nanofilamentary Resistive Switching in Binary Oxide System; A Review on the Present Status and Outlook. *Nanotechnology* **2011**, *22*, 254002.
31. Wong, H. S. P.; Lee, H. Metal–oxide RRAM. *Proc IEEE* **2012**, *100* (6), 1951–1970.
32. Valov, R. W. Electrochemical Metallization Memories—Fundamentals, Applications, Prospects. *Nanotechnology* **2011**, *22* (25), 254003.
33. Flocke, T. G. Nano-Crossbar Arrays for Nonvolatile Resistive RAM (RRAM) Applications. 2008 8th IEEE Conference on Nanotechnology, 2008; p. 319.
34. Chua, L. O. Memristor-The Missing Circuit Element. *IEEE Trans. Circuit Theory* **1971**, *18* (5), 507–519.
35. Strukov, B.; Snider G. S. The Fourth Circuit Element. *Nature* **2008**, 80–83.
36. Akinaga, H. S. Resistive Switching Effect in Metal/Insulator/Metal Heterostructures and Its Application for Non-volatile Memory. *IEEE Transactions,* 2007.
37. Loertscher, J. W. Reversible and Controllable Switching of a Single-molecule Junction. *Materials Science, Medicine, Small,* 2006.
38. Evgeny, Y.; Tsymbal, H. K. Tunneling Across a Ferroelectric. *Science* **2006**.
39. Wuttig, M.; Yamada, N. Phase-change Materials for Rewriteable Data Storage. *Nat. Mater.* **2007**, 824–832.

40. Kozicki, M. N.; Yun, M. Applications of Programmable Resistance Changes in Metal-Doped Chalcogenides. *J. Electrochem.* **1999**.
41. Dearnaley, A. M. Electrical Phenomena in Amorphous Oxide Films. *Reports on Progress in Physics, IOP Science,* 1970.
42. Oxley, D. P. Electroforming, Switching and Memory Effects in Oxide Thin Films, 1977.
43. Sotnik, H. P. A Regeneration Model for Conducting Filaments in MIM Diodes. *Physica Status Solidi,* 1985.
44. Asamitsu, Y. T. Current Switching of Resistive States in Magnetoresistive Manganites. *Nature,* 1997.
45. Beck, J. G. Reproducible Switching Effect in Thin Oxide Films for Memory Applications. *Appl. Phys. Lett.* **2000.**
46. Aono, R. W. Nanoionics-based Resistive Switching Memories. *Nat. Mater.* **2007**, 833–840.
47. Sawa, A. Resistive Switching in Transition Metal Oxides. *Mater. Today* **2008**.
48. Kozicki, M. N.; Barnaby, H. Conductive Bridging Random Access Memory—Materials, Devices and Applications. *Semicond. Sci. Technol.* **2016.**
49. Goux, L., Valov, I. Electrochemical Processes and Device Improvement in Conductive Bridge RAM Cells. *Phys. Stat Solidi*, **2016.**
50. Huang, Y., Shen, Z. Amorphous ZnO Based Resistive Random Access Memory. *RSC Adv.* **2016**, 17867–17872.
51. Hsu, C. W.; Wang, I. Self-rectifying Bipolar TaOx/TiO_2 RRAM with Superior Endurance over 1012 Cycles for3D High-density Storage-class Memory 3D High-density Storage-class Memory. Symposium on VLSI Technology, IEEE, 2013; pp. 166–167.
52. Waser, R.; Aono, M. Nanoionics-based Resistive Switching Memories. *Nat. Mater.* **2007**, 833–840.
53. Pan, F.; Chen, C. Nonvolatile Resistive Switching Memories-Characteristics, Mechanisms and Challenges. *Prog. Nat. Sci. Mater.* **2010.**
54. Zhu, D.; Li, Y. Resistive Random Access Memory and Its Applications in Storage and Nonvolatile Logic. *J. Semiconductors* **2017**, *38* (7), 071002.
55. Prakash, A.; Hwang, H. Multilevel Cell Storage and Resistance Variability in Resistive Random Access Memory. *Phys. Sci. Rev.* **2016.**
56. Lee, H.; Chen, P. Low Power and High Speed Bipolar Switching with a Thin Reactive Ti Buffer Layer in Robust HfO_2 Based RRAM. *IEEE Int. Electron Devices* **2008**.
57. Wu, Y.; Lee, B. Al_2O_3-Based RRAM Using Atomic Layer Deposition (ALD) with 1-μA RESET Current. *IEEE Electron Device Lett.* **2010**, 1449–1451.
58. Chen, Y. C.; Chang, Y. Dynamic Conductance Characteristics in HfO_2 Based Resistive Random Access Memory. *RSC Adv.* **2017**, *7 (21)*, 12984–12989.
59. Chiu, F. C.; Li, P. Reliability Characteristics and Conduction Mechanisms in Resistive Switching Memory Devices Using ZnO Thin Films. *Nanoscale Res Lett.* **2012.**
60. Goux, L.; Sankaran, K. Field-driven Ultrafast Sub-NS Programming in W Al_2O_3 Ti CuTe-based. *1T1R CBRAM system. In: 2012 Symposium on VLSI Technology (VLSIT), IEEE,* 2012.
61. Park, M. R.; Abbas, Y. Resistive Switching Characteristics of Tantalum Oxide Thin Film and Titanium Oxide Nanoparticles Hybrid Structure. *J. Nanosci. Nanotechnol.* **2015**, 8613–8616.

62. Woo, J.; Belmonte, A. Introduction of WO_3 Layer in a Cu-based Al_2O_3 Conductive Bridge RAM System for Robust Cycling and Large Memory Window. *IEEE J. Electron Devices* **2016**, 163–166.
63. Lim, S.; Yoo, J. Excellent Threshold Switching Device (IOFF~ 1 pA) with Atom-scale Metal Filament for Steep Slope (< 5 mV/dec), Ultra Low Voltage (VDD = 0.25 V) FET Applications. *IEEE International Electron Devices Meeting (IEDM). IEEE,* 2016.
64. Chang, T. C.; Chang, K. Resistance Random. *Mater Today* **2016**, 254–264.
65. Su, Y. T.; Liu, H. A Method to Reduce Forming Voltage without Degrading Device Performance in Hafnium Oxide-based 1T1R Resistive Random Access Memory. *IEEE J. Electron Devices Soc.* **2018**, *6*, 341–345.
66. La Torre, C. L.; Fleck, K. Dependence of the SET Switching Variability on the Initial State in HfOx-based ReRAM. *Phys Status Solidi* **2016**, 316–319.
67. Feng, W.; Shima, H. Investigation of Switching Mechanism in HfO x-ReRAM under Low Power and Conventional Operation Modes. *Sci. Rep.* **2016**, 1–8.
68. Long, S.; Lian, X. Cycle-to-cycle Intrinsic RESET Statistics in HfO_2-based Unipolar RRAM Devices. *IEEE Electron Device Lett.* **2013**, 623–625.
69. Chen, W.; Lu, W. Switching Characteristics of W/Zr/HfO_2/TiN ReRAM Devices for Multi-level Cell Non-volatile Memory Applications. *Semicond. Sci. Technol.* n.d., *30* (7), 075002.
70. Zhao, H.; Chen, H.-Y.; Wu, S. C. Multi-level Control of Conductive Nano-filament Evolution in HfO_2 ReRAM by Pulse-train Operations. *Nanoscale* **2014**, 5698–5702.
71. Huang, Y. J.; Shen, T. Low-power Resistive Random Access Memory by Confining the Formation of Conducting Filaments. *AIP Adv.* **2016**, *6*, 065022, 065022-1-065022-7.
72. Acharyya, D.; Hazra, A. A Journey towards Reliability Improvement of TiO_2-based Resistive Random Access Memory: A Review. *Microelectron. Reliab.* **2014**, *54* (3), 541–560.
73. Lu, W.; Chen, W. Self Current Limiting MgO ReRAM Devices for Low-power Non-volatile Memory Applications. *IEEE J. Emerg. Sel. Topics Circuits Syst.* **2016**, 163–170.
74. Tsigkourakos, M.; Bousoulas, P. Ultra-low Power Multilevel Switching with Enhanced Uniformity in Forming Free TiO_2–x-based RRAM with Embedded Pt Nanocrystals. *Phys. Status Solidi A* **2017**, *214* (12), 700570.
75. Bousoulas, P.; Tsoukalas, D. Understanding the Formation of Conducting Filaments in RRAM through the Design of Experiments. *Int. J. High Speed Electron. Syst.* **2016**, *25* (01n02), 640007.
76. Yang, L.; Kuegeler, C. The Influence of Copper Top Electrodes on the Resistive Switching Effect in TiO_2 Thin Films Studied by Conductive Atomic Force Microscopy. *Appl. Phys. Lett.* **2009.**
77. Chen, S. X.; Chang, S. Highly Stable ITO/Zn_2TiO_4/Pt Resistive Random Access Memory and Its Application in Two-bit-per-cell. *RSC Adv.* **2018,** 17622–17628.
78. Chen, S. X.; Chang, S. Highly Stable Ultrathin TiO_2-based Resistive Random Access Memory with Low Operation Voltage. *ECS J. Solid State Sci. Technol.* **2018.**
79. Maikap, J.; Jana, D. P. Self-compliance RRAM Characteristics Using a Novel W/TaOx/TiN Structure. *Nanoscale Res. Lett.* **2014**, *9* (1), 292.
80. Prakash, A.; Jana, D. TaO x-based Resistive Switching Memories: Prospective and Challenges. *Nanoscale Res. Lett.* **2013**, *8* (1), 418.

81. Prakash, A.; Jana, D. Resistance Controllability and Variability Improvement in a TaOx-based Resistive Memory for Multilevel Storage Application. *Appl. Phys. Lett.* **2015**, *106* (23), 233104.
82. Chung, Y. L.; Cheng, W. Joint Contributions of Ag Ions and Oxygen Vacancies to Conducting Filament Evolution of Ag/TaOx/Pt Memory Device. *J. Appl. Phys.* **2014**, *116* (16), 164502.
83. Prakash, A.; Park, J. Demonstration of Low Power 3-bit Multilevel Cell Characteristics in a TaO x-based RRAM by Stack Engineering. *IEEE Electron Device Lett.* **2014**, *36* (1), 32–34.
84. Ma, G.; Tang, X. Effects of Stress on Resistive Switching Property of the NiO RRAM Device. *Microelectron. Eng.* **2015**, 43–47.
85. Long, S.; Cagli, C. Reset Statistics of NiO-based Resistive Switching Memories. *IEEE Electron. Device Lett.* **2011**, 1570–1572.
86. Lee, S.; Chae, S. Scaling Behaviors of Reset Voltages and Currents in Unipolar Resistance Switching. *Appl. Phys. Lett.* **2008.**
87. Nardi, F.; Ielmini, D. Sub-10 μA Reset in NiO-based Resistive Switching Memory (RRAM) Cells. *In: 2010 IEEE International Memory Workshop. IEEE,* 2010.
88. Seo, J. W.; Baik, S. Characteristics of ZnO Thin Film for the Resistive Random Access Memory. *MRS Online Proc.* **2010.**
89. Han, Y.; Cho, K. Characteristics of Multilevel Bipolar Resistive Switching in Au/ZnO/ITO Devices on Glass. *Microelectron. Eng.* **2011**, 2608–2610.
90. Chang, W. Y.; Lai, Y. Unipolar Resistive Switching Characteristics of ZnO Thin Films for Nonvolatile Memory Applications. *Appl. Phys. Lett.* **2008.**
91. Zhang, S.; Long, S. Resistive Switching Characteristics of MnOx-based ReRAM. *J. Phys. D Appl. Phys.* **2009.**
92. Yang, M. K.; Park, J. Bipolar Resistive Switching Behavior in Ti/MnO_2/Pt Structure for Nonvolatile Memory Devices. *Appl. Phys. Lett.* **2009.**
93. Chiu, F. C.; Shih, W. Conduction Mechanism of Resistive Switching Films in MgO Memory Devices. *J. Appl. Phys.* **2012.**
94. Lin, C. Y.; Lee, D. Effect of Thermal Treatment on Resistive Switching Characteristics in Pt/Ti/Al_2O_3/Pt Devices. *Surf. Coat Technol.* **2008**, 628–631.
95. Banerjee, W.; Xu, X. Occurrence of Resistive Switching and Threshold Switching in Atomic Layer Deposited Ultrathin (2 nm) Aluminium Oxide Crossbar Resistive Random Access Memory. *IEEE Electron Device Lett.* **2015**, 333–335.
96. Wu, M. C.; Jang, W. A Study on Low-power Nanosecond Operation and Multilevel Bipolar Resistance Switching in Ti/ZrO_2/Pt Nonvolatile Memory with 1T1R Architecture. *Semicond. Sci. Technol.* **2012.**
97. Lin, C. Y.; Wu, C. Effect of Top Electrode Material on Resistive Switching Properties of ZrO_2 Film Memory Devices. *IEEE Electron Device Lett.* **2007**, 366–368.
98. Sun, B.; Liu, Y. Highly Uniform Resistive Switching Characteristics of TiN/ZrO_2/Pt Memory Devices. *J. Appl. Phys.* **2009.**
99. Wang, S. Y.; Lee, D. Effects of Ti Top Electrode Thickness on the Resistive Switching Behaviors of rf sputtered ZrO_2 Memory Films. *Appl. Phys. Lett.* **2009.**
100. Gao, S.; Song, C. Dynamic Processes of Resistive Switching in Metallic Filament-based Organic Memory Devices. *J. Phys. Chem. C* **2012**, 17955–17959.

101. Son, J. Y.; Shin, Y. NiO Resistive Random Access Memory Nanocapacitor Array on Graphene. *ACS Nano* **2010**, *4* (5), 2655–2658.
102. Tsai, C. L.; Xiong, F. Resistive Random Access Memory Enabled by Carbon Nanotube Crossbar Electrodes. *Acs Nano* **2013**, 5360–5366.
103. Tang, G.; Zeng, F. Resistive Switching Behaviour of a Tantalum Oxide Nanolayer Fabricated by Plasma Oxidation. *Phys. Status Solidi (RRL)–Rapid Res. Lett.* **2013**, *7* (4), 282–284.
104. Tang, G.; Zeng, F. Resistive Switching with Self-rectifying Behavior in Cu/SiOx/Si Structure Fabricated by Plasma-oxidation. *J. Appl. Phys.* **2013,** *113* (24), 244502.
105. Huang, Y. C.; Chou, C. High-performance Resistive Switching Characteristics of Programmable Metallization Cell with Oxidized Cu-Ti Electrodes. *Appl. Phys. Lett.* **2013**, *103* (14), 142905.
106. Goux, L.; Opsomer, K. Influence of the Cu-Te Composition and Microstructure on the Resistive Switching of Cu-Te/Al_2O_3/Si Cells. *Appl. Phys. Lett.* **2011**, *99* (5), 053502.
107. Wang, J. C.; Jian, D. Platinum–aluminum Alloy Electrode for Retention Improvement of Gadolinium Oxide Resistive Switching Memory. *Appl. Phys. A* **2013**, *113* (1), 37–40.
108. Lee, M. J.; Lee, D. A Plasma-treated Chalcogenide Switch Device for Stackable Scalable 3D Nanoscale Memory. *Nat. Commun.* **2013**, 1–8.
109. Cao, X.; Li, X. All-ZnO-based Transparent Resistance Random Access Memory Device Fully Fabricated at Room Temperature. *J. Phys. D Appl. Phys.* **2011**, *44* (25), 255104.
110. Zheng, K.; Sun, X. An Indium-free Transparent Resistive Switching Random Access Memory. *IEEE Electron Device Lett.* **2011**, *32* (6), 797–799.
111. Kim, H. D.; An, H. Unipolar Resistive Switching Phenomena in Fully Transparent SiN-based Memory Cells. *Semicond. Sci. Technol.* **2012.**
112. Hickmott, T. W. Low Frequency Negative Resistance in Thin Anodic Oxide Films. *J. Appl. Phys.* September **1962**, *33* (9), 2669–2682.
113. Gibbons, F.; Beadle, W. E. Switching Properties of Thin NiO Films. *Solid. State. Electron.* **1964**, *7* (11), 785–790.
114. Baek, G. et al. Highly Scalable Non-volatile Resistive Memory Using Simple Binary Oxide Driven by Asymmetric Unipolar Voltage Pulses. *IEDM Technical Digest. IEEE International Electron Devices Meeting*, 2004; pp. 587–590.
115. Villena, A.; Roldan, J. B.-M. SIM2 RRAM: A Physical Model for RRAM Devices Simulation. *J. Comput. Electron.* **2017**, *16* (4), 1095–1120.
116. Ye, C.; Wu, J-J.; Panb, C-H.; Tsai, T-M.; Chang; K-C., Wu; H., Deng, N.; Qian, H. Boosting the Performance of Resistive Switching Memory with a Transparent ITO Electrode Using Supercritical Fluid Nitridation. *RSC Adv.* **2017,** *7,* 11585–11590.
117. Sawa, T. F. Colossal Electro-resistance Memory Effect at Metal/La_2CuO_4 Interfaces. *Jpn. J. Appl. Phys.* **2005**, *44* (pL), L1241.
118. Chang, W. Z.; Chu, J. P. Resistive Switching Behavior of a Thin Amorphous Rare-earth Scandate: Effects of Oxygen Content. *Appl. Phys. Lett.* n.d., *101* (1).
119. Baek, S.; Lee, D. L.-H. Novel Digital Nonvolatile Memory Devices Based on Semi-conducting Polymer Thin Films. *Adv. Funct. Mater.* **2007**, *17* (15), 2637–2644.

120. Ha, H.; Kim, O. Bipolar Switching Characteristics of Nonvolatile Memory Devices Based on Poly(3,4-ethylenedioxythiophene):Poly(styrenesulfonate). *Appl. Phys. Lett.* **2008**, *93* (3), 033309.
121. Puglisi, M.; Larcher, L. 2D h-BN Based RRAM Devices. *2016 IEEE International Electron Devices Meeting (IEDM)*, 2016; pp. 34.8.1–4.
122. Hu, B. Electrically Controlled Electron Transfer and Resistance Switching in Reduced Graphene Oxide Noncovalently Functionalized with Thionine. *J. Mater. Chem.* **2012**, *22* (32), 16422–16430.
123. Xu, X.-Y.; Yin, Z.-Y. Resistive Switching Memories in MoS_2 Nanosphere Assemblies. *Appl. Phys. Lett.* **2014**, *104* (3), 033504.
124. Li, Y.; Long, S.; Liu, Q.; Lv, H.; Liu, M. Resistive Switching Performance Improvement via Modulating Nanoscale Conductive Filament, Involving the Application of Two-Dimensional Layered Materials, Small, Volume 13, Issue 35, Special Issue: Beyond Graphene—Booming Development of Novel 2D Nanomaterials, 2017, 1604306.
125. Hu, Q.; Park, M. Forming-free Resistive Switching Characteristics in Tantalum Oxide and Manganese Oxide Based Crossbar Array Structure. *Microelectron. Eng.* **2018.**
126. Wu, Q.; Banerjee, W. Improvement of Durability and Switching Speed by Incorporating Nanocrystals in the HfOx Based Resistive Random Access Memory Devices. *Appl. Phys. Lett.* **2018.**
127. Bousoulas, P.; Giannopoulos, I. Investigating the Origins of High Multilevel Resistive Switching in Forming Free Ti/TiO_2−x-based Memory Devices through Experiments and Simulations. *J. Appl. Phys.* **2017.**
128. Bousoulas, P.; Stathopoulis, S. Low-power and Highly Uniform 3-b Multilevel Switching in Forming free TiO_2−x-based RRAM with Embedded Pt Nanocrystals. *IEEE Electron Device Lett.* **2016**, *37* (7), 874–877.
129. Chand, U.; Huang, C. Suppression of Endurance Degradation by Utilizing Oxygen Plasma Treatment in HfO_2 Resistive Switching Memory. *Appl. Phys. Lett.* **2015.**
130. Philip Wong, H. S. Multi-level Control of Conductive Nano-filament Evolution in HfO_2 ReRAM by Pulse-train Operations. *Nanoscale* **2014**, *6* (11), 5698–5702.
131. Yu, S.; Wu, Y. Characterization of Switching Parameters and Multilevel Capability in HfOx/AlOx Bi-layer RRAM Devices. *Proceedings of 2011 International Symposium on VLSI Technology, Systems and Applications. IEEE,* 2011.
132. Yu, S.; Wu, Y. Investigating the Switching Dynamics and Multilevel Capability of Bipolar Metal Oxide Resistive Switching Memory. *Appl. Phys. Lett.* **2011**, *98* (10).
133. Chen, L.; Gou, H. Enhancement of Resistive Switching Characteristics in Al_2O_3-based RRAM with Embedded Ruthenium Nanocrystals. *IEEE Electron Device Lett.* **2011**, 794–796.
134. Li, Y.; Long, S. Resistive Switching Properties of Au/ZrO_2/Ag Structure for Low-voltage Nonvolatile Memory Applications. *IEEE Electron Device Lett.* **2009**, 117–119.
135. Kim, S.; Zhou, J. Crossbar RRAM Arrays: Selector Device Requirements during Write Operation. *IEEE Trans. Electron Devices* **2014**, 2820–2826.
136. Park, S. G.; Yang, M. A Non-linear ReRAM Cell with Sub-μA Ultralow Operating Current for High Density Vertical Resistive Memory (VRRAM). *2012 International Electron Devices Meeting. IEEE,* 2012.

137. Russo, U.; Kamalanathan, D. Study of Multilevel Programming in Programmable Metallization Cell (PMC) Memory. *IEEE Trans. Electron Devices* n.d., *56* (5), 1040–1047.
138. Chen, Z.; Huang, W. Ultrafast Multilevel Switching in Au/YIG/n-Si RRAM. *Adv. Electron Mater.* **2019,** *5* (2), 1800418.
139. Liu, J. C.; Hsu, C. Categorization of Multilevel-cell Storage-class Memory: An RRAM. *IEEE Trans. Electron Devices* **2015**, *62* (8), 2510–2516.
140. Kim, W.; Menzel, S. 3-Bit Multilevel Switching by Deep Reset Phenomenon in Pt/W/TaOx/Pt-ReRAM Devices. *IEEE Electron Device Lett.* **2016**, *37* (5), 564–567.
141. Balatti, S.; Ambrogio, S. Set Variability and Failure Induced by Complementary Switching in Bipolar RRAM. *IEEE Electron Device Lett.* **2013**, *34* (7), 861–863.
142. Sedghi, N.; Li, H. The Role of Nitrogen Doping in ALD Ta_2O_5 and Its Influence on Multilevel Cell Switching in RRAM. *Appl. Phys. Lett.* **2017,** *110* (10).
143. Li, H.; Gao, B. A Learnable Parallel Processing Architecture towards Unity of Memory and Computing. *Sci. Rep.* **2015.**
144. Yang, J. J.; Strukov, D. Memristive Devices for Computing. *Nat. Nanotechnol.* **2013**.
145. Gao, S.; Zeng, F. Implementation of Complete Boolean Logic Functions in Single Complementary Resistive Switch. *Sci. Rep.* **2015.**
146. Rosezin, R.; Linn, E. Crossbar Logic Using Bipolar and Complementary Resistive Switches. *IEEE Electron Device Lett.* **2011**, *32* (6), 710–712.
147. Prezioso, M.; Merrikh Bayat, F. Training and Operation of an Integrated Neuromorphic Network Based on Metal-oxide Memristors. *Nature* **2015**, 61–64.
148. Chen, Z.; Gao, B. Optimized Learning Scheme for Grayscale Image Recognition in a RRAM Based Analog Neuromorphic System. *IEEE International Electron Devices Meeting (IEDM), IEEE,* 2015; pp. 7–17.
149. Wu, H.; Wang, X. Resistive Random Access Memory for Future Information Processing System. *Proc. IEEE* **2017**, *105* (9).
150. Du, C.; Ma, W. Biorealistic Implementation of Synaptic Functions with Oxide Memristors through Internal Ionic Dynamics. *Adv. Funct. Mater.* **2015**, *25* (27), 4290–4299.
151. He, W.; Huang, K. Enabling an Integrated Rate-temporal Learning Scheme on Memristor. *Sci. Rep.* **2014.**
152. Chen, A. Comprehensive Assessment of RRAM-based PUF for Hardware Security Applications. *2015 IEEE International Electron Devices Meeting (IEDM). IEEE,* 2015.
153. Gao, Y.; Ranasinghe, D.-S. Memristive Crypto Primitive for Building Highly Secure Physical Unclonable Functions. *Sci. Rep.* **2015.**
154. Nakayama, M. ReRAM Technologies: Applications and Outlook. I. *2017 IEEE International Memory Workshop (IMW). IEEE,* 2017.
155. Liu, R.; Wu, H. Experimental Characterization of Physical Unclonable Function Based on 1 kb Resistive Random Access Memory Arrays. *IEEE Electron Device Lett.* **2015**, 1380–1383.
156. Chiu, P. F.; Chang, M. Low Store Energy, Low VDDmin, 8T2R Nonvolatile Latch and SRAM with Vertical-stacked Resistive Memory (memristor) Devices for Low Powermobile Applications. *IEEE J. Solid State Circ.* **2012**, *47* (6), 1483–1496.
157. Huang, L. Y.; Chang, M. ReRAM-based 4T2R Nonvolatile TCAM with 7x NVM-stress Reduction, and 4x Improvement in Speed-word length-capacity for Normally-off

Instant-on Filter-based Search Engines Used in Big-data Processing. *2014 Symposium on VLSI Circuits Digest of Technical Papers, IEEE,* 2014.

158. Lo, C. P.; Chen, W. A ReRAM-based Single-NVM Nonvolatile Flip-flop with Reduced Stress-time and Write-power against Wide Distribution in Write-time by Using Self-write-termination Scheme for Nonvolatile Processors in IoT era. *2016 IEEE International Electron Devices Meeting (IEDM). IEEE*, 2016.
159. International Technology Roadmap for Semiconductors, 2011 Edition, August. Retrieved from International Technology Roadmap for Semiconductors, 2011 Edition, http://www.itrs.net/Links/2011ITRS/Home2011.html
160. Pan, C. et al. Coexistence of Grain-boundaries-assisted Bipolar and Threshold Resistive Switching in Multilayer Hexagonal Boron Nitride. *Adv. Funct. Mater.* **2017**, *27* (10).
161. Shi, Y. et al. Coexistence of Volatile and Non-volatile Resistive Switching in 2D h-BN Based Electronic Synapses. *2017 IEEE International Electron Devices Meeting (IEDM),* 2017.
162. Schindler, C.; Meier, M.; Waser, R.; Kozicki, M. N. Resistive Switching in Ag-Ge-Se with Extremely Low Write Currents. Proc. IEEE Non-Volatile Memory Technology Symp, 2007; pp. 82–85.
163. 163. Kahng, S. M. A Floating Gate and Its Application to Memory Devices. *Bell. Syst. Tech.* May 16, **1967**, 1288–1295.
164. 164. Jameson, J. R.; Blanchard, P.; Cheng, C.; Dinh, J.; Gallo, A.; Gopalakrishnan, V. et al. Conductive-bridge Memory (CBRAM) with Excellent High-temperature Retention. *IEDM*, 2013; pp. 30.1.1–30.1.4.
165. Sills, S.; Yasuda, S. A Copper ReRAM Cell for Storage Class Memory Applications. *Symposium on VLSI Technology Digest of Technical Papers*, 2014; pp. 1–2.
166. Liu, T. Y. et al. A 130.7 mm2 2-Layer 32Gb ReRAM Memory Device in 24nm Technology. *2013 IEEE International Solid-State Circuits Conference*, 2013; pp. 210–212.

CHAPTER 3

Analog and RF Performance Analysis and Its Sensitivity to Critical Geometrical Parameters in Junctionless Accumulation Mode Bulk FinFET

KALYAN BISWAS[1*] and ANGSUMAN SARKAR[2]

[1]*ECE Department, MCKV Institute of Engineering, Howrah, West Bengal, India*

[2]*ECE Department, Kalyani Government Engineering College, Kalyani, Nadia, West Bengal, India*

[*]*Corresponding author. E-mail: bkalyan.ece@gmail.com*

ABSTRACT

Junctionless MOSFET is a device where same type of doping is used for source to drain. FinFETs are considered as the most popular structure in the area of 22-nm technology and beyond due to the lower Short Channel Effects (SCEs) and superior gate control, compared to standard MOSFETs. Analog/RF performance of a junctionless accumulation-mode (JAM) bulk FinFET and its performance in digital circuits for variation of different geometrical parameters has been investigated through intensive TCAD simulations and presented in this chapter. On current (I_{ON}), off-current(I_{OFF}), I_{ON}/I_{OFF}, transconductance (g_m), transconductance generation factor (g_m/I_d), gate capacitances, cut-off frequency (f_T), maximum oscillation frequency (f_{max}) etc. are simulated and analyzed for variation of different structural parameters such as materials for spacer, length of spacer, fin width, fin shape etc. These results will be supportive to design junctionless FinFET with better performance for its specific applications.

3.1 INTRODUCTION

Metal-oxide-semiconductor field-effect transistor (MOSFET) using the theory of field-effect conductivity modulation is one of the most significant inventions of the 20th century, which was first proposed by J. E. Lilienfeld in 1928.[1] Determined by the Moore's Law, the size of a MOSFET is sinking by a factor of two in every 18 months, without changing the other basic structures. This basic structure of MOSFETs continued its successful geometric scaling without any specific changes, following the ideal scaling principles "constant-field scaling." However, as the supply voltage (V_{dd}) reaches 1V, the "constant-field scaling" deviates due to difficulty in reducing threshold voltage (V_{Th}). Further reduction in threshold voltage causes increased leakage current, which in turn prevents supply-voltage scaling. Also, high electric field inside the transistor decreases device reliability. Moreover, high channel doping creates problems such as increased "band-to-band-tunnelling" (BTBT), degradation of carrier mobility, "gate-induced-drain-leakage (GIDL)," and fluctuations in threshold voltage owing to random dopant variation. These are known as "short channel effects" (SCEs) and these effects increase with decrease in device dimensions. SCEs affect the performance of the designed circuit by using those MOSFETs having lower dimensions. High-frequency performance and consumption of power by the circuits also largely depends on SCEs. At the level of nano-dimension, the impact of SCEs on common MOSFET characteristics should be noted.[2] Researchers are trying to solve the aforementioned problems mainly in two ways: (i) structural variations such as structures using multiple gates for greater drive current keeping the over-drive voltage constant and (ii) use of high-mobility channel materials to improve carrier mobility.

To avoid SCEs, several device structures, that is, silicon-on-insulator (SOI) MOSFETs, double-gates MOSFETs, junction-less MOSFET, and multigate MOSFET, etc. are proposed.[3] In comparison to the design of a transistor using single gate, the use of three gates around the Fin ensures tremendous electrostatic control on the channel.[3] From the range of multi-gate MOSFETs, FinFETs are considered as the utmost popular structure in the area of 22-nm technology and above due to the lower short-channel effects and superior gate control compared to standard MOSFETs. These devices also provide the drain current ON/OFF rating and indicate the higher current drive compared to devices with planar structures. Because of these benefits, the industry has embraced FinFET technology as a mature technology.[4] Junctionless MOSFET is another device with the same doping type for the semiconductor all the way starting from the source to the drain, which behaves as a resistor. Junctionless FETs have tougher protection against short-channel effects in comparison to FETs working in inversion mode. It is evident that there are many advantages of JL MOSFETs in comparison to standard devices that includes better protection from short-channel effects, greater scalability, improved DIBL, and so on.[5, 6] The high reliability of a junctionless FET compared to inversion-mode FETs is shown by researchers. Unlike standard short-channel MOSFETs, formation of junctions is not required in JL MOSFETs thus reducing number of process steps. In addition, JL MOSFET can overcome many of the challenges of implementation such as controling doping level and also thermal budget. Acculation mode (JAM) FinFET-based FinFETs are developed and deployed.[7, 8] The junctionless field-effect transistor was considered by lots of investigators[8–12] as a suitable replacement for "highly scaled" devices. Recently, the p-channel junctionless accumulation-mode (JAM) MOSFET has been used successfully and has been reported.[13] Ultra low-power (ULP) junctionless MOSFET performance in achieving advanced analog/RF metrics is also reported.[11] The literature reveals that many of junctionless FinFETs show a good ON/OFF current ratio and better characteristics of short-channels effects.[14] It is also shown that spacer usage can improve device performance. Recently, the research reports on the "junctionless-accumulation-bulk FinFETs" revealed that for devices using high-κ gate spacers, efficiency of the device such as better DIBL, SS, and I_{ON}/I_{OFF} ratio are obtained.[8] For the application in wireless communication domain, there is a great need of high-power amplifier circuits, which should be capable of handling high power and high frequency. So, it is very much essential to evaluate the performance of the evolving new CMOS devices for its applicability in high-frequency analog/RF domain.[15–21]

Analog/RF performance investigation of a junctionless accumulation-mode (JAM) bulk FinFET and its performance in digital circuits for different geometrical parameters variation are the main objectives of this chapter. Researchers have shown that FinFET performance depends essentially on device parameters and Fin structural parameters.[22–24] Since FinFET performance depends mostly on the geometry of the Fin, it is anticipated that the Fin cross-section must influence its RF/analog performance. Importance is given to reviewing the performance for application in analog/RF circuit by computing transconductance (g_m), transconductance generation factor (g_m/I_d), gate capacitances, cut-off frequency (f_T), maximum oscillation frequency (f_{max}), etc. and its digital performance. Therefore, an investigation of the variability of device performance by device parameters such as material for spacer, length of spacer, Fin width, cross-section of fin, etc, for its use in analog/RF circuits have been carried out in this work.

3.2 DEVICE STRUCTURE UNDER STUDY

3.2.1 JUNCTIONLESS FINFET

In a junctionless transistor, source region, channel region, and drain region have the same doping type as shown in Figure 3.1. Normally, junctionless transistor requires lower doping level in source, drain, or channel region but that may increase the parasitic resistances. To diminish the parasitic resistance for the region of source and drain extension, junctionless MOSFET with extra source/drain-doping concentrations is considered by the researchers. This device is called junctionless accumulation mode MOSFET. The junctionless accumulation mode (JAM) bulk FinFET using spacers on gate is considered here for operation with n-channel. Simulation model is calibrated with the published results by varying the model parameters. Once the matching is achieved, the same parameters for the model are applied for other investigations. By using extensive 3D TCAD simulations of n-channel junctionless transistors, it is demonstrated that the high-k gate spacers enhance the electrostatic integrity of MOSFETs with sub-20 nm gate lengths. To reduce the SCEs, gate spaces are employed on both sides of the gate. Various materials with different dielectric constants (k) value for spacer are considered to validate the effect of spacer materials on performance of the FinFET. Application of high-k gate spacers improve the digital and analog functionality of the device. However, employing high-k spacer material, this device shows a smaller f_T and f_{max}. The length of the device spacer varies to determine its effect. It is also witnessed that the increment in space length improves the SS and DIBL and also increases V_{th} at the same time. The results obtained from ssimulation also recommend that the RF/Aanalog performance with a high-k-dielectric spacer can be upgraded by decreasing the length of the spacer.[25]

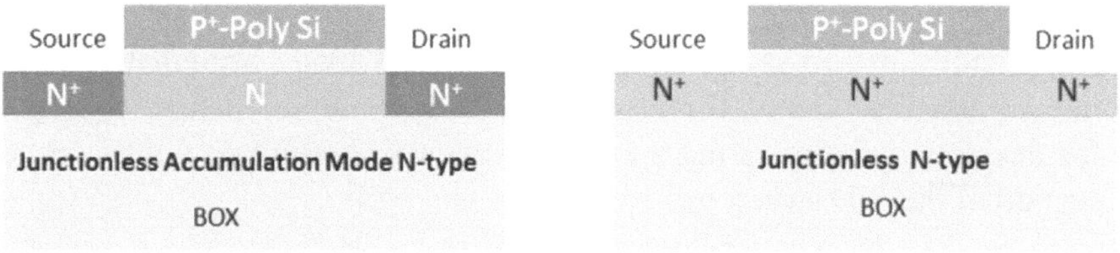

FIGURE 3.1 A schematic digram of SOI FinFET and Bulk FinFET structures.

3.2.2 BULK FINFET

FinFETs can be classified as SOI FinFET and Bulk FinFET. Initially, FinFETs were fabricated on SOI wafers and were familiar as SOI MOSFETs. Structure of SOI MOSFETs helped in overcoming the short-channel effects-related problems. SOI FinFETs are also easy to fabricate and possess good scalability. Thesse devices also have advantages of less parasitic capacitances. But these devices suffer from "floating body" problems and heat dissipation issues because of thick "buried oxide." SOI wafers are also expensive in comparison to silicon wafers because of additional oxide layer. To overcome these challenges, bulk FinFETs are proposed, that is, body is directly fabricated on the Si wafers. These devices have very good heat dissipation path through the bulk Si wafers.[26] Fabrication steps of bulk FinFET are also compatible with those of planner 2D devices. They also provide almost the same scalability as SOI FinFET. A schematic cross-sectional view of SOI FinFET and bulk FinFET are shown in Figure 3.2.

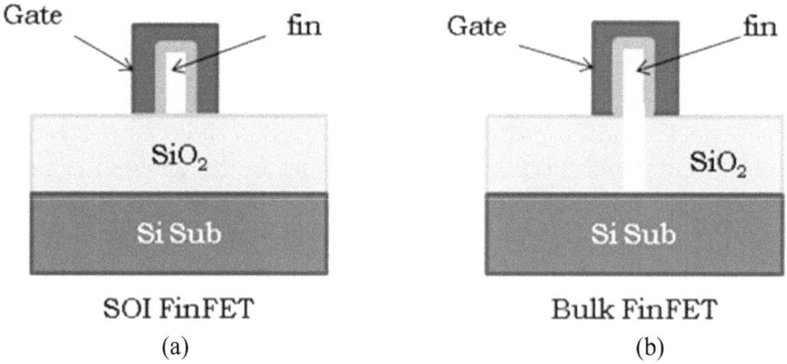

FIGURE 3.2 A schematic view of Fin cross-section for SOI FinFET and Bulk FinFET.

Figure 3.3a displays the structure (3-D) of the FinFET device under study and 3.3b shows the view of vertical cut plane hrough the center line of the Fin. Generally, researchers adopt simplified methodologies for device simulation using a two-dimensional model by taking a cross-section of the 3-D structure.

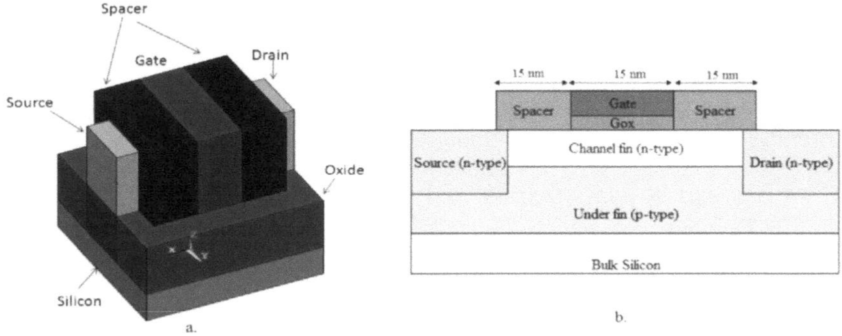

FIGURE 3.3 Structure of the junctionless accumulation-mode-bulk FinFET. (a) 3-D view. (b) Vertical cut-plane view.

Source: Reprinted with permission from Ref. [25]. © The Institution of Engineering and Technology

3.2.3 SIMULATION MODEL

To simulate the JAM bulk FinFET under study, TCAD simulation tool for device simulation is used. As the device has a 3-D structure, 2-D analysis may provide inaccurate simulation results[27] as a 2-D model may not account for all the interactions coming from those portions of the model outside the considered plane. In a tri-gate FinFET device, there are two vertical gates on both sides of the "Fin"-shaped channel and a horizontal gate situated on top. Moreover, it is important to consider 3-D electric field for accurate capacitance evaluation of the nanoscale FinFET devices.[27] The n-type junctionless transistor studied here has higher doping of the source and drain (10^{21}/cm^3) and moderate n-type doping in the channel region (10^{18}/cm^3). The doping concentrations and types of different regions are listed in Table 3.1. The higher doping in source and drain regions are required to avoid parasite resistance in the source region and drain region. A metal having a work function value equal to 4.65 eV is considered as gate. Generally, polysilicon is the preferred material for the gate in MOSFETs as its work-function (WF) can be changed to obtain different flat-band voltage, which is responsible for a threshold voltage (V_{th}) variation. But due to its requirement of high thermal budget and degration because of doped polysilicon[28] depletion, it is not as popular as a material for thegate in nano-scale MOSFETS. In this respect, metal gate is more popular these days than polysilicon.[29, 30] This goal may be achieved by Molybdenum having variable work function in the range of 4.5– 4.9 eV through the application of nitrogen in molybdenum following a hot anneal,[29] the gate work-function value equal to 4.65 eV has been considered because it is compatible with molybdenum. To reduce the current leakage of the gate, HfO_2 gate dielecters with "equivalent oxide thickness" (EOT) of 1.25 nm are used. The device has a gate length (L_g) of 20 nm, a height of Fin (H_{fin}) 20 nm, and a Fin width (W_{fin}) of 20 nm. Among other parameters of the device, the spacer length is considered to be 15 nm having a source/drain to spacer overlap region of length of 3 nm. A 60 nm under Fin thickness is used and a field oxide size of 35 nm is taken. Because of this design, the channel is very well controlled by the gate as a large part of the channel region is below the spacer. To estimate the effect of the length of the spacer, the lengths of both spaces varied from 10 nm to 20 nm.

In the simulation model, the carrier movement is modeled employing the field-dependent mobility model (FLDMOB). Also, the "Shockley-Read-Hall (SRH)" and "Auger" recombination model are included in calculating the active carrier lifetime with an accurate measurement of performance parameters. Foriterative calculations, Newton, and Gummel's numerical methods were used and a temperature of 300 K is considered. Some of the device parameters applied in model for this simulation are put in Table 3.1.[8] A few key device-related parameters used for the modeling of our device are set according to ITRS 2013 guidelines.[31]

TABLE 3.1 Different Device Parameters Used.

Parameters	**Value**
Channel-Fin doping	n-type, 1×10^{18} cm^{-3}
Source/Drain doping	n-type, 1×10^{21} cm^{-3}
Under-Fin doping	p-type, 5×10^{17} cm^{-3}
Fin width	15 nm
Gate length	20 nm
Gate oxide material	HfO_2
Drain supply voltage	1 V
Gate work function	4.65 eV

3.3 RESULT AND ANALYSIS

3.3.1 EFFECT OF SPACER

It is well established that the electrostatic potential for lightly doped channel situated under the underlap portion is mostly delicate to the fringing field generated at the gate region and passes through the spacer as it is not straightforwardly underneath the gate electric field. An exploration of act of the device with spacer and without using spacer has been delivered to establish the importance of using spacer. The performance of the MOSFET with spacer and without using spacer is figured and charted in the Table 3.2.

All output values are reported in Table 3.2 with V_{ds} = 1 V, gate length 20 nm, and Fin width 20 nm. A spacer length equal to 15 nm and k-value 22 were considered to understand the impact of the spacer created on the device performance. From Table 3.2, it is evident that the key operating parameters considered for the device lime the ON-state I_{ON}, OFF-state current, I_{OFF}, Subthreshold Slope, SS, DIBL, and transconductance (g_m) have improved considerably using the spacer compared to the device without using gate spacer. It is fair to say that the obtained results are consistent with the results previously reported by Pradhan et al.[32]

TABLE 3.2 Performance Parameters of the Device Using Spacer and Without Using Spacer.

Gate length 20 nm	I_{ON} current (mA)	I_{OFF} current (nA)	SS (mV/dec)	DIBL (mV/V)	Transconductance (mmoh)
No spacer used	0.0858	4.5	85	48	0.136
With high-k spacer	0.198	1.2	78	43	0.371

All output values are reported in Table 3.2 with V_{ds} = 1 V, gate length 20 nm, and Fin width 20 nm. A spacer length equal to 15 nm and k-value 22 were used to understand the influence of the spacer created on the device performance. It is evident from Table 3.2 that the key operating parameters considered for the device lime the ON-state I_{ON}, OFF-state current, I_{OFF}, subthreshold slope, SS, DIBL, and transconductance (g_m) have improved considerably using the spacer compared to the device without using gate spacer. It is fair to say that the obtained results are consistent with the results previously reported by Pradhan et al.[32]

3.3.2 EFFECT OF SPACER LENGTH

In case of the FinFET device design, high-k spacer posseses a major role in charge transport dynamics. Therefore, spacer length defines the performance of the device. Since the current drive and fringe capacitance play a vital role in predicting the performance of performance in circuit, the impact of spacer on the maximum drain current (I_d) is assessed. For the simulations here, spacer length has been varied from 10 nm to 20 nm and a HfO_2 ($k = 22$) spacer material is employed.

Figure 3.4 shows the variation of I_{on} and I_{on}/I_{off} for three separate spacer lengths 10 nm, 15 nm, and 20 nm. From the Figure 3.4, it is detected that maximum drain current, I_{on}, of the device rises with reduction of length of spacer. This outcome includes a decrease in device resistance due to a decrease in space length. With the increase in the length of the spacer, the I_{On}/I_{Off} ratio increases in the device under investigation. As the length of the spacer increases, the ON-current of the device decreases due to increased resistance to the channel, but OFF-current also decreases. Overall, the I_{On}/I_{Off} ratio of drain current for the device rises with increase in spacer length.

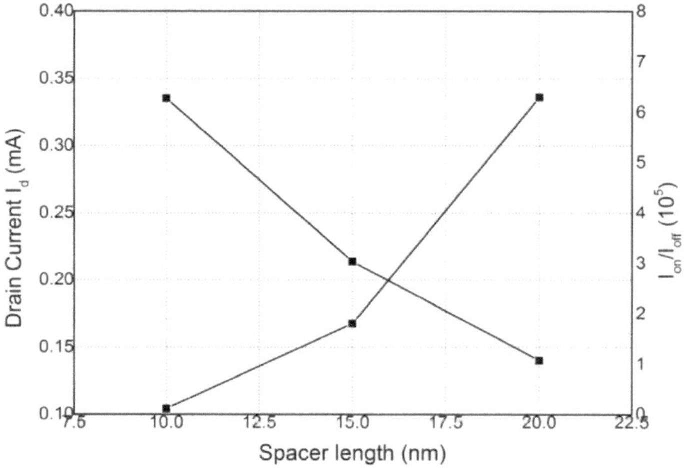

FIGURE 3.4 Variation of I_{on} and I_{on}/I_{off} for different spacer length.

Variation of transconductance (g_m) and TGF (g_m/I_{ds}) with gate overdrive voltage (V_{gt}) are shown in Figure 3.5. From Figure 3.5, it's obviously agreed that device transconductance surge for lower spacer length. An increase within the spacer length consequently helps to boost effective length of the channel. As an end result, g_m declines with the growing length of the spacer. The parameter g_m/I_{ds} stands for the obtainable gain as per-unit value of power dissipation. g_m/I_{ds} is in upper limit in the weak inversion region and it drops with the growing drain current.[33, 34] Moreover, there is a fall in the spacer length origin's drop in g_m/I_{ds} due to rise in short-channel effects (SCEs). It happens because the reduced spacer length reduces the effective channel length, which causes enhancement in the detrimental SCEs. In addition, an increase in the spacer length also reduces the I_{ds}.

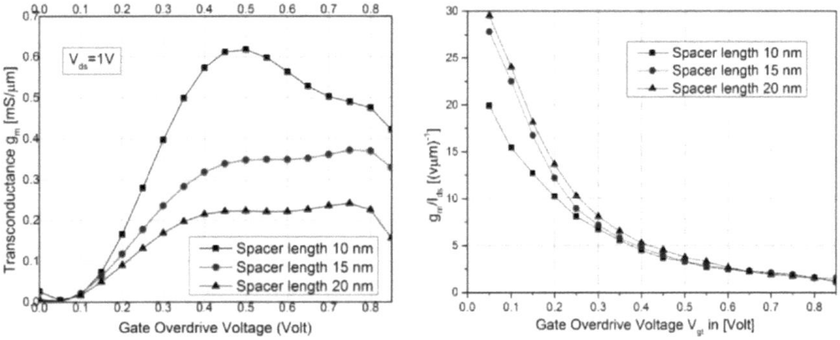

(a) (b)

FIGURE 3.5 (a) Variation of g_m with V_{gt} for different spacer length at V_{ds} = 1.0 V. (b) Variation of g_m/I_{ds} with V_{gt} for different spacer length at V_{ds} = 1.0 V.

Source: Reprinted with permission from Ref. [25]. © The Institution of Engineering and Technology

The performance of the device under study in RF circuits was tested by estimating f_T (cut-off frequency) and f_{max} (maximum frequency of oscillation), which are important factors in the suitability of RF circuits. The f_T and f_{max} are considered for various width of Fin top. The f_T is taken out as the frequency for which the short-circuit current gain is unity. f_{max} is the very important characteristic of the device, which consists of source resistance, gate resistance, and drain resistance. The ability of the device to deliver power gains at high frequencies is guaranteed by f_{max}. It usually signifies the frequency at which power gain is unity. The simple expressions for f_T and f_{max} are reported in Refs. 35 and 36 and used for this calculation.

$$f_T = \frac{g_m}{2\pi(C_{gd} + C_{gs})} \tag{3.1}$$

$$f_{\max} = \frac{g_m}{2\pi C_{gs}\sqrt{4*(R_s + R_i + R_g)(g_{ds} + g_m \frac{C_{gd}}{C_{gs}})}} \tag{3.2}$$

where C_{gs}, C_{gd}, R_s, R_g, and R_i are the the gate-to-source capacitance, the gate-to-drain capacitance, the source resistance, the gate resistance, and the intrinsic-channel resistances, respectively.

Figure 3.6 shows the deviation of f_T and f_{max} with variation in gate overdrive voltage (V_{gt}) for various spacer length. With the reduction in the length of the spacer, f_T rises following the inclination of transconductance (g_m), as shown in Figure 3.6a. From the application consideration in analog circuits, and to get higher f_{max}, not only a better g_m is wanted, but also elevated g_m/C_{gs} and C_{gs}/C_{gd} ratios are also desired. Obtained results shows that with the reduction in length of spacer, g_m increases, which it turn surges the f_{max} as a collective effect. Consequently, f_T and f_{max} of the device tend to surge for smaller spacer length signifying that it is possible to enhance RF performances by scaling down the length of spacer. Figure 3.6b plots the variation of f_{max} with gate-overdrive voltage (V_{gt}) for different spacer length.

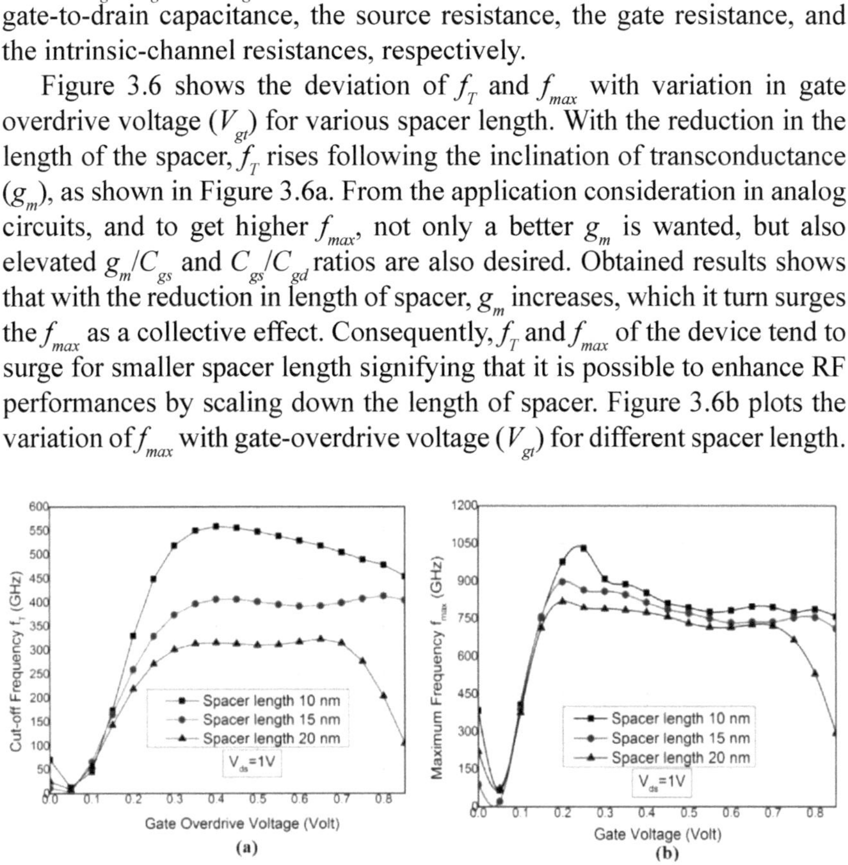

FIGURE 3.6 (a) Variation of f_T with V_{gt} for different spacer length. (b) Variation of f_{max} with V_{gt} for different spacer length.

Source: Reprinted with permission from Ref. [25]. © The Institution of Engineering and Technology

3.3.3 EFFECT OF FIN WIDTH

The dependency of significant performance on geometrical dimensions of the device like gate length, Fin width, and length of spacer have been summarized. Figure 3.7a–c plots the change of SS, DIBL, and threshold voltage (V_{th}) with gate length (15–25 nm) for different spacer lengths (10–20 nm). Figure 3.8a–c plots the deviation of SS, DIBL, and threshold voltage (V_{th}) on Fin width (15–25 nm) for different spacer dimensions (10–20 nm).

For a fixed length of spacer, the SS and DIBL of the FinFET decline with the rise in length of the gate as shown in Figure 3.7a and b. However, Figure 3.7c reveals that the threshold voltage (V_{th}) of the device rises with increase in gate length. Moreover, it is understood that the surge in spacer length is linked to the enhancement in performance in terms of SS and DIBL and increase in V_{th} (lower V_{th} roll-off). The findings are very much in agreement with the results reported earlier.[37, 38]

In Figure 3.8a–c, SS, DIBL, and V_{th} are plotted for different fin widths and spacer lengths. For a fixed gate length, if Fin width increases, the increment in SS and DIBL and a reduction in threshold voltage are noticed. These findings may be largely attributed to the reduced gate control because of increased Fin width. It is also understood that by increasing the spacer length, the various digital performance-related parameters such as SS, DIBL, and V_{th} roll-off may be improved. These findings match well with that earlier reported in Refs 39 and 40.

3.3.4 EFFECT OF ASPECT RATIO OF FIN

Effect of the aspect ratio for the fin structure of junctionless FinFET is verified and discussed in this section. For the determination of the channel resistance and ON state current, effective channel cross-sectional area is very significant; therefore, only the variation in aspect ratio of Fin is considered while the channel cross-sectional area is kept fixed so that the effect of Fin aspect ratio can be found out. In literatures, many investigators have fixed aspect ratio as W_{fin}/H_{fin}. However, the ratio defined as height/width (H_{fin}/W_{fin}) is considered appropriate and is used here. Four different combinations of Fin width and Fin height defining separate Fin aspect ratios of the Fin structure are used here. The height and width of the Fin are considered in such a way that the the total Fin cross-sectional area remains unchanged. Fin height and width for the channel are listed in Table 3.3. It is appropriate to use a rectangular cross-section of the channel region to estimate the effect of Fin aspect ratio.

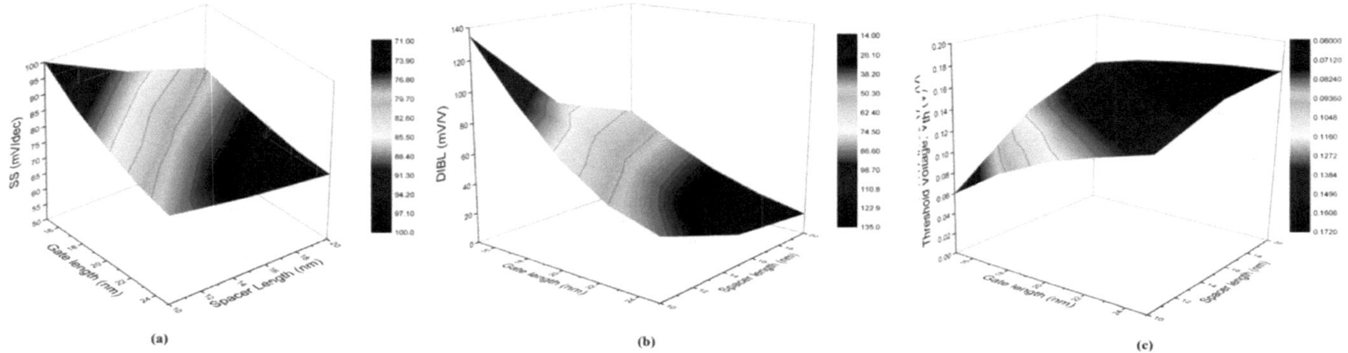

FIGURE 3.7 Performance parameters of JAM bulk FinFET (a). Variation of SS with gate length for different spacer lengths. (b). Variation of DIBL with gate length for different spacer lengths. (c). Variation of threshold voltage, V_{th} with gate length for different spacer lengths.

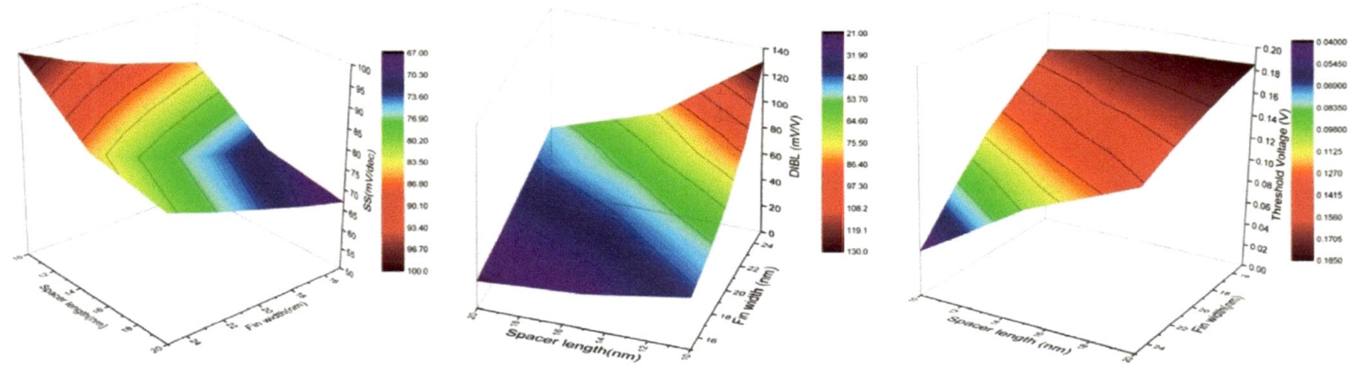

(a) (b) (c)

FIGURE 3.8 Short-channel performance of JAM bulk FinFET devices. (a) Variation of SS with Fin width for different spacer lengths. (b) Variation of DIBL with Fin width for different spacer lengths. (c) Variation of threshold voltage, V_{th} with Fin width for different spacer lengths.

TABLE 3.3 Different Designs Considered.

Design no.	Fin width (W_{fin}) in nm	Fin height (H_{fin}) in nm	Aspect ratio (Fin height/ Fin width) = H_{fin}/W_{fin}
C1	20	20	1
C2	16	25	1.56
C3	10	40	4
C4	8	50	6.25

The Fin width (W_{fin}) and Fin height (H_{fin}) determine the electrostatic integrity for FinFET. Therefore, these are considered as the most noteworthy parameters for FinFET scaling. The aattained drain current for V_{ds} = 0.9 V for the simulated model for several aspect ratio (AR) of the Fin is shown in Table 3.4. It is understood that the shape of the device having a greater aspect ratio for Fins keeps a higher ON-state current (I_{ON}) as the structure with fin aspect ratio 6.25 delivers 1.2 times greater I_{ON} than the device with aspect ratio 1.

TABLE 3.4 Variation of I_{On} with I_{on}/I_{off} for Different Fin Aspect Ratio.

Fin aspect ratio (AR)	I_{on} (mA)	I_{On}/I_{Off}
1	1.77×10^{-04}	1.71
1.56	1.80×10^{-04}	3.25
4	2.00×10^{-04}	6.13
6.25	2.12×10^{-04}	6.28

Table 3.4 also displays the I_{ON}/I_{OFF} current ratio for various Fin aspect ratio. I_{OFF} corresponds to V_{gs} of 0 V. I_{ON}/I_{OFF} is considered as one of the most significant parameters to be considered for use in digital circuit as a switch. Figure 3.9a plots the change of transconductance ($g_m = \delta I_{ds}/\delta V_{gs}$) with V_{gs} for four separate aspect ratios considered for the Fin structure. From Figure 3.9a, it is clear that g_m is increased for a device that has a Fin structure with higher aspect ratio. Higher g_m/I_{ds} of the device is wanted to obtain greater parameters for analog circuit applications. The plot of g_m/I_{ds} with vate voltage for different Fin aspect ratios is shown in Figure 3.9b. Higher g_m/I_{ds} is found for structure with higher aspect ratio of the Fin.

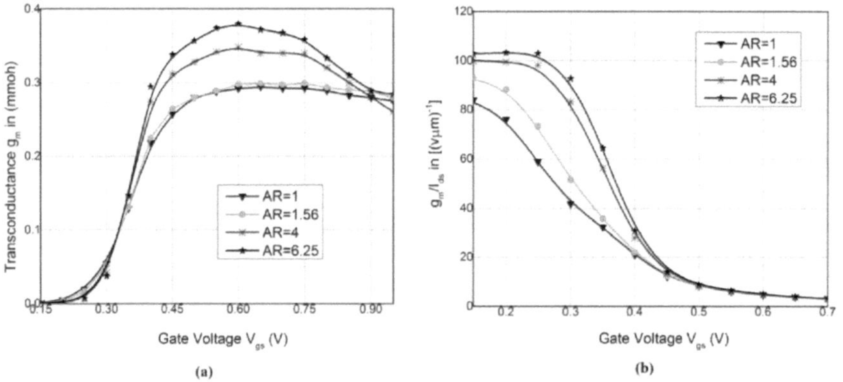

FIGURE 3.9 (a) Variation of g_m with of V_{gs} for different Fin aspect ratio at $V_{ds} = 0.9$ V. (b) Variation of g_m/I_{ds} for a range of V_{gs} for different Fin aspect ratio at $V_{ds} = 0.9$ V.

For applications in RF circuit, the f_T and f_{max} are the key performance factors of the FinFET. Both the factors may be evaluated by employing simple analysis of S-parameters as reported in literature. It is detected that for smaller value of AR, the total gate capacitance (C_{gg}) becomes low, which, in sequence, improves the f_T. The higher value of C_{gg} with a higher aspect ratio not only develops channel controllability, but it also affects mostly the speed of the MOSFET. Figure 3.10 illustrates the highest value of f_T for value of Fin aspect ratio considered. It may be assumed from Figure 3.10 that the value of f_T decreases as aspect ratio of device Fin becomes higher. Maximum frequency of oscillation f_{max} is a more beneficial image of suitability for analog circuit applications. Figure 3.10 shows a decrease in the f_{max} with an increase of Fin aspect ratio.

3.3.5 EFFECT OF FIN SHAPE

In previous section, the effect of Fin aspect ratio on device paerformance is evaluated. Effect of Fin cross-sectional shapes is considered in this section. To evaluate the influence of Fin shapes, the width of fin top is varied from 15 nm (rectangular shape) to 3 nm while the Fin bottom width is kept constant (15 nm). Different analog/RF performance parameters were evaluated for the non-rectangular channel structure.

V_{gs}-I_{ds} curve for the FinFET device considered here is presented in Figure 3.11 for various shapes of the Fin. It is found that I_{ON} decreases when Fin top-width decrease, because of reduction in effective Fin width (W_{eff}). The effective Fin width (W_{eff}) is defined as boundary of the Fin cross-section neighboring to the high-k gate oxide (HfO_2), estimated using the simple Pythagorean theorem[22] as per equation (3.3).

$$W_{eff} = 2\sqrt{\left(\frac{W_{bottom} - W_{top}}{2}\right)^2 + \left(H - \frac{W_{top}}{2}\right)^2} + \pi\frac{W_{top}}{2} \tag{3.3}$$

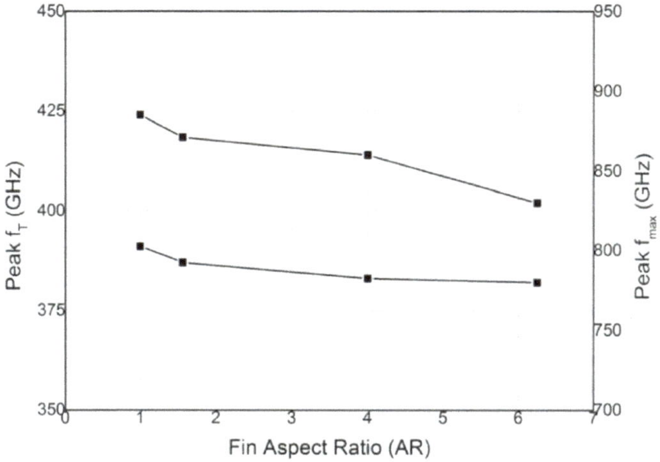

FIGURE 3.10 Fluctuations of f_T and f_{max} for different designs.

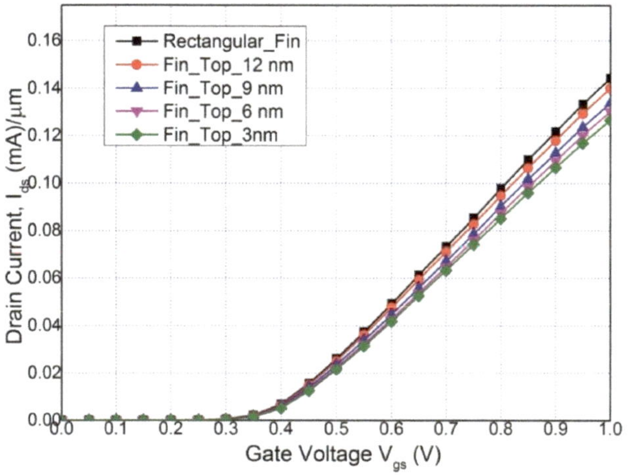

FIGURE 3.11 Variation of I_{ds} with V_{gs} for various Fin top width.

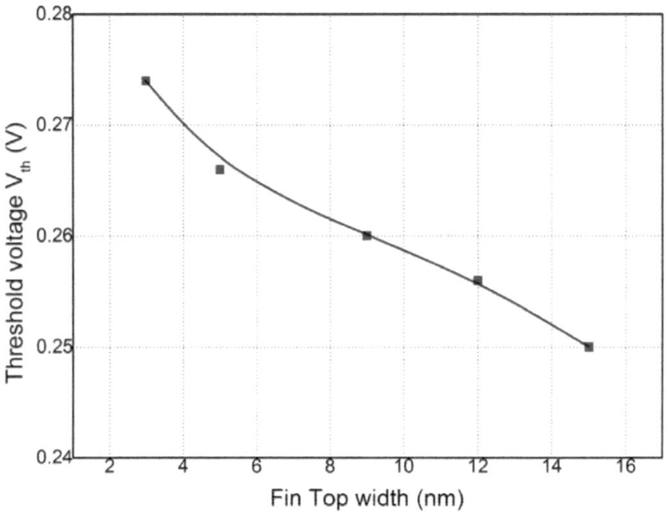

FIGURE 3.12 Change in threshold voltage, V_{th} with Fin top width.

Source: Reprinted with permission from Ref. [24]. © 2018 Springer.

In any FinFET device, the decrease of the width of Fin top actually makes the effective Fin width less and rises the series resistance of the channel, which agrees with results previously published by Rezali et al.[23] For Fin top width with smaller value, higher series resistance is expected. This is the reason for lower I_{ds} when Fin top width reduces.

Threshold voltage, V_{th}, of the device was estimated using the method of constant current ($I_{ds} = 10^{-7} \times$W/L A) using the plotting of I_{ds} vs. V_{gs} curves. Here, W denotes the effective width of fin and L denotes the length of channel. Significant deviation for threshold voltage (V_{th}) obtained for variation of Fin top width is shown in Figure 3.12. Results clearly display that V_{th} becomes higher for smaller Fin top width. For FinFET with reduced Fin width, V_{th} rises because of the narrow Fin width effect. V_{th} becomes higher for triangular FinFET when compared to rectangular Fin because of charge density increase across the Fin for narrower fin width. The results signifies that multi-threshold FinFETs may be fabricated on a sibgle chip by controlling the fin shapes. However, many fabrication issues may be there to be solved.

ON-current, I_{on}, OFF-current I_{off}, and I_{on}/I_{Off} of the device are tabulated in Table 3.5. Results reveals that when Fin top width becomes smaller, ON current (I_{ON}) and also leakage current (I_{OFF}) of the device decreases. Improvement in terms of lower OFF current (I_{OFF}) are attained with the lower I_{ON}, which is caused by higher parasitic series resistance. From results, it is agreed that ON-current (I_{ON}) is worsened up to 12% (for Fin top = 3 nm) as the shape of the Fin modifies to trapezoidal form from rectangular shape. Also, for reduced Fin top width, decline in leakage current is obtained for the devices. The usage of tapered-shaped Fin structure permits stronger influence of the of the gate in the middle portion of the Fin, which in turn results in I_{OFF} reduction. Almost 7% reduction is seen in OFF-current by sinking the Fin top width from 15 nm to 3 nm.

TABLE 3.5 I_{ON}, I_{OFF} and Ratio of ON Current and OFF Current for Different Fin Top Width.

Fin top width	On current, I_{On} (A)	Off current, I_{off}(A)	On current/off current ratio I_{On}/I_{Off}(×10^{6})
15	1.44×10^{-04}	6.73×10^{-11}	2.15
12	1.39×10^{-04}	6.61×10^{-11}	2.12
9	1.34×10^{-04}	6.42×10^{-11}	2.09
5	1.30×10^{-04}	6.31×10^{-11}	2.06
3	1.26×10^{-04}	6.28×10^{-11}	2.03

Changes to the I_{ON}/I_{OFF} values for different Fin top widths are also estimated and revealed in Figure 3.13. It indicates the distinction between the ON state current and OFF state current of the device under investigation. In digital systems, the device must switch in between ON state and OFF state at a very high speed. I_{ON}/I_{OFF} ratio is calculated by taking the ON state (I_{ds} in V_{gs} = 1.0 V and V_{ds} = 0.9 V) and OFF state current (I_{ds} in V_{gs} = 0.0 V and V_{ds} = 0.9 V). The higher I_{ON}/I_{OFF} rating is desirable. Proper I_{ON}/I_{OFF} value of the device increases device speed and reduces leakage. With a reduction in the top width of Fin, I_{ON}/I_{OFF} decreases because I_{OFF} decreases for smaller Fin top width but, at the same time, I_{ON} also decreases, indicating a decrease in I_{O}/I_{OFF}. The study of device vulnerability in SCEs, SS, and DIBL is widely used as it a measure. From the trend of DIBL values as shown in Figure 3.13, it is understood that controlling of SCEs is possible by controlling the Fin-width variation. It is significant to note that DIBL improves by reducing Fin top width. The variation of SS for various Fin top width is also plotted in Figure 3.13. From the investigation, it is agreed that SCEs for the device improves for the structure, which has smaller width of Fin top. It's worth revealing that the obtained results are in agreement with the results reported by earlier researchers.[23]

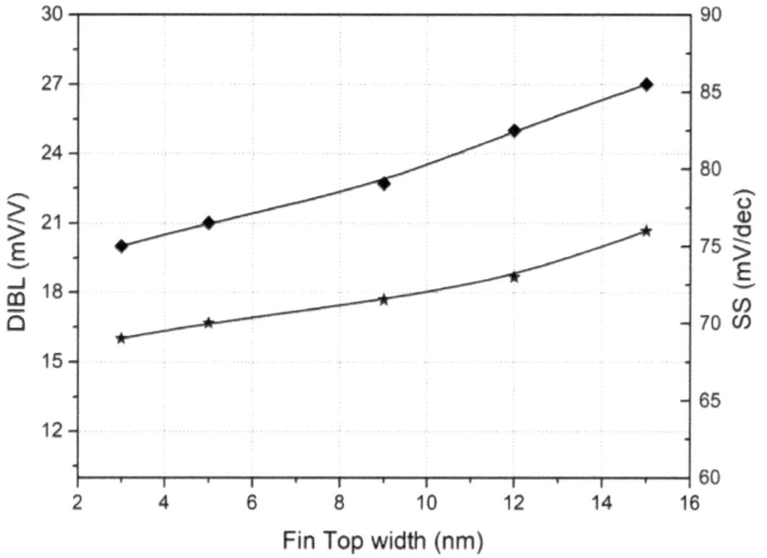

FIGURE 3.13 Variation of SS and DIBL for different shapes of Fin.

Source: Reprinted with permission from Ref. [24]. © 2018 Springer.

Different analog FoMs for the device were also calculated by varting the Fin top width. The superior gate control of FinFET for different Fin shapes is realized through the analysis of transconductance values. In Figure 3.14, the variation of transconductance (g_m) and transconductance to drain current ratio (g_m/I_d) with gate voltage (V_{gs}) is shown for different Fin top width. It is apparent from the Figure 3.14 that the variation of g_m/I_d is insignificant for different Fin top width.

Figure 3.15 plots the total gate capacitance ($C_{gg} = C_{gs} + C_{gd}$) with change in gate voltage (V_{gs}) for different values of Fin top widths. It is detected from Figure 3.15 that for a deviation of Fin shape, which changes the fin shape from rectangular to a trapezoidal one, the total gate capacitance (C_{gg}) reduces. Figure 3.15 tells that the gate capacitance for the Fin with nearly triangular shape decreases nearly 6.5% when the width of top changes from 15 nm to 3 nm. At the same time, the transconductance (g_m) value of the device also falls with the reduction in the width for the Fin top. As a consequence, the cut-off frequency, f_T, rests almost unaffected to Fin top width variation. Figure 3.15 plots the variation of cut-off frequency (f_T) with change of V_{gs} for different top widths also.

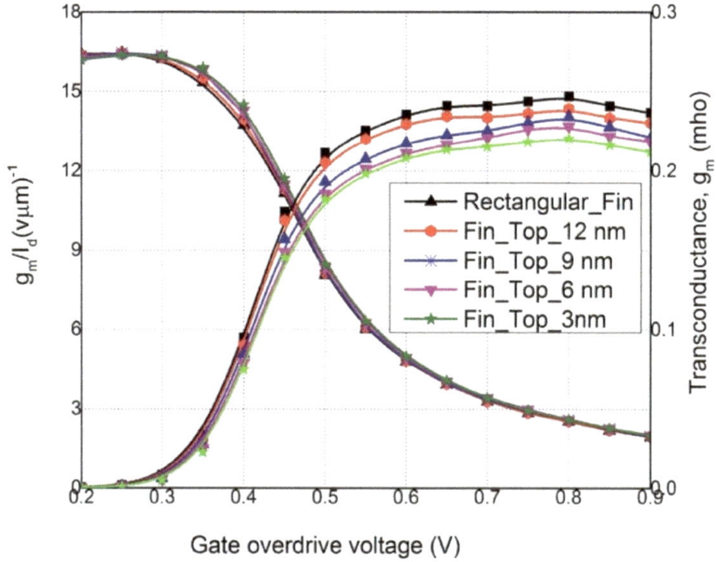

FIGURE 3.14 Variation of g_m and g_m/I_{ds} with V_{gs} for different Fin top width at $V_{ds} = 0.9$ V. *Source*: Reprinted with permission from Ref. [24]. © 2018 Springer.

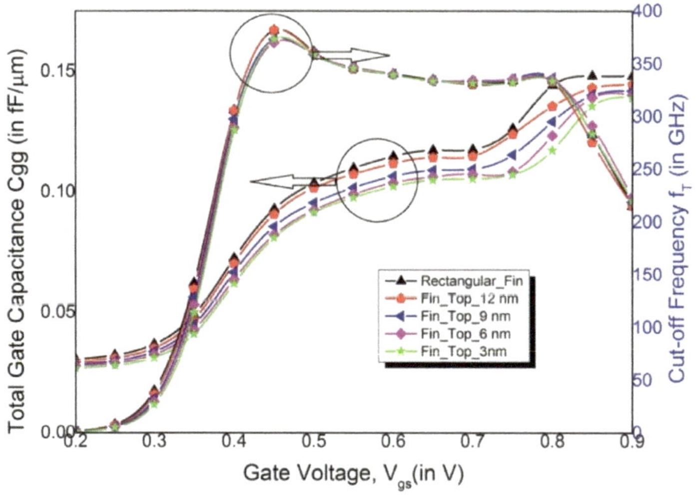

FIGURE 3.15 Variation of gate capacitance, C_{gg} ($=C_{gd} + C_{gs}$) and cut-off frequency f_T as a function of V_{gs} for different Fin top widths.

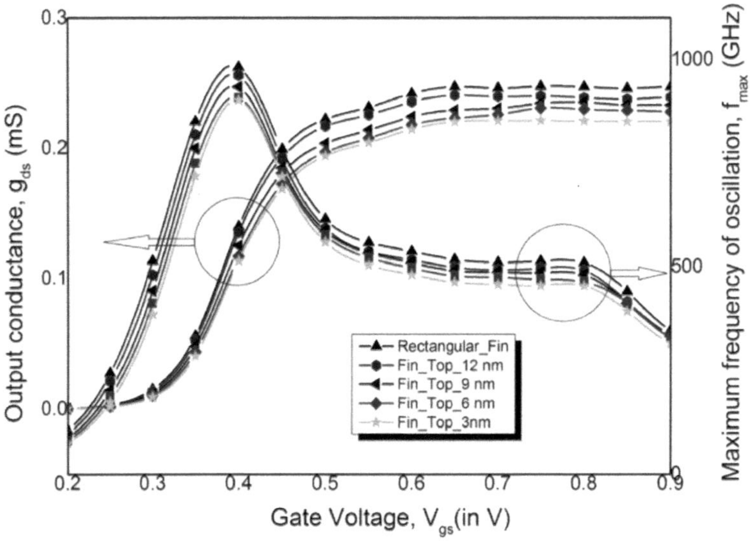

FIGURE 3.16 Variation of output conductance (g_{ds}) and f_{max} with V_{gs} for different Fin top widths.

The reduction in the width of Fin top increases the channel resistance, which has a straight effect on the maximum oscillation (f_{max}) frequency of the device. Figure 3.16 displays the change of g_{ds} with V_{gs} of different widths of Fin. Output conductance (g_{ds}), shown in Figure 3.16, depends on Fin status. FinFET with a 3 nm Fin top width has a minimum value of $g_{ds.}$ The variation of f_{max} of the different values of the Fin top width is also shown in Figure 3.16. Figure 3.16 indicates that the f_{max} decreases with a decrease in the Fin top width. Closely 10% drop in f_{max} was detected when the Fin top width reduced from 15 nm to 3 nm. The main reason behind it is the increase of channel resistance with a fall in effective Fin width. The obtained results may be compared well with the model published recently by Kim et al.[41]

3.4 CONCLUSION

The performance of a junctionless accumulation-mode (JAM) bulk FinFET and its digital performance is presented. New viewpoints for the design and optimization of junctionless transistors for analog/RF applications are studied to continue the historical cadence of CMOS scaling. This research for emerging nanoscale devices, such as junctionless accumulation mode bulk through intensive TCAD simulations, suggests the optimal device parameters and structures. Therefore, it is established that understanding of device physics insight with the design of new structures and material compositions are very crucial for the enhancement of next-generation device performances. Performance variation with many critical geometrical parameters is shown to have a design guideline for the junctionless accumulation mode bulk FinFET device.

KEYWORDS

- **junctionless MOSFET**
- **MOSFET structures**
- **spacer**
- **gate length**
- **fin shape**

REFERENCES

1. Lilienfeld, J. E. Device for Controlling Electric Current. U.S. Patent 1,900,018 filed on 1928-03-28, a thin film MOSFET.
2. Kim, Y. B. Challenges for Nanoscale MOSFETs and Emerging Nanoelectronics. *Trans. Electr. Electron. Mater*. June **2010**, *11* (3), 93–105.
3. Colinge, J. P. *FinFETs and Other Multi-Gate Transistors*; Springer, 2008.
4. www.intel.com/content/www/us/en/silicon-innovations/intel-22nm-technology.html
5. Lee, C. W.; Nazarov, A. N.; Ferain, I.; Akhavan, N. D.;, Yan, R.; Razavi, P.; Yu, R.; Doria, R. T.; Colinge, J. P. Low Subthreshold Slope in Junctionless Multigate Transistors. *Appl. Phys. Lett.* March **2010**, *96* (10), 102106.
6. Lee, C. W.; Ferain, I.; Afzalian, A.; Yan, R.; Akhavan, N. D.; Razavi, P.; Colinge, J. P. Performance Estimation of Junctionless Multigate Transistors. *Solid-State Electron.* **2010**, *54* (2), 97–103.
7. Choi, S. J.; Moon, D. I.; Kim, S.; Duarte, J. P.; Choi, Y. K. Sensitivity of Threshold Voltage to Nanowire Width Variation in Junctionless Transistors. *IEEE Electron. Device Lett.* February **2011**, *32* (2), 125–127.
8. Choi, J. H.; Kim, T. K.; Moon, J. M.; Yoon, Y. G.; Hwang, B. W.; Kim, D. H.; Lee, S. H. Origin of Device Performance Enhancement of Junctionless Accumulation-Mode (JAM) Bulk FinFETs With High-κ Gate Spacers. *IEEE Electr. Device Lett.* December **2014**, *35* (12).

9. Toledano-Luque, M.; Matagne, P.; Sibaja-Hernández, A.; Chiarella, T.; Ragnarsson, L.A.; Sorée, B.; Cho, M.; Mocuta, A.; Thean, A. Superior Reliability of Junctionless pFinFETs by Reduced Oxide Electric Field. *IEEE Electr. Device Lett.* December **2014**, *35* (12).
10. Gundapaneni, S.; Ganguly, S.; Kottantharayil, A. Bulk Planar Junctionless Transistor (BPJLT): An Attractive Device Alternative for Scaling. *Electron Device Lett., IEEE* **2011**, *32* (3), 261–263.
11. Colinge, J. P. et al. Nanowire Transistors Without Junctions. *Nat. Nanotechnol.* **2010**, *5*, 225–229.
12. Ghosh, D.; Parihar, M. S.; Armstrong, G. A.; Kranti, A. High-Performance Junctionless MOSFETs for Ultralow-Power Analog/RF Applications. *IEEE Electr. Device Lett.* October **2012**, *33* (10).
13. Kim, T. K. et al. First Demonstration of Junctionless Accumulation-Mode Bulk FinFETs with Robust Junction Isolation. *IEEE Electr. Device Lett.* December **2013**, *34* (12).
14. Han, M. H.; Chang, C. Y.; Chen, H. B.; Wu, J. J.; Cheng, Y. C.; Wu, Y. C. Performance Comparison Between Bulk and SOI Junctionless Transistors. *IEEE Electr. Device Lett.* February **2013**, *34* (2), 169–171.
15. Sarkar, A. Study of RF Performance of Surrounding Gate MOSFET with Gate Overlap and Underlap. *Adv. Nat. Sci.: Nanosci. Nanotechnol.* *5* (3), 035006.
16. Chakraborty, A.; Sarkar, A. Investigation of Analog/RF Performance of Staggered Heterojunctions Based Nanowire Tunneling Field-effect Transistors. *Superlattices Microstruct.* *80*, 125–135.
17. Biswas, K.; Sarkar, A.; Sarkar, C. K. Impact of barrier thickness on Analog, RF and linearity performance of nanoscale DG heterostructure MOSFET. *Superlattices Microstruct.* *86*, 95–104.
18. Sarkar, A.; De, S.; Dey, A.; Sarkar, C. K. A New Analytical Subthreshold Model of SRG MOSFET with Analogue Performance Investigation. *Int. J. Electron.* *99* (2), 267–283.
19. Biswal, S. M.; Baral, B.; De, D.; Sarkar, A. Study of Effect of Gate-Length Downscaling on the Analog/RF Performance and Linearity Investigation of InAs-based Nanowire Tunnel FET. *Superlattices Microstruct.* *91*, 319–330.
20. Sarkar, A.; Sarkar, C. K. RF and Analogue Performance Investigation of DG Tunnel FET. *Int. J. Electron. Lett.* *1* (4), 210–217.
21. Datta, E.; Chattopadhyay, A.; Mallik, A. Relative Study of Analog Performance, Linearity, and Harmonic Distortion Between Junctionless and Conventional SOI FinFETs at Elevated Temperatures. *J. Elect. Mater.* **2020**, *49*, 3309–3316.
22. Gaynor, B. D.; Hassoun, S. Fin Shape Impact on FinFET Leakage With Application to Multithreshold and Ultralow-Leakage FinFET Design. IEEE Trans. Electr. Devices **2014**, *61* (8), 2738–2744.
23. Md Rezali, F. A.; Othman, N. A. F.; Mazhar, M.; Wan Muhamad Hatta, S.; Soin, N. Performance and Device Design Based on Geometry and Process Considerations for 14/16-nm Strained FinFETs. *IEEE Trans. Electr. Devices* **2016**, *63* (3), 974–981.
24. Biswas, K.; Sarkar, A.; Sarkar, C. K. Fin Shape Influence on Analog and RF Performance of Junctionless Accumulation-mode Bulk FinFETs. *Microsyst. Technol.* *24* (5), 2317–2324.
25. Biswas, K.; Sarkar, A.; Sarkar, C. K. Spacer Engineering for Performance Enhancement of Junctionless Accumulation-mode Bulk FinFETs. *IET Circ. Devices Syst.* **2017**, *11* (1), 80–88.

26. Lee, J. H. Bulk FinFETs: Design at 14 nm Node and Key Characteristics. In *Nano Devices and Circuit Techniques for Low-Energy Applications and Energy Harvesting*; Kyung, C. M., Eds.; KAIST Research Series. Springer: Dordrecht, 2016.
27. Zhao, H.; Yeo, Y. C.; Rustagi, S. C.; Samudra, G. S. Analysis of the Effects of Fringing Electric Field on FinFET Device Performance and Structural Optimization Using 3-D Simulation. *IEEE Trans. Electr. Devices* May **2008**, *55* (5), 1177–1184.
28. Wong, B.; Mittal, A.; Cao, Y.; Starr, G. W. *Nano-CMOS Circuit and Physical Design*; Wiley-IEEE Press, ISBN: 978-0-471-46610-9, November 2004.
29. Ha, D.; Takeuchi, H.; Choi, Y. K.; King, T. J. Molybdenum Gate Technology for Ultrathin-Body MOSFETs and FinFETs. *Trans. Electr. Devices* December **2004**, *51* (12), 1989–2004.
30. Ranade, P.; Choi, Y. K.; Ha, D.; Agarwal, A.; Ameen, M.; King, T. J. In *Tunable Work Function Molybdenum Gate Technoo*, CA, USA, pp. 363–366, 2002.
31. The International Technology Roadmap for Semiconductors, [Online]. Available: http://www.itrs.net.
32. Pradhan, K. P.; Mallikarjunarao, P.; Sahu, P. K. Exploration of Symmetric High-k Spacer (SHS) Hybrid FinFET for High Performance Application. *Superlatt. Microstruct.* February **2016**, *90*, 191–197.
33. Murmann, B.; Nikaeen, P.; Connelly, D. J.; Dutton, R. W. Impact of Scaling on Analog Performance and Associated Modeling Needs. *IEEE Trans. Electr. Devices* September **2006**, *53* (9), 2160–2167.
34. Sarkar, A.; De, S.; Dey, A.; Kumar Sarkar, C. 1/f Noise and Analogue Performance Study of Short-channel Cylindrical Surrounding gate MOSFET Using a New Subthreshold Analytical Pseudo-two-dimensional Model. *IET Circ. Devices Syst.* **2012**, *6* (1), 28.
35. Mohankumar, N.; Syamal, B.; Sarkar, C. K. Influence of Channel and Gate Engineering on the Analog and RF Performance of DG MOSFETs. *IEEE Trans. Electr. Devices* **2010**, *57* (4), 820–826.
36. Sarkar, A.; Das, A. K.; De, S.; Sarkar, C. K. Effect of Gate Engineering in Double-gate MOSFETs for Analog/RF Applications. *Microelectron. J.* **2012**, *43*, 873–882.
37. Sachid, B. et al. In *Sub-20 nm Gate Length FinFET Design: Can High-κ Spacers Make a Difference*?, 2008 IEEE International Electron Devices Meeting, San Francisco, CA, pp. 1–4, 2008.
38. Saini, G.; Rana, A. K. Physical Scaling Limits of FinFET Structure: A Simulation Study. *Int. J. VLSI Des. Commun. Syst. (VLSICS)* March **2011**, *2* (1), 26–35.
39. Poljak, M.; Jovanović, V.; Suligoj, T. Improving Bulk FinFET DC Performance in Comparison to SOI FinFET. *Microelectron. Eng.* October **2009**, 86 (10), 2078–2085.
40. Subramanian, V.; Mercha, A.; Parvais, B.; Loo, J.; Gustin, C.; Dehan, M.; Collaert, N.; Jurczak, M.; Groeseneken, G.; Sansen, W.; Decoutere, S. Impact of Fin Width on Digital and Analog Performances of n-FinFETs. *Solid-State Electron.* April **2007**, *51* (4), 551–559.
41. Kim, J.; Huynh, H. A.; Kim, S. Modelling of FinFET Parasitic Source/Drain Resistance with Polygonal Epitaxy. *IEEE Trans. Electr. Devices* **2017**, *64* (5), 2072–2079.

CHAPTER 4

Charge Storage Mechanism in Proteotronic Capacitors

SHRABANI GUHATHAKURATA[1], NABIN BARAN MANIK[2], SHEETIKANTA MOHANTY[1], DIPUN JENA[1], SUBHASIS SATAPATHY[1], DEBASISH PANDA[1], AJIT DASH[1], ROHINI MAHATO[3], GUFRAN AHMAD[4], PRASHANT KUMAR SINGH[5], and SANDIPAN MALLIK[1*]

[1]*Department of Electronics and Communication Engineering, National Institute of Science and Technology, Berhampur, Odisha 761008, India*

[2]*Department of Physics, Condensed Matter Physics Research Centre, Jadavpur University, Kolkata 700032, India*

[3]*Department of Life Science, Central University of South Bihar, Gaya, Bihar 824236, India*

[4]*Department of Electrical Engineering, Dayalbagh Educational Institute, Agra 282005, India*

[5]*Department of Electronics and Communication Engineering, University College of Engineering and Technology, V.B.U., Hazaribag, Jharkhand 825301, India*

[*]*Corresponding author. E-mail: sandi.iitkgp@gmail.com*

ABSTRACT

Flexible electronics has evolved as the most enthralled subject in the domain of modern electronic devices owing to its user-friendly and versatile ease of fabrication properties. Various studies on the fabrication of active devices over flexible platforms have shown numerous promising applications. Paper, with its incredible properties of lightweight, flexibility, low cost, mass productivity, disposability, porosity, and ease of processing, is showing great potential as a future flexible substrate. Organic materials are getting attention for the evolution of green electronics that will be environment-friendly and biodegradable. Protein as an organic material has been widely used in fabrication of biosensors, organic field-effect transistors (OFETs), memristors, etc. The coupling of protein and electronics popularly known as proteotronics will emerge as the breakthrough technology in future electronic industry.

In the first section of this chapter, a brief review has been carried out on the importance of energy-saving fabrication processes using interdigitated pencil-on-paper (PoP) strategies in the area of low-cost flexible devices applications, along with the use of protein as an emerging dielectric material. In the second section of this chapter, the electrical properties of bovine serum albumin (BSA)-based capacitors with interdigitated graphite PoP substrate has been studied and discussed in detail. The number of dipoles and the corresponding dipole moment has been evaluated using an analytical approach. In the last section of this chapter, the charge storage mechanism of proteotronics capacitors using hydrated BSA will be discussed in detail with corresponding models.

4.1 INTRODUCTION

The word "*Proteotronic*" was derived from the amalgamation of two words: "proteomics," instilled by Marc Wilkins in 1996 to explain the extensive study of proteins, its structures, and functions, and "electronics," instilled by Thomas Alva Edison in 1883, to stipulate the evolution of devices for manipulation of electric current to execute effective work.[1] Precisely, after convergence, a completely new branch of electronics is formulated to explore the electrical characteristics of protein, mainly focusing on the development of bio-electronic devices such as bio-sensors,[2] bio-capacitors,[3] bio-memristors,[4] bio-hybrid solar cells,[5,6] etc. Particularly, these devices have gained distinctive attention due to the key element, protein, which shows excellent electrical and optical properties,[7] due to which the precision and sensitivity of the device has increased to an unimaginable level, thereby, allowing the fabrication of fast and sensitive devices.[8]

In the field of proteotronic, one of the most abundantly used proteins to make bio-devices is bovine serum albumin (BSA or Fraction V). BSA is a serum albumin, globular protein, derived from cows, which consists of 583 amino acids to make up a single chain of polypeptides, no carbohydrates, having the molecular weight of ~66kDa, having an intrinsic viscosity 0.041 and a pH of 5.2–7 for 1% solution.[9] BSA is widely available with high purity and low cost. Along with that, the organic material has unique attributes, such as, it allows simplistic fabrication methods and large-scale production.[10] BSA is one of the most reliable protein models for interaction studies due to its water-soluble nature,[11,12] which gives us a good time in merging and blending with polymers and other solvents, which leads to biodegradable and biocompatible electronics,[13,14] which are abstruse for inorganic materials. Besides, BSA is reported as a highly stable protein complex.[15] It is accepted that to assess the work of proteins, its dielectric properties play a vital role. It is reported that proteins have a high dielectric constant,[16] due to which they might replace traditional materials for proteotronic device applications and can be used as an organic dielectric medium in a metal-insulator-metal (MIM) capacitors. In the semiconductor industry, MIM technology is very well established, its implementation to energy storage devices is an attractive option, especially in terms of device performance, process simplicity, and cost; and due to the excellent advantage of their rapid-moving charge and high burst power, they are used widely for energy storage application. Since the energy is stored as dielectric polarization of the insulating film, charge discharge rates in these devices are determined by RC time constants of external circuits, which can be in the milliseconds range much faster than conventional batteries. The fast charge-discharge rates are useful in applications that require high power for a short duration of time.

Recently, MIM structures has been successfully fabricated and characterized over different types of flexible substrates like polyethylene terephthalate (PET), Kapton, etc.[17–20] In the present time, paper-based flexible electronics also have gathered a remarkable popularity due to its versatile attributes to fabricate thin, light weight, portable, low-priced, and biocompatible devices.[21–24] Paper as a substrate is much more appealing compared to other metallic/plastic flexible substrates—titanium or stainless-steel foils, PET, polyethylene naphthalate (PEN), polydimethylsiloxane (PDMS), etc.—for low-cost electronics.[25] Commonly used electrical components such as thermochromic displays,[26] cellulose-based batteries,[27] foldable printed circuit boards,[28] etc., have recently been produced on paper substrates.

4.1.1 PENCIL ON PAPER (POP) AS A FLEXIBLE SUBSTRATE

Flexible electronics have aggregated substantial attention due to their promising potentials in making future wearable, miniature, and portable electronics. Many electronic and optoelectronic devices are built on the flexible substrate for sublime applications such as flexible displays,[29] wearable devices,[30–32] implantable devices,[33,34] microfluidic devices,[35] electronic skin devices for detection and treatment,[36–38] and many more. The fabricated devices become more unique in their way when paper—a day-to-day used material for writing and drawing purpose—is selected as the organic flexible substrate. In the recent time, paper-based electronics is giving tough combat to already-established conventional electronics based on silicon wafers, glass, and plastic substrates.[39]

Paper is an insulating and heat-resistant material; hence, various approaches and techniques have been used to fabricate conducting devices and structures such as resistors, supercapacitors, capacitors, and diodes by the pencil-on-paper (PoP) method,[40] pen-on-paper method,[41] inkjet printing, photolithography,[42] spray coating,[43] and screen printing.[44] The performance and the outcomes of paper-based electronics depend on the electrical characteristics of the conducting electrodes like silver and copper inks, graphene powder, carbon nanotubes, carbon fibers, etc. However, using carbon-based materials for depositing conducting electrodes through thermal evaporation and sputtering is complex and demands treatment, besides being an expensive procedure to be employed..[45] Therefore, pencil writing on paper, PoP, a day-to-day paradigm since the 16th century, has been trending and gaining the attention of the researchers to fabricate flexible, portable, and conformable electronic circuits on a paper, which is providing a unique outlook to the field of pencil-used paper-based electronics.

Recently, an electronic capacitor and an RC circuit were constructed by hand-printing on a paper substrate using a pentalic-brand type-B pencil, having an aim of measuring the capacitive sensing by Thompson et al.[46] Down et al. have reported a supercapacitor, which used an inter-digitated design of electrodes by PoP.[47] The fluid-flow electrokinetics has also been studied with paper and pencil devices. Paper-based analytical devices for glucose biosensing using graphite pencil electrodes have been developed and reported.[48,49] Paper is also considered as a suitable platform for sensors due to its porous nature; which is caused by the three-dimensional arrangement of the cellulose fibers, making it convenient for the solvents or the liquids to cross through its thickness.[50] A thin film of ZnO on pencil-drawn interdigitated electrodes on the paper was used to construct an UV detector.[51] A low-cost device is fabricated considering the hygroscopic nature of the paper, and a full pencil-drawn capacitive sensor for measuring the breath rate and patterns has been reported.[52] Using PoP technique, two different types of functional sensing devices like strain gauge and chemiresistors constructed by tracing pencil on the paper has been reported.[53]

4.1.2 INTERDIGITATED ELECTRODES AND PHYSICS

The word "*interdigitated*" refers to a comb-like or a multiple finger-like periodic arrangement of parallel, in-plane, and co-planar electrodes on a rigid surface. One of the major applications of the interdigitated electrodes (IDEs) is modeling numerous capacitive-based circuits; due to natural positioning of the multiple pair of electrodes. Specifically, whenever the IDEs are having multiple dielectric layers sandwiched between the individual finger-like projections, which ultimately builds up a capacitance for the multi-layered structure; these particular kinds of structures are named to be interdigitated capacitors (IDCs), which provides an appreciable value of capacitance comparing to flat, cylindrical traditional capacitors, along with higher sensitivity and less signal-to-noise ratio (SNR), which is preferred for high-frequency applications and sensor device fabrication such as lumped elements in microwave-integrated circuits, radio-frequency circuits, humidity sensors, supercapacitors, chemical sensors, gas sensors, etc. IDC/IDE structures can be engraved by few methods like lithography,[54] direct laser scribing,[55] PoP method,[56] and inkjet printing by using conducting materials such as gold (Au), silver (Ag), copper (Cu), aluminium (Al), and graphitic species, etc. Nowadays, electrodes made up of graphitic species (primary source is pencil graphite lead) are of utter importance due to their impressive sensitivity, viability, and sustainability. Besides being low-cost and easily available, it allows easy designing of IDEs/IDCs and it also allows avoiding preliminary steps for cleaning and polishing of electrode surfaces.

A low-cost UV sensor was fabricated using hands-free pencil-drawn graphite IDEs and further screen printing the active ZnO nanocrystal on the surface of the paper substrate, as most of the area of the IDE was exposed to UV light, there was an increase in optical sensitivity and responsivity of the device as reported.[57] Similarly, sensitivity of the NO_2 gas sensor constructed with Ag-IDEs on paper has increased due to decreasing the finger width of the IDEs and was reported.[58] Many researchers have tried to implement the capacitance of IDEs in biosensing applications. A capacitive humidity sensor was fabricated on silicon substrate with Au-IDE electrodes by deposition technique with stable and linear response over a wide range of relative humidity, proving that interdigitated capacitive humidity sensors are preferred over parallel plate electrodes and reported.[59] Recently, a supercapacitor was reported with IDEs of graphite with PoP substrate, which exhibits a reliable range of capacitance and capable of storing the charge within it.[47]

An analytical expression for the capacitance between the two comb electrodes of a periodic interdigital capacitive sensor, based on conformal mapping techniques has been reported by *Igreja et al.*[60] This application independent proposed general model is not only suitable to apply for any space and finger width, but also applicable for any number of layers with different thickness and permittivity. The capacitance of a particular sensor configuration is a function of the dielectric permittivity of the materials used and the length of the fingers. It also depends on the ratio between the space and finger widths; and the ratio between the thickness of the sensitive layer and the spatial sensor wavelength. The total capacitance of the IDC-S is given by

$$C_{IDC} = (N-3)\frac{C_{I,IDC}}{2} + 2\frac{C_{I,IDC}C_{E,IDC}}{C_{I,IDC} + C_{E,IDC}} \tag{4.1}$$

where $C_{E,IDC}$ is the exterior half-capacitance, $C_{I,IDC}$ is the interior half-capacitance, and N is the number of electrode figures ($N > 3$). Using eq 4.1, the corresponding capacitance of the BSA-based PoP capacitors were measured with IDE electrodes.

4.1.3 USE OF PROTEIN AS A DIELECTRIC

Proteins are natural polymers composed of one or more long chains of amino acids, having a specific sequence, linked to each other by peptide bonds. Proteins are considered as polarizable materials[61] because it contains neutral, polar, and charged side chains,[62] and possesses polar recurring units of amino acids[63] due to which proteins are sensitive to dielectric operating procedure and put up a distinct dielectric constant in a particular condition. Its significantly acknowledged dielectric properties play a very vital role to access the electrical characteristics of the proteins. Several studies have been reported to investigate the dielectric properties of proteins,[16] its high dielectric constant,[64] electron transport mechanism, and dipole-moment study.[65] The main tool by which a material's dielectric property can be analyzed is by its electron transport mechanism. Primarily, the backbone of protein is made up of phosphodiester bonds; due to water loss, phosphodiester bonds are converted into zwitterion for stabilizing the charge at the isoelectric point by resonance, ensuring the transport of electrons through proteins. The Ramachandran Plot elucidates the electron flow in protein through continuous redox reactions, defined as electron transport in proteins (ETp). More particularly, the ETp ascertains that the electron transport in organic dielectrics is quite similar to that of the charge transport within inorganic dielectrics.[66] The protein, as dielectric, has many applications such as organic thin-film transistors (OTFT), MIM, organic field-effect transistors (OFET), bio-memristors, and bio-sensors and organic light-emitting diodes (OLED). A transparent bio-memristor was fabricated with natural silk-fibroin protein to investigate the bipolar memristive switching behavior and a model on charge trapping and de-trapping in silk protein and reported.[4] Proteins are ideal candidates for bio-sensing applications due to their high specificity. In many bio-sensors, proteins are immobilized on the solid substrate and act as transducers for the physical signals and convert them into electrical signals.[67] Currently, many protein-based bio-sensors are being used in toxin and disease detection[68–70] and drug monitoring.[71] A low-cost bio-Sensor was fabricated with gold-interdigitated electrodes with C-reactive protein (CRP; a plasma protein), which was reported to be left immobilized with CRP antibodies to detect the presence of CRP antigens.[72] In a report, OTFTs were fabricated using Keratin protein (protein extracted from poultry chicken feathers) to make biodegradable gates to establish a semiconductor-dielectric interface.[73]

Stimulated by the earlier-explained accomplishments with flexible substrate and organic materials, in this report we have fabricated a BSA-based biodegradable, portable, light weight, and flexible proteotronic MIM capacitors on flexible trace-on paper substrate with pencil-drawn interdigitated graphite electrodes. Electrical characterization of the proteotronic capacitor has been investigated through capacitance-voltage (C-V) characteristics. The charge storage mechanism is also explained with corresponding models.

4.2 EXPERIMENTAL

A MIM capacitor using BSA as the active insulating layer was fabricated on a trace-on paper substrate. Prior to the fabrication of devices, the paper substrates were cleaned with methanol for 10 min to remove the residual surface contaminants. On the cleaned substrate, thick lines were drawn with different pencils (HB, 2B, and 4B) to measure electrode resistance. The line drawn with 4B pencil shows the lowest resistance, hence it was considered to be used in drawing the conducting electrodes for further device fabrication. IDEs were drawn on the clean substrate in 1.5 cm × 1.5 cm area with length of the electrodes strand to be 1.3 cm and an inter-electrode distance of 0.1 cm. The interdigitated electrodes were drawn very close to each other, each of equal length and thickness to achieve high capacitance. High-speed N_2 flash was done repeatedly to remove the loosely bonded graphite particles. A 2 mg/mL aqueous solution of BSA (Sigma Aldrich A2058) was prepared, followed by an ultrasonic bath for 10 min. The prepared solution was drop casted on the electrode drawn on paper substrate and was left to dry for 6 h under room temperature. Further, the Agilent E4980A precision LCR meter was deployed for measuring the capacitance-volage (CV) characteristics of the fabricated devices.Figure 4.1 shows the schematic of different steps of device fabrication methodology and the device characterization setup.

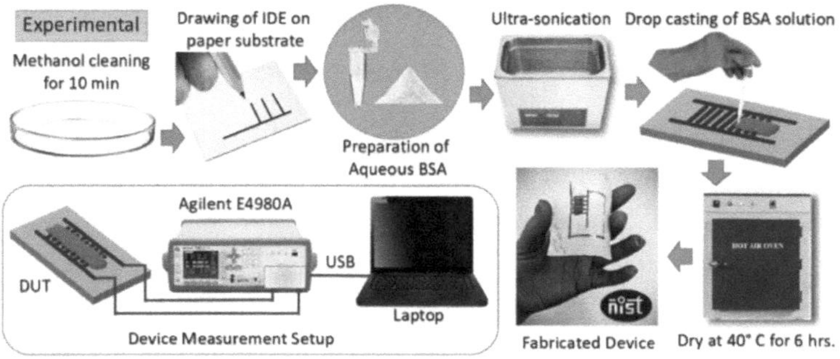

FIGURE 4.1 Process schematic of device fabrication methodology and characterization setup.

4.3 RESULTS AND DISCUSSION

The capacitance-voltage characteristics of the BSA-based MIM structures were measured in different frequency ranges from 50 kHz to 650 kHz, with an applied bias of -12 to +12 V. The normalized capacitance can be expressed as second-order polynomial equation,[74]

$$\frac{\Delta C}{C_0} = \alpha V^2 + \beta V \tag{4.2}$$

where $\Delta C/C_0$ is the ratio of the change in capacitance, to the capacitance at zero bias (C_0), α and β are the quadratic voltage coefficient and linear voltage coefficients of capacitance, respectively, expressed in parts per million (ppm), and V is the applied voltage.

Normalized capacitance versus voltage plot of MIM capacitor, using BSA as the active insulating layer is shown in Figure 4.2a. The experimental data is fitted with theoretically computed normalized capacitance and is shown in Figure 4.2b for 650 kHz frequency, where a positive curvature is observed with a VCC-α value of 0.005×10^4 ppm/V^2. It can be note that the CV characteristics are not symmetrical in nature and having different voltage coefficients found for different bias values. To recognize the origin of capacitance variation with applied bias, several models have been reported, including charge trapping,[75] field-enhanced bond polarizability,[76] electrode polarization,[77] field-induced entropy variation,[78] electrostriction and nonlinear electronic polarization,[79] and ion oscillation in double wells.[80] To explain nonlinearities in MIM capacitors, Gonon et al.[81] have proposed a physical model based on electron hopping through oxygen vacancy defects.

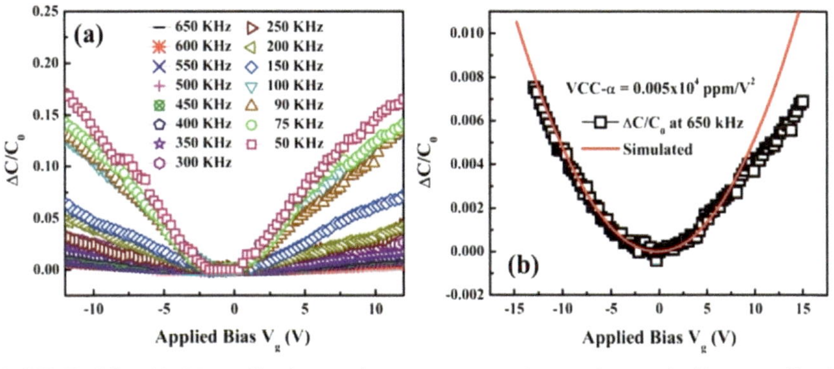

FIGURE 4.2 (a) Normalized capacitance versus voltage plot; and (b) normalized capacitance at 650 kHz with simulated value for the BSA-based MIM structure.

As per their proposed model, oxygen vacancies are considered as the possible reason of nonlinearity and there are two probable cases. In first case, the dipole polarization is provided due to the isolated pairs of oxygen vacancies localized electron hopping, and in second case, the connected pairs, extending in between electrode, provide a percolation path, where electrons drift by hopping under the influence of the dc bias, giving rise to electrode polarization. In presence of only a.c. bias electrons hop between two traps that can be compared with oscillating dipole. On the other hand, under d.c. bias, the a.c. hoping mechanism is negligible and electrons drift through the percolation path provided by connected pairs that may be present at grain boundaries. Accumulation layer may be formed at the anode due to tunneling mechanism from trap to anode fermi level in case of non-ohmic contact. The electron flux that is modulated by the a.c. signal through barrier energy variation along the conduction path leads to alteration in the space charge region close to the anode, corresponding to an electrode polarization mechanism. Electrode polarization is equivalent to a macroscopic dipole, whose length is one of the defect conductionpaths. Considering a frequency dependence of the capacitance, one can get that the relative capacitance variation with voltage follows an exponential law, which can ultimately written as,

$$\frac{\Delta C}{C_0} = C\omega^{-2n}\left[\exp(enER / kT) - 1\right] \tag{4.3}$$

where $C = A(\varepsilon / \delta_0)^{-2n}$, δ_0 is the low field conductivity, and E is the applied DC field.

Figure 4.3a shows the variation in capacitance density as a function of applied frequency at a bias of 1 V. Trap-induced charges attribute to such frequency dispersion as the traps fail to follow applied signal at high frequency.[15] A value of 3.2×10^{10} F/μm^2 capacitance density was obtained at 50 kHz frequency, which gradually degraded non-linearly to 2.2×10^{10} F/μm^2 at 650 MHz applied frequency. The inset of Figure 4.3a shows quadratic voltage coefficient of capacitance of the MIM capacitor versus applied frequency,which is an essential factor for dynamic analog circuits,[82] has a tradeoff with capacitance density. International Technology Roadmap for Semiconductors (ITRS) suggests the VCC-α value less than 100 as a standard for MIM structures.[83] At 50 KHz applied frequency, the value of VCC-α is 0.15×10^4 ppm/V^2, which is well below the maximum standard and decreases gradually with frequency. A positive VCC-α was obtained for the BSA capacitor resulting due to induced dipole in the dielectric due to the ionic polarization.[84] Positive sign in VCC-α denotes dominance of negative charge traps,[85] which continuously trap and de-trap the electrons originating from the electrodes resulting in upward bent of the CV curve. Moreover, it has a profound influence on the relaxation time as carrier mobility decreases with increase in frequency a longer relaxation time is observed at a higher frequency and can result in electrode polarization.[86] On the contrary, voltage linearity is highly dependent on dielectric constant, carrier concentration, and dielectric thickness. VCC-β, which depends on trap distribution within the dielectric, work function difference between the electrodes, and symmetry of the depletion layer in the electrodes,[87] also decreases with increase in frequency (figure not shown).

The dissipation factor of the dielectric quantifies the dielectric loss in the material and its frequency dispersion can be attributed to high roughness at the interfaces, formation of interfacial layer in between dielectric and electrode, and electrode resistance.[88] Variation of dissipation factor versusapplied frequency is shown in the inset of Figure 4.3(a), where the dissipation factor decreases with increment in frequency and reaches a value of 145 at 650 MHz.

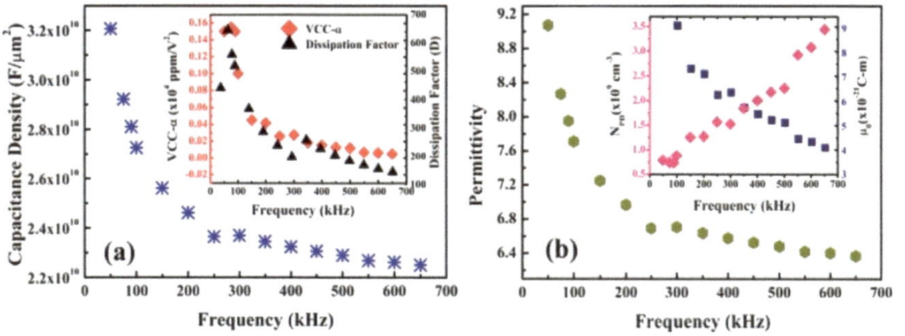

FIGURE 4.3 (a) Normalized capacitance density versus frequency plot, VCC-α and dissipation factor plotted with frequency (inset); and (b) variation of permittivity of BSA filmwith frequency, no. of dipoles and induced dipole moment variation with frequency (inset).

The values of the relative permittivity of BSA in a frequency range of 50 kHz to 650 kHz is represented in Figure 4.3b. The value of relative permittivity was evaluated as 9.1 at applied frequency of 50 MHz, which is close to the reported value.[84,89] The relationship of quadratic voltage coefficient with the number of dipoles (N_{PD}) and the dipole moment (μ_0) can be established following an analytical model described by Kannadassan et al.[90]

Understanding this model requires an idea of basic concept of polarization and dipole moment. The relation between induced dipole moment μ with applied electric field E is given as $\mu = \alpha_{ie} E$, where, α_{ie} denotes the internal electronic polarizability. E_{loc} is termed as local electric field which is the field within the spherical dipole and its position is dependent on the applied field. It is related to the applied field as $E_{loc} = \lambda E$, where $\lambda = 3\varepsilon_r/(2\varepsilon_r + 1)$ is field correction factor. The average induced dipole moment can be expressed as,

$$\overline{\mathrm{M}} = \alpha_{ie}\,\overline{\cos^2\theta}.E_{loc} \tag{4.4}$$

Using these models of ionic polarization, the total polarization can be expressed as,

$$P = \left(\varepsilon_0 \chi_e + N_{di}\left[\frac{3\varepsilon_0}{N_{di}}\left(\frac{\varepsilon_r - 1}{\varepsilon_r + 2}\right)\right] L_2(\beta)\right) E \tag{4.5}$$

Considering the second-order Langevin function, the field-dependent permittivity of dielectric layer can be expressed in term of static permittivity ε_r as,

$$\varepsilon(V) = \varepsilon_0 + \varepsilon_0 \chi_e + N_{pd}\left[\frac{3\lambda\varepsilon_0}{N_{pd}}\left(\frac{\varepsilon_r - 1}{\varepsilon_r + 2}\right)\right]^3 \frac{E^2}{15(kT)^2} \tag{4.6}$$

This equation can be compared with the empirical relation, $C(V) = C_0(\alpha V^2 + \beta V + C)$, where C_0 is the capacitance at bias voltage of zero. Therefore, the quadratic (α) coefficient of capacitance is,

$$\alpha \approx 1.8\left(\frac{\varepsilon_0}{N_{pd} kT}\right)^2 \left[\lambda\left(\frac{\varepsilon_r - 1}{\varepsilon_r + 2}\right)\right]^3 \frac{1}{d^2} \tag{4.7}$$

Thus, the no. of dipoles can be obtained from a known α value from eq 4.7,

$$N_{pd} \approx \frac{\varepsilon_0}{kTd}\left(\frac{1.8}{\alpha}\right)^{\frac{1}{2}} \left[\lambda\left(\frac{\varepsilon_r - 1}{\varepsilon_r + 2}\right)\right]^{\frac{3}{2}} \tag{4.8}$$

The relation between induced dipole moment and no. of dipoles can be established using the equations, $\chi/\varepsilon_0 = (\varepsilon - 1)$ and $\chi = \mu^2 N_{pd}/3KT$. Thus, the induced dipole moment can be given as,

$$\mu_0 = \sqrt{\frac{3KT\varepsilon_0(\varepsilon_r - 1)}{N_{pd}}} \tag{4.9}$$

Using the known value of VCC-α and the permittivity ε_r for different frequencies, the variation of N_{pd} and μ_0 with frequency can be estimated using eq 4.8 and 4.9 and is shown in the inset of Figure 4.3b. It is observed that the value of μ_0 increases with frequency while the induced dipole moment μ_0 value decreases non-linearly. The values of N_{pd} and μ_0 are found 3.4×10^9 cm^{-3} and 1×10^{-21} C-m, respectively, at 650 kHz applied frequency for the BSA MIM capacitor.

4.4 CHARGE STORAGE MECHANISM OF BSA-BASED MIM STRUCTURE

In the biological world, the most abundant and novel macromolecules present in all parts of cell are proteins, whose structural units are amino acids. There are 20 different amino acids commonly found in proteins.[91] The main elements of an amino acid are carbon (C), hydrogen (H), oxygen (O), and nitrogen (N), although other elements are found in the side chains of some amino acids. Amino acid consists of a central carbon atom (the α carbon), which is attached to an amine ($-NN_2$) group, a carboxyl (–COOH) group, and a side chain (R group).[92,93] These four different groups amid α-carbon atom can engage two unique spatial configurations due to tetrahedral arrangement of bonding orbitals and hence giving rise to two possible stereoisomers of the amino acid (L & D) and the α- carbon, which is called as chiral center.[94] These arrangements are contrasting in their side chains (R group), which vary with certain parameters such as size, structure, and electrical charge due to which it influences the solubility of amino acids in water. Depending on the side chains, amino acids are categorized into hydrophobic, aromatic, polar uncharged, basic, and acidic. When an amino acid is dissolved in water, it exists in solution as the dipolar ion, that is, the carboxyl group (COO) and the amino group (NH_3^+). Because the resulting amino acid contains one positive and one negative charge, it is a neutral molecule called a zwitterion.[94] Figure 4.4a and b shows both the non-ionic and zwitter-ionic forms of amino acids, respectively. Therefore, in aqueous solution, there is an existence of equilibrium between the molecular form and the zwitterion form of amino acids. Amino acids are also referred as amphiprotic (both donate and accept protons) in nature, as it contains both amine and carboxylic acid functional groups. A conjugate acid–base pair is formed up of proton donor and its corresponding proton acceptor.[95] The tendency of any acid (HA) to lose a proton and form its conjugate base

(A) is defined by the equilibrium constant (K_{eq}). For ionization reactions, equilibrium constants are called as ionization or dissociation constants (K_a). Its negative logarithm (pK_a) can be defined as $pK_a = \log(1/K_a) = -\log K_a$,[96] which is a tool to measure the tendency of groups to liberate a proton and it is evident from the pK_a expression that for tenfold decreasing tendency,pK_a value increases by one unit.[97] When $pH < pK_a$, protonation occurs and similarly when pH > pKa, deprotonation occurs.

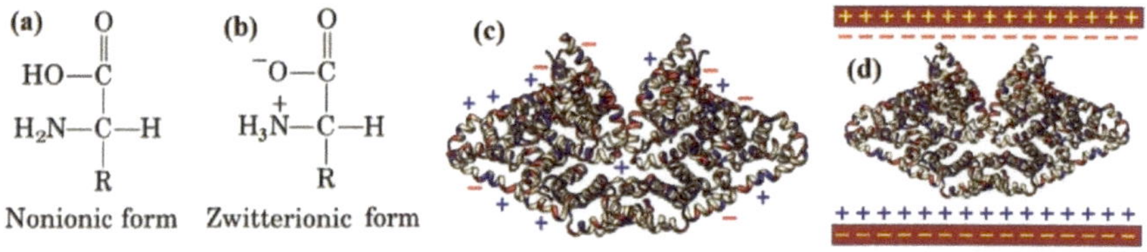

FIGURE 4.4 (a) Non-ionic and (b) zwitterionic forms of amino acids. (c) Side chains (positive and negative) of hydrated BSA with mobile ions, where H_3O^+ mobile ions are marked with "+" sign and OH^- mobile ions are marked with "–" sign. (d) Electric double-layer capacitors formation by hydrated BSA, on applying bias.

At $pH = pK_1$, molecule of $NH_3^+CH(R)CO_2H$ (cation) and $NH_3^+CH(R)CO_2^-$ (zwitterion) and similarly at $pH = pK_2$, molecules of $NH_3^+CH(R)CO_2^-$ (zwitterion) and $NH_2CH(R)CO_2^-$ (anion) are of equal concentrations. At the exact center of the two pK_a values, there is a balance between the net positive charge and the negative charge[98]; as a result of which the average charge of all forms become null,[99], this point is known as isoelectric point (pI); where the mobility of electrons becomes zero. Generally, compared with the known pK values of amines and carboxylic acid, very often it is assumed that concentration of zwitterion is much higher as compared to the concentration of neutral molecules. The zwitterion formation and pI of an amino acid greatly depends upon its R group. Amino acid can yield a protein, if the molecules of amino acids are linked covalently with a substituted amide, the bond formed is named as peptide bond. These bonds are formed by removing elements of water (dehydration) from the amino group and carboxyl group of amino acids.[100].

Theodor von Grotthuss, a German chemist, has come up with a mechanism for transport of protons by tunneling or hopping across chains of hydrogen bonds in water from one molecule to another. This transport mechanism upshots higher mobility (3.6×10^{-3} cm^2 sec^{-1} V^{-1}) of H^+ ions than K^+ ions with alike hydrodynamic radius.[98] In this mechanism, by hopping a proton moves from one molecule to another by exchanging a hydrogen bond with a covalent bond. However, there is no result of unit charge transfer, as it leaves the water in polarized state by moving through a single water chain. Hence, the water chain needs to rearrange itself, to accept other protons by transporting a Bjerrum D orientation defect for the orientation of water molecules in the chain. Ultimately, there will be a transfer of unit charge by the combined motion of H^+ and a D defect. Similarly, a negatively charged OH^- (known as proton hole) can be transferred through the chain, by the same two-step processes[101]. Hence, it is evident that protonic conductivity possesses two different charge carriers, which is similar to semiconductors. Stimulated by the Grotthuss mechanism and proton channel, a protein-based electronic device has been fabricated and demonstrated in this chapter, where the electric field controls the flow of H^+ and OH^- ions in biopolymers, which is quite analogous to electrons and holes in semiconductors.[102,103]

The charge storage mechanism of BSA (hydrated) as the gate dielectric material has been proposed by Lee et al. earlier for organic field-effect transistors applications.[104] The generation of mobile ions in the hydrated BSA thin film and the formation of ionic double layers (IDL) are propounded to describe the electrical characteristics. In hydrated BSA, it is observed that mobile ions are generated when acidic/basic amino acid residues interact with H_2O.[105] It is found that the acidic and basic responses of a protein is unaffected, when the α-amino and α-carboxyl group of amino acids get converted into peptide bonds in proteins as there are only one α-amino group (at the N terminus) and one α-carboxyl group (at the C terminus). When the protein is in hydrated condition, the R group of the acidicand basic amino acids interacts with water and acts as an acid or base.[106] Table 4.1 shows the properties of basic and acidic amino acids found in BSA with both the negatively and positively charged R group. The pK_R values of glutamic acid and aspartic acid, found in BSA, are 4.25 and 3.65, respectively, which are less than the pH value of water (pH = 7 for normal water).

Due to this, carboxyl groups of side chains of these amino acids are deprotonated when interact with water (H_2O) and produce H_3O^+ mobile ions and negatively charged side chains $(CH_2)_2COO^-$ & CH_2COO^-. Although Histidine has a positively charged R group, its pK_R value (pK_R = 6) is less than the pH value of water, acts as acid and deprotonates when interact with water. Similarly, the pK_Rvalues of basic amino acids lysine and arginine are 10.53 and 12.48, respectively, greater than pH value of water. Therefore, amino groups of side chains of these amino acids are protonated when react with water and separate into OH^- mobile ions and their respective positively charged $(CH_2)_4^-NH_3^+$ side chains.

TABLE 4.1 Properties of Basic and Acidic Amino Acids Found in BSA.

Amino acid	Abbreviation	Symbol	pK_a values		
			pK_1 (-COOH)	pK_2 ($-NH_3^+$)	pK_R (R group)
Positively charged R groups					
Lysine	Lys	K	2.18	8.95	10.53
Histidine	His	H	1.82	9.17	6.00
Arginine	Arg	R	2.17	9.04	12.48
Negatively charged R groups					
Aspartate	Asp	D	1.88	9.60	3.65
Glutamate	Glu	E	2.19	9.67	4.25

Figure 4.4c shows the molecular structure of BSA (Homo dimer), highlighting acidic amino acid residues with blue and basic amino acid residues with red, along with an elementary picture showing the negatively and positively charged side chains with mobile ions of hydrated BSA placed between two electrodes. The ionic double layers (IDL) are considered to be formed when proteotronic capacitors are biased at a voltage, is shown in Figure 4.4d. Since the positively and negatively charged side chains in BSA are taken to be static to give response to biased voltages as it has a long chain structure, the H_3O^+ and OH^- ions will accumulate near negative and positive electrodes, respectively, leading to the formation of ionic double layers. By drifting directly or by Grotthuss transfer of H_3O^+ and OH^- ions, there will be a free proton (H^+) mechanism to further electrodes. Emergence of this ionic double layer is supposed to increase the capacitance of the capacitor in agreement with the experimental result.

4.5 CONCLUSION

In summary, a BSA-based biodegradable, portable, light weight, and flexible proteotronic MIM capacitors on flexible trace-on paper substrate with pencil-drawn interdigitated graphite electrodes were fabricated. Electrical characterization of the proteotronic capacitor has been investigated through capacitance-voltage(C-V) characteristics. The charge storage mechanism is also explained with corresponding models. However, it is a challenge to further reduce the separation of the fingers of the IDE electrodes by adopting various alternatives procedures by not elevating the dissipation factor, which might improve the capacitance density for high-frequency device applications and to meet the requirements of ITRS roadmap.

ACKNOWLEDGMENT

Sandipan Mallik and Shrabani Guhathakurata would like to sincerely thank the Department of Science and Technology-Science and Engineering Research Board (DST-SERB), New Delhi, India, for providing research funding [project no: ECR/2018/002477] under the DST-ECR scheme.

KEYWORDS

- **bovine serum albumin (BSA)**
- **interdigitated graphite electrode**
- **pencil-on-paper (PoP) substrate**
- **charge storage**
- **flexible capacitor**

REFERENCES

1. Alfinito, E.; Pousset, J.; Reggiani, L. PROTEOTRONICS: The Emerging Science of Protein-Based Electronic Devices. *J. Phys. Conf. Ser.* **2015**, *647*, 12002.
2. Bontidean, I.; Berggren, C.; Johansson, G.; Csöregi, E.; Mattiasson, B.; Lloyd, J. R.; Jakeman, K. J.; Brown, N. L. Detection of Heavy Metal Ions at Femtomolar Levels Using Protein-Based Biosensors. *Anal. Chem.* **1998**, *70* (19), 4162–4169.
3. San, B. H.; Kim, J. A.; Kulkarni, A.; Moh, S. H.; Dugasani, S. R.; Subramani, V. K.; Thorat, N. D.; Lee, H. H.; Park, S. H.; Kim, T. Combining Protein-Shelled Platinum Nanoparticles with Graphene to Build a Bionanohybrid Capacitor. *ACS Nano* **2014**, *8* (12), 12120–12129.

4. Hota, M. K.; Bera, M. K.; Kundu, B.; Kundu, S. C.; Maiti, C. K. A Natural Silk Fibroin Protein-Based Transparent Bio-Memristor. *Adv. Funct. Mater.* **2012**, *22* (21), 4493–4499.
5. Trihutomo, P.; Soeparman, S.; Widhiyanuriyawan, D.; Yuliati, L. Performance Improvement of Dye-Sensitized Solar Cell-(DSSC-) Based Natural Dyes by Clathrin Protein. *Int. J. Photoenergy* **2019**, *2019*.
6. Zeynali, A.; Ghiasi, T. S.; Riazi, G.; Ajeian, R. Organic Solar Cell Based on Photosystem I Pigment-Protein Complex, Fabrication and Optimization. *Org. Electron.* **2017**, *51*, 341–348.
7. Zotti, L. A.; Cuevas, J. C. Electron Transport through Homopeptides: Are They Really Good Conductors? *ACS Omega* **2018**, *3* (4), 3778–3785.
8. Alfinito, E.; Reggiani, L. In *Nanodevices Based on Proteins: The Emerging Science of Proteotronics*, 2015 1st Workshop on Nanotechnology in Instrumentation and Measurement (NANOFIM); IEEE, 2015, pp. 47–50.
9. Nosrati, H.; Sefidi, N.; Sharafi, A.; Danafar, H.; Manjili, H. K. Bovine Serum Albumin (BSA) Coated Iron Oxide Magnetic Nanoparticles as Biocompatible Carriers for Curcumin-Anticancer Drug. *Bioorg. Chem.* **2018**, *76*, 501–509.
10. Campana, A.; Cramer, T.; Greco, P.; Foschi, G.; Murgia, M.; Biscarini, F.; Facile M. Fabrication of Organic Field Effect Transistors on Biodegradable Substrates. *Appl. Phys. Lett.* **2013**, *103* (7), 153_1.
11. Valanciunaite, J.; Bagdonas, S.; Streckyte, G.; Rotomskis, R. Spectroscopic Study of TPPS 4 Nanostructures in the Presence of Bovine Serum Albumin. *Photochem. Photobiol. Sci.* **2006**, *5* (4), 381–388.
12. Kragh-Hansen, U. Relations between High-Affinity Binding Sites of Markers for Binding Regions on Human Serum Albumin. *Biochem. J.* **1985**, *225* (3), 629–638.
13. Zhuang, X.; Zhang, D.; Wang, X.; Yu, X.; Yu, J. Biocompatible and Degradable Gelatin Dielectric Based Low-Operating Voltage Organic Transistors for Ultra-High Sensitivity NH_3 Detection. *Appl. Phys. Lett.* **2018**, *113* (26), 263301.
14. Klimovich, I. V; Leshanskaya, L. I.; Troyanov, S. I.; Anokhin, D. V; Novikov, D. V; Piryazev, A. A.; Ivanov, D. A.; Dremova, N. N.; Troshin, P. A. Design of Indigo Derivatives as Environment-Friendly Organic Semiconductors for Sustainable Organic Electronics. *J. Mater. Chem. C* **2014**, *2* (36), 7621–7631.
15. Mohamad, A. S.; Hamzah, R.; Hoettges, K. F.; Hughes, M. P. A Dielectrophoresis-Impedance Method for Protein Detection and Analysis. *AIP Adv.* **2017**, *7* (1), 15202.
16. Bibi, F.; Villain, M.; Guillaume, C.; Sorli, B.; Gontard, N. A Review: Origins of the Dielectric Properties of Proteins and Potential Development as Bio-Sensors. *Sensors.* **2016**. https://doi.org/10.3390/s16081232.
17. Meena, J. S.; Chu, M.-C.; Kuo, S.-W.; Chang, F.-C.; Ko, F.-H. Improved Reliability from a Plasma-Assisted Metal-Insulator-Metal Capacitor Comprising a High-k HfO_2 Film on a Flexible Polyimide Substrate. *Phys. Chem. Chem. Phys.* **2010**, *12* (11), 2582–2589.
18. Meena, J. S.; Chu, M.-C.; Singh, R.; Shieh, H.-P. D.; Liu, P.-T.; Ko, F.-H. Controlled Deposition of New Organic Ultrathin Film as a Gate Dielectric Layer for Advanced Flexible Capacitor Devices. *J. Mater. Sci. Mater. Electron.* **2013**, *24* (6), 1807–1812.
19. Numan-Al-Mobin, A. M.; Petersen, J.; Liu, M.; Cross, W. M.; Kellar, J. J.; Crawford, G. A.; Jordan, J.; Ponchak, G. E. Additive Manufacturing of RF Metal-Insulator-Metal (MIM) Capacitors on Flexible Substrate. In 2019 IEEE Aerospace Conference; IEEE, 2019; pp. 1–10.

20. Hota, M. K.; Bera, M. K.; Maiti, C. K. Flexible Metal–Insulator–Metal Capacitors on Polyethylene Terephthalate Plastic Substrates. *Semicond. Sci. Technol.* **2012**, *27* (10), 105001.
21. Sun, S.; Duan, Z.; Wang, X.; Lai, G.; Zhang, D.; Wei, H.; Liu, L.; Ma, N. Cheap, Flexible, and Thermal-Sensitive Paper Sensor through Writing with Ionic Liquids Containing Pencil Leads. *ACS Appl. Mater. Interfaces* **2017**, *9* (34), 29140–29146.
22. Koga, H.; Nagashima, K.; Huang, Y.; Zhang, G.; Wang, C.; Takahashi, T.; Inoue, A.; Yan, H.; Kanai, M.; He, Y. Based Disposable Molecular Sensor Constructed from Oxide Nanowires, Cellulose Nanofibers, and Pencil-Drawn Electrodes. *ACS Appl. Mater. Interfaces* **2019**, *11* (16), 15044–15050.
23. Kanaparthi, S.; Badhulika, S. Low Cost, Flexible and Biodegradable Touch Sensor Fabricated by Solvent-Free Processing of Graphite on Cellulose Paper. *Sensors Actuators B Chem.* **2017**, *242*, 857–864.
24. Anagnostou, D. E.; Gheethan, A. A.; Amert, A. K.; Whites, K. W. A Direct-Write Printed Antenna on Paper-Based Organic Substrate for Flexible Displays and WLAN Applications. *J. Disp. Technol.* **2010**, *6* (11), 558–564.
25. Prebianto, N. F.; Futra, A. D. Paper as a Substrate for Sensor Applications: A Review. *In 2018 International Conference on Applied Engineering (ICAE)*; IEEE, 2018; pp. 1–5.
26. Siegel, A. C.; Phillips, S. T.; Wiley, B. J.; Whitesides, G. M. Thin, Lightweight, Foldable Thermochromic Displays on Paper. *Lab Chip* **2009**, *9* (19), 2775–2781.
27. Nystrom, G.; Razaq, A.; Strømme, M.; Nyholm, L.; Mihranyan, A. Ultrafast All-Polymer Paper-Based Batteries. *Nano Lett.* **2009**, *9* (10), 3635–3639.
28. Siegel, A. C.; Phillips, S. T.; Dickey, M. D.; Lu, N.; Suo, Z.; Whitesides, G. M. Foldable Printed Circuit Boards on Paper Substrates. *Adv. Funct. Mater.* **2010**, *20* (1), 28–35.
29. Sekitani, T.; Nakajima, H.; Maeda, H.; Fukushima, T.; Aida, T.; Hata, K.; Someya, T. Stretchable Active-Matrix Organic Light-Emitting Diode Display Using Printable Elastic Conductors. *Nat. Mater.* **2009**, *8* (6), 494–499.
30. Yokota, T.; Zalar, P.; Kaltenbrunner, M.; Jinno, H.; Matsuhisa, N.; Kitanosako, H.; Tachibana, Y.; Yukita, W.; Koizumi, M.; Someya, T. Ultraflexible Organic Photonic Skin. *Sci. Adv.* **2016**, *2* (4), 1–9. https://doi.org/10.1126/sciadv.1501856.
31. Jang, K. I.; Han, S. Y.; Xu, S.; Mathewson, K. E.; Zhang, Y.; Jeong, J. W.; Kim, G. T.; Webb, R. C.; Lee, J. W.; Dawidczyk, T. J.; Kim, R. H.; Song, Y. M.; Yeo, W. H.; Kim, S.; Cheng, H.; Rhee, S. Il; Chung, J.; Kim, B.; Chung, H. U.; Lee, D.; Yang, Y.; Cho, M.; Gaspar, J. G.; Carbonari, R.; Fabiani, M.; Gratton, G.; Huang, Y.; Rogers, J. A. Rugged and Breathable Forms of Stretchable Electronics with Adherent Composite Substrates for Transcutaneous Monitoring. *Nat. Commun.* **2014**, *5*, 1–10.
32. Someya, T.; Sekitani, T.; Iba, S.; Kato, Y.; Kawaguchi, H.; Sakurai, T.; Whitesides, G. M. A Large-Area, Flexible Organic Skin Applications Pressure Transistors Sensor Matrix for Artificial With. *Proc. Natl. Acad. Sci. U. S. A.* **2010**, *101* (27), 9966–9970.
33. Kim, D. H.; Lu, N.; Ghaffari, R.; Kim, Y. S.; Lee, S. P.; Xu, L.; Wu, J.; Kim, R. H.; Song, J.; Liu, Z.; Viventi, J.; De Graff, B.; Elolampi, B.; Mansour, M.; Slepian, M. J.; Hwang, S.; Moss, J. D.; Won, S. M.; Huang, Y.; Litt, B.; Rogers, J. A. Materials for Multifunctional Balloon Catheters with Capabilities in Cardiac Electrophysiological Mapping and Ablation Therapy. *Nat. Mater.* **2011**, *10* (4), 316–323.
34. Lee, C. H.; Kim, H.; Harburg, D. V.; Park, G.; Ma, Y.; Pan, T.; Kim, J. S.; Lee, N. Y.; Kim, B. H.; Jang, K. I.; Kang, S. K.; Huang, Y.; Kim, J.; Lee, K. M.; Leal, C.;

Rogers, J. A. Biological Lipid Membranes for On-Demand, Wireless Drug Delivery from Thin, Bioresorbable Electronic Implants. *NPG Asia Mater.* **2015**, *7* (11), e227-9.

35. Ko, H.; Lee, J.; Kim, Y.; Lee, B.; Jung, C. H.; Choi, J. H.; Kwon, O. S.; Shin, K. Active Digital Microfluidic Paper Chips with Inkjet-Printed Patterned Electrodes. *Adv. Mater.* **2014**, *26* (15), 2335–2340.
36. Gao, W.; Emaminejad, S.; Nyein, H. Y. Y.; Challa, S.; Chen, K.; Peck, A.; Fahad, H. M.; Ota, H.; Shiraki, H.; Kiriya, D.; Lien, D. H.; Brooks, G. A.; Davis, R. W.; Javey, A. Fully Integrated Wearable Sensor Arrays for Multiplexed in Situ Perspiration Analysis. *Nature* **2016**, *529* (7587), 509–514.
37. Lee, H.; Choi, T. K.; Lee, Y. B.; Cho, H. R.; Ghaffari, R.; Wang, L.; Choi, H. J.; Chung, T. D.; Lu, N.; Hyeon, T.; Choi, S. H.; Kim, D. H. A Graphene-Based Electrochemical Device with Thermoresponsive Microneedles for Diabetes Monitoring and Therapy. *Nat. Nanotechnol.* **2016**, *11* (6), 566–572.
38. Imani, S.; Bandodkar, A. J.; Mohan, A. M. V.; Kumar, R.; Yu, S.; Wang, J.; Mercier, P. P. A Wearable Chemical-Electrophysiological Hybrid Biosensing System for Real-Time Health and Fitness Monitoring. *Nat. Commun.* **2016**, *7* (May), 1–7.
39. Nery, E. W.; Kubota, L. T. Sensing Approaches on Paper-Based Devices: A Review. *Anal. Bioanal. Chem.* **2013**, *405* (24), 7573–7595.
40. Weng, M.; Zhou, P.; Chen, L.; Zhang, L.; Zhang, W.; Huang, Z.; Liu, C.; Fan, S. Multiresponsive Bidirectional Bending Actuators Fabricated by a Pencil-on-Paper Method. *Adv. Funct. Mater.* **2016**, *26* (40), 7244–7253.
41. Russo, A.; Ahn, B. Y.; Adams, J. J.; Duoss, E. B.; Bernhard, J. T.; Lewis, J. A. Pen-on-Paper Flexible Electronics. *Adv. Mater.* **2011**, *23* (30), 3426–3430.
42. Yu, L.; Shi, Z. Z. Microfluidic Paper-Based Analytical Devices Fabricated by Low-Cost Photolithography and Embossing of Parafilm®. *Lab Chip* **2015**, *15* (7), 1642–1645.
43. Li, Y.; Liu, C.; Xu, Y.; Minari, T.; Darmawan, P.; Tsukagoshi, K. Solution-Processed Organic Crystals for Field-Effect Transistor Arrays with Smooth Semiconductor/Dielectric Interface on Paper Substrates. *Org. Electron.* **2012**, *13* (5), 815–819.
44. Dungchai, W.; Chailapakul, O.; Henry, C. S. A Low-Cost, Simple, and Rapid Fabrication Method for Paper-Based Microfluidics Using Wax Screen-Printing. *Analyst* **2011**, *136* (1), 77–82.
45. Kawde, A. N.; Aziz, M.; Baig, N.; Temerk, Y. A Facile Fabrication of Platinum Nanoparticle-Modified Graphite Pencil Electrode for Highly Sensitive Detection of Hydrogen Peroxide. *J. Electroanal. Chem.* **2015**, *740*, 68–74.
46. Thompson, J. E. Pencil-on-Paper Capacitors for Hand-Drawn RC Circuits and Capacitive Sensing. *J. Chem.* **2017**, *2017*, 1–5.
47. Down, M. P.; Foster, C. W.; Ji, X.; Banks, C. E. Pencil Drawn Paper Based Supercapacitors. *RSC Adv.* **2016**, *6* (84), 81130–81141.
48. Sałlam, Ö.; KIzIlkaya, B.; Uysal, H.; Dilgin, Y. Biosensing of Glucose in Flow Injection Analysis System Based on Glucose Oxidase-Quantum Dot Modified Pencil Graphite Electrode. *Talanta* **2016**, *147*, 315–321.
49. Santhiago, M.; Kubota, L. T. A New Approach for Paper-Based Analytical Devices with Electrochemical Detection Based on Graphite Pencil Electrodes. *Sensors Actuators, B Chem.* **2013**, *177*, 224–230.
50. Bracher, P. J.; Gupta, M.; Whitesides, G. M. Patterning Precipitates of Reactions in Paper. *J. Mater. Chem.* **2010**, *20* (24), 5117–5122.

51. Ansari, M. A. Paper Based ZnO Thin Film UV Light Detector Using Graphite Pencil Based Electrodes. *Adv. Mater. Proc.* **2017**, *2* (7), 428–432.
52. Kanaparthi, S. Pencil-Drawn Paper-Based Non-Invasive and Wearable Capacitive Respiration Sensor. *Electroanalysis* **2017**, *29* (12), 2680–2684.
53. Lin, C. W.; Zhao, Z.; Kim, J.; Huang, J. Pencil Drawn Strain Gauges and Chemiresistors on Paper. *Sci. Rep.* **2014**, *4*, 1–6.
54. Niu, Z.; Zhang, L.; Liu, L.; Zhu, B.; Dong, H.; Chen, X. All-Solid-State Flexible Ultrathin Micro-Supercapacitors Based on Graphene. *Adv. Mater.* **2013**, *25* (29), 4035–4042.
55. Gao, W.; Singh, N.; Song, L.; Liu, Z.; Reddy, A. L. M.; Ci, L.; Vajtai, R.; Zhang, Q.; Wei, B.; Ajayan, P. M. Direct Laser Writing of Micro-Supercapacitors on Hydrated Graphite Oxide Films. *Nat. Nanotechnol.* **2011**, *6* (8), 496–500.
56. Li, J.; Mishukova, V.; Östling, M. All-Solid-State Micro-Supercapacitors Based on Inkjet Printed Graphene Electrodes. **2016**, *123901*.
57. Ul Hasan, K.; Nur, O.; Willander, M. Screen Printed ZnO Ultraviolet Photoconductive Sensor on Pencil Drawn Circuitry over Paper. *Appl. Phys. Lett.* **2012**, *100* (21).
58. Zhang, J.; Huang, L.; Lin, Y.; Chen, L.; Zeng, Z.; Shen, L.; Chen, Q. Sensors Pencil-Trace on Printed Silver Interdigitated Electrodes for Paper-Based NO_2 Gas Sensors. **2015**, *143101* (2), 1–6.
59. Hong, J. K. B. M. S. Capacitive Humidity Sensors Based on a Newly Designed Interdigitated Electrode Structure. **2012**, 31–35.
60. Igreja, R.; Dias, C. J. Analytical Evaluation of the Interdigital Electrodes Capacitance for a Multi-Layered Structure. *Sensors Actuators A Phys.* **2004**, *112* (2–3), 291–301.
61. Bonincontro, A.; Mari, C.; Mengoni, M.; Risuleo, G. A Study of the Dielectric Properties of E. Coli Ribosomal RNA and Proteins in Solution. *Biophys. Chem.* **1997**, *67* (1–3), 43–50. https://doi.org/10.1016/S0301-4622(97)00014-8.
62. Pitera, J. W.; Falta, M.; van Gunsteren, W. F. Dielectric Properties of Proteins from Simulation: The Effects of Solvent, Ligands, PH, and Temperature. *Biophys. J.* **2001**, *80* (6), 2546–2555.
63. Figueiró, S. D.; Macêdo, A. A. M.; Melo, M. R. S.; Freitas, A. L. P.; Moreira, R. A.; De Oliveira, R. S.; Góes, J. C.; Sombra, A. S. B. On the Dielectric Behaviour of Collagen-Algal Sulfated Polysaccharide Blends: Effect of Glutaraldehyde Crosslinking. *Biophys. Chem.* **2006**, *120* (2), 154–159.
64. Lund, M.; Jönsson, B.; Woodward, C. E. Implications of a High Dielectric Constant in Proteins. *J. Chem. Phys.* **2007**, *126* (22), 225103.
65. Scheider, W.; Dintzis, H. M.; Oncley, J. L. Changes in the Electric Dipole Vector of Human Serum Albumin Due to Complexing with Fatty Acids. *Biophys. J.* **1976**, *16* (5), 417–431.
66. Gannett, P. M.; Bostick, C. D.; Mukhopadhyay, S.; Pecht, I. Protein Bioelectronics : A Review of What We Do and Do Not Know. **2018**.
67. Wang, X. D.; Wolfbeis, O. S. Fiber-Optic Chemical Sensors and Biosensors (2015-2019). *Anal. Chem.* **2020**, *92* (1), 397–430.
68. Ligler, F. S.; Taitt, C. R.; Shriver-Lake, L. C.; Sapsford, K. E.; Shubin, Y.; Golden, J. P. Array Biosensor for Detection of Toxins. *Anal. Bioanal. Chem.* **2003**, *377* (3), 469–477.
69. Palchetti, I.; Mascini, M. Electroanalytical Biosensors and Their Potential for Food Pathogen and Toxin Detection. *Anal. Bioanal. Chem.* **2008**, *391* (2), 455–471.

70. Baj-Rossi, C.; de Micheli, G.; Carrara, S. Electrochemical Detection of Anti-Breast-Cancer Agents in Human Serum by Cytochrome P450-Coated Carbon Nanotubes. *Sensors (Switzerland)* **2012**, *12* (5), 6520–6537.
71. Baj-rossi, C.; Micheli, G. De; Carrara, S. For Personalized Medicine. **2005**.
72. Quershi, A.; Gurbuz, Y.; Kang, W. P.; Davidson, J. L. Novel Interdigitated Capacitor Based Biosensor for Detection of Cardiovascular Risk Marker. *Biosensors Bioelectron. A* **2009**, *25*, 877–882.
73. Singh, R.; Lin, Y.; Chuang, W.; Ko, F. AC SC. *Org. Electron.* **2017**.
74. Ding, S.-J.; Zhu, C.; Li, M.-F.; Zhang, D. W. Atomic-Layer-Deposited Al_2O_3–HfO_2–Al_2 O_3 Dielectrics for Metal-Insulator-Metal Capacitor Applications. *Appl. Phys. Lett.* **2005**, *87* (5), 53501.
75. Van Huylenbroeck, S.; Decoutere, S.; Venegas, R.; Jenei, S.; Winderickx, G. Investigation of PECVD Dielectrics for Nondispersive Metal-Insulator-Metal Capacitors. *IEEE Electron Device Lett.* **2002**, *23* (4), 191–193.
76. Bécu, S.; Crémer, S.; Autran, J.-L. Microscopic Model for Dielectric Constant in Metal-Insulator-Metal Capacitors with High-Permittivity Metallic Oxides. *Appl. Phys. Lett.* **2006**, *88* (5), 52902.
77. Gonon, P.; Vallée, C. Modeling of Nonlinearities in the Capacitance-Voltage Characteristics of High-k Metal-Insulator-Metal Capacitors. *Appl. Phys. Lett.* **2007**, *90* (14), 142906.
78. Blonkowski, S. Nonlinear Capacitance Variations in Amorphous Oxide Metal-Insulator-Metal Structures. *Appl. Phys. Lett.* **2007**, *91* (17), 172903.
79. Wenger, C.; Lupina, G.; Lukosius, M.; Seifarth, O.; Müssig, H.-J.; Pasko, S.; Lohe, C. Microscopic Model for the Nonlinear Behavior of High-k Metal-Insulator-Metal Capacitors. *J. Appl. Phys.* **2008**, *103* (10), 104103.
80. Kliem, H. Non-Linear Dielectric Response by Protons in Amorphous Insulators. In [Proceedings] 1992 Annual Report: Conference on Electrical Insulation and Dielectric Phenomena; IEEE, 1992, pp. 81–86.
81. Gonon, P.; Vallée, C. Understanding Capacitance-Voltage Nonlinearities in Microelectronic Metal-Insulator-Metal (MIM) Capacitors. In 2015 IEEE 11th International Conference on the Properties and Applications of Dielectric Materials (ICPADM); IEEE, 2015, pp. 636–639.
82. Tan, K.-S.; Kiriaki, S.; De Wit, M.; Fattaruso, J. W.; Tsay, C.-Y.; Matthews, W. E.; Hester, R. K. Error Correction Techniques for High-Performance Differential A/D Converters. *IEEE J. Solid-State Circuits* **1990**, *25* (6), 1318–1327.
83. Arden, W. M. The International Technology Roadmap for Semiconductors—Perspectives and Challenges for the next 15 Years. *Curr. Opin. Solid State Mater. Sci.* **2002**, *6* (5), 371–377.
84. Eden, J.; Gascoyne, P. R. C.; Pethig, R. Dielectric and Electrical Properties of Hydrated Bovine Serum Albumin. *J. Chem. Soc. Faraday Trans. 1 Phys. Chem. Condens. Phases* **1980**, *76* (0), 426–434.
85. Babcock, J. A.; Balster, S. G.; Pinto, A.; Dirnecker, C.; Steinmann, P.; Jumpertz, R.; El-Kareh, B. Analog Characteristics of Metal-Insulator-Metal Capacitors Using PECVD Nitride Dielectrics. *IEEE Electron Device Lett.* **2001**, *22* (5), 230–232.
86. Fang, C.; Wang, M.; Han, P.; Cao, Y.-Q.; Wu, D.; Li, A.-D. High-Performance MIM Capacitors Using Zr-Sn-Ti-O Dielectrics Derived from Atomic Layer Deposition. *IEEE Electron Device Lett.* **2019**, *40* (5), 682–685.

87. Besset, C.; Bruyere, S.; Blonkowski, S.; Cremer, S.; Vincent, E. MIM Capacitance Variation under Electrical Stress. *Microelectron. Reliab.* **2003**, *43* (8), 1237–1240.
88. Thomas, R.; Dube, D. C.; Kamalasanan, M. N.; Chandra, S.; Bhalla, A. S. Structural, Electrical, and Low-Temperature Dielectric Properties of Sol–Gel Derived $SrTiO_3$ Thin Films. *J. Appl. Phys.* **1997**, *82* (9), 4484–4488.
89. Lin, C.-M.; Wu, P.-C.; Lee, M.-J.; Lee, W. Label-Free Protein Quantitation by Dielectric Spectroscopy of Dual-Frequency Liquid Crystal. *Sensors Actuators B Chem.* **2019**, *282*, 158–163.
90. Kannadassan, D.; Karthik, R.; Baghini, M. S.; Mallick, P. S. Modeling the Voltage Nonlinearity of High-k MIM Capacitors. *Solid. State. Electron.* **2014**, *91*, 112–117.
91. Guyton, A. C.; Hall, J. E. Medical Physiology. *Gökhan N, Çavuşoğlu H* **2006**, *3*.
92. Lehninger, A. L. *Lehninger Principles of Biochemistry: David L. Nelson, Michael M. Cox*; Recording for the Blind & Dyslexic, 2004.
93. Westbrook, J. D.; Shao, C.; Feng, Z.; Zhuravleva, M.; Velankar, S.; Young, J. The Chemical Component Dictionary: Complete Descriptions of Constituent Molecules in Experimentally Determined 3D Macromolecules in the Protein Data Bank. *Bioinformatics* **2015**, *31* (8), 1274–1278.
94. Creighton, T. E. *Proteins: Structures and Molecular Properties*; Macmillan, 1993.
95. Petrucci, R. H.; Herring, F. G.; Bissonnette, C.; Madura, J. D. *General Chemistry: Principles and Modern Applications*; Pearson, 2017.
96. Rossotti, F. J. C.; Rossotti, H. *The Determination of Stability Constants: And Other Equilibrium Constants in Solution*; McGraw-Hill, 1961.
97. Perrin, D. D.; Dempsey, B.; Serjeant, E. P. *PKa Prediction for Organic Acids and Bases*; Springer, 1981; Vol. 1.
98. Bard, A. J.; Faulkner, L. R.; Fundamentals and Applications. *Electrochem. Methods* **2001**, *2* (482), 580–632.
99. McNaught, A. D.; Wilkinson, A. *Compendium of Chemical Terminology*; Blackwell Science Oxford, 1997; Vol. 1669.
100. Cornishbowden, A. Nomenclature and Symbolism for Amino-Acids and Peptides Recommendations 1983. *Eur J Biochem.* **1984**, *138*, 9–37.
101. Meisenberg, G.; Simmons, W. H. *Principles of Medical Biochemistry*; Elsevier Health Sciences, 2006.
102. Noy, A. Bionanoelectronics. *Adv. Mater.* **2011**, *23* (7), 807–820.
103. Svennersten, K.; Larsson, K. C.; Berggren, M.; Richter-Dahlfors, A. Organic Bioelectronics in Nanomedicine. *Biochim. Biophys. Acta (BBA)-General Subj.* **2011,** *1810* (3), 276–285.
104. Lee, C. Y.; Hwang, J. C.; Chueh, Y. N.; Chang, T. H.; Cheng, Y. Y.; Lyu, P. C. Hydrated Bovine Serum Albumin as the Gate Dielectric Materialfor Organic Field-effect Transistors. *Organic Electron.* **2013**, *14*, 2645–2651.
105. Nielsen, J. E.; Vriend, G. Optimizing the Hydrogen-bond Network in Poisson–Boltzmann Equation-based PKa Calculations. *Proteins Struct. Funct. Bioinforma.* **2001**, *43* (4), 403–412.
106. Aftabuddin, M.; Kundu, S. Hydrophobic, Hydrophilic, and Charged Amino Acid Networks within Protein. *Biophys. J.* **2007**, *93* (1), 225–231.

CHAPTER 5

Recent Advancement in Graphene-Based Metasurface Structures

SAMBIT KUMAR GHOSH[1], SANTANU DAS[2], and
SOMAK BHATTACHARYYA[1*]

[1]*Department of Electronics Engineering, Indian Institute of Technology (Banaras Hindu University), Varanasi, Uttar Pradesh 221005, India*

[2]*Department of Ceramic Engineering, Indian Institute of Technology (Banaras Hindu University), Varanasi, Uttar Pradesh 221005, India*

[*]*Corresponding author. E-mail: sambitkrghosh.rs.ece17@iitbhu.ac.in*

ABSTRACT

Surface plasmon polaritons (SPP) formed in graphene exhibits promising properties to become one of the mainstream research topics in nanophotonics. Graphene surface plasmons (GSPs) are excited most efficiently in periodic graphene-based devices, including diffraction gratings, graphene plasmonic crystals, corrugated graphene and conductivity gratings, etc. The metasurface is a strong candidate for electromagnetic wave modulation in optical systems because EM wave can be manipulated in terms of amplitude and phase. Metallic metasurfaces suffer from unwanted correlation between amplitude and phase control of the EM wave as one meta-atom has only one resonance condition. On the other hand, graphene metasurface can achieve tunable conductivity under the exposure of external biasing, chemical doping or stretching the graphene surface mechanically. Therefore, the graphene meta-atoms can achieve multiple resonance conditions within a single design, which further allows them to complete the wavefront's complete control. Therefore, periodic graphene ribbons have been used independently on top of the dielectric, working as tunable terahertz (THz) metasurfaces. Graphene has been integrated with the metallic nanostructures as a hybrid combination to enhance the performance of existing metallic metasurface devices. Later, only graphene patterns have been employed to enhance SPP of the electromagnetic wave at the graphene-dielectric interface for several applications like cross-polarization converters, absorbers, and spatial filters etc. Periodic single-layer graphene patterns have been used in the metadevices for various applications in the THz domain. Recently, two separately patterned layers of graphene have been employed on the two opposite sides of a dielectric substrate for achieving maximum absorption possible of EM wave within the THz gap. The tunable surface conductivity of graphene makes it possible to attain versatile characteristics by altering the Fermi level in graphene.

5.1 INTRODUCTION TO METASURFACE

Electromagnetic metasurface, the two-dimensional equivalent of metamaterial, exhibits exotic electromagnetic (EM) properties under the incident electromagnetic wave.[1] Metasurface generates great interest to the scientific community for achieving unusual properties under the incidence of electromagnetic wave because of its sub-wavelength thickness and periodicity.[2–6] Metasurface manipulates the wavefront of the incoming EM wave approaching it by imposing gradient phase shift, thereby resulting in the generalized Snell's law of light propagation.[2] Electromagnetic (EM) wave analysis through the two-dimensional (2D) metasurface structures can be assumed to have a wide range of values of effective medium parameters instead of the bulk parameters. Metasurfaces have been reported for various applications including absorbers, polarization converters, antennas, filters, etc. over a broader electromagnetic spectrum.[7–15]

5.2 INTRODUCTION TO GRAPHENE

Graphene, a two-dimensional (2D) material, has attracted immense growing interest to the research community due to its outstanding electrical, mechanical, thermal, and optical properties. Graphene is a 2D allotrope of carbon.[16, 17] Apart from graphene, carbon has several allotropes, viz., graphite, diamond, carbon nanotubes, fullerene, etc.[18] Graphene can be demonstrated chemically by defining it as a single-layered atomically thick slice of graphite with hexagonal lattice with extraordinarily high tensile strength of 130 gigapascals.[19] Carbon has four electrons in its outermost cell while taking part in the covalent bonding process. On the graphene lattice, the neighboring electrons contribute in the sigma bond formation along the horizontal plane, whereas pi-electrons are placed along the vertical plane.[18] These delocalized pi-electrons are moving in nature and play a key role in achieving high mobility of the electrons on the 2D graphene surface, which leads to very high electrical conductivity (σ). Pristine graphene is generally a zero-bandgap semiconductor, and it is very promising for the nanotechnology domain as the Dirac points are the junctions of the conduction and valance bands.[18] The conductivity of graphene is tunable under external factors, such as magnetostatic gating, electrostatic gating, and chemical doping.[16, 20] These external factors alter the Fermi level in graphene, which is essential for forming the surface plasmon polariton (SPP).[21, 22] The versatile nature of graphene forms a useful platform for futuristic applications for its integration in the metasurface domain, including absorbers, polarizers, filters, cloaking, nonlinear optical applications, and antenna applications, etc.[23–35] (Figure 5.1).

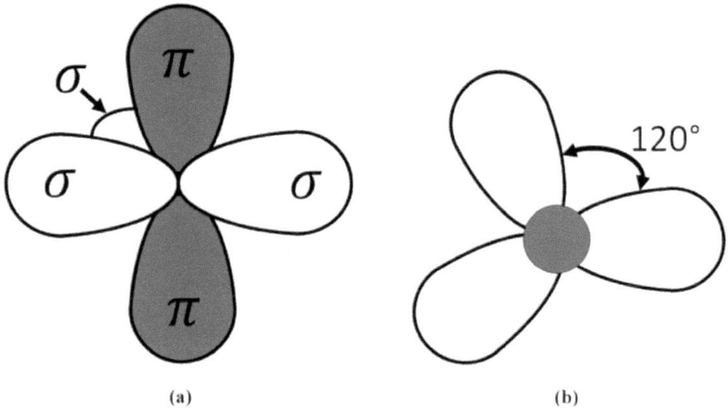

FIGURE 5.1 Schematic representation of formation of covalent bonds in graphene.

5.3 MATHEMATICAL MODELING OF GRAPHENE

Monolayer graphene can be considered as an infinitesimal thin surface with zero bandgap. The 2D graphene surface's conductivity is assumed to be the most suitable parameter to define the electromagnetic properties of the graphene-based devices. Therefore, the graphene layer is modeled in terms of surface conductivity formulated by Kubo.[36] Graphene has an unconventional electrical conductivity, which is the crucial factor for the absorption of the electromagnetic (EM) wave within the mid-infrared (MIR) spectrum. The complex surface conductivity of graphene has two components viz., interband conductivity and intraband conductivity as proposed by Kubo and provided in eq 5.1,[36] where μ = chemical potential of graphene produced via chemical doping and mechanical straining of the graphene layer, k_B = Boltzmann's constant, e = electronic charge, E_f = Fermi energy of graphene, ω = angular frequency of the incident EM wave, T = operating temperature considered as 300 K, Γ = scattering rate of the Dirac electrons considered as inversely proportional to relaxation time (τ), $\hbar$ = reduced Plank's constant, and $f_d(\varepsilon) = \left(e^{\frac{(\varepsilon-\mu)}{k_B T}} + 1\right)^{-1}$, the Fermi Dirac distribution function, ε = the associated energy. The above-said conductivity components of graphene are provided in eqs 5.2 and 5.3; the frequency-dependent transfer function $H(\Omega)$ is defined in eq 5.4.[36]

$$\sigma(\omega, E_f, \Gamma, T) = \sigma_{inter} + \sigma_{intra} \tag{5.1}$$

$$\sigma_{intra} = \frac{2k_b T e^2}{\pi\hbar^2} ln\left(2\cos h\frac{E_f}{2k_b T}\right)\frac{i}{(\omega + i\Gamma)} \tag{5.2}$$

$$\sigma_{inter} = \frac{e^2}{4\hbar}\left[H\left(\frac{\omega}{2}\right) + i\frac{4\omega}{\pi}\int_0^\infty \frac{H(\Omega) - H\left(\frac{\omega}{2}\right)}{\omega^2 - 4\Omega^2} d\Omega\right] \tag{5.3}$$

$$H(\Omega) = sinh\left(\frac{\hbar\Omega}{k_b T}\right) \Big/ \left[\cos h\left(\frac{\hbar\Omega}{k_b T}\right) + \cos h\left(\frac{E_f}{k_b T}\right)\right] \tag{5.4}$$

The intraband conductivity rules over the interband one when the Fermi level is more than half of the photon energy. In this condition, the latter one is insignificant due to Pauli blocking.[37] The Fermi level (E_f) can be modulated by applying external electrical biasing, chemical doping, or mechanical stretching, as evident from eq 5.5. As per the Drude mathematics, the surface conductivity of graphene (σ_g) is directly dependent on the Fermi level (E_f) within the infrared region as verified from eq 5.6.[30] The relative permittivity $\varepsilon(\omega)$is directly dependent on σ_g, where t_g stands for the thickness of the 2D graphene layer as experienced from eq 5.7. The single-layer graphene is usually considered as a 1 nm thick ultrathin film.[38–40]

$$E_f = \hbar v_f \sqrt{\frac{\pi \varepsilon_p \varepsilon_o \mu}{et^2}} \tag{5.5}$$

$$\sigma_g = \frac{e^2 E_f}{\pi \hbar^2} \cdot \frac{i}{\omega + i\tau^{-1}} \tag{5.6}$$

$$\varepsilon(\omega) = 1 + i\frac{\sigma_g(\omega)}{t_g \varepsilon_o(\omega)} \tag{5.7}$$

$$\sigma_{xx} = \sigma_{yy} = \sigma(\mu(E_o)) \tag{5.8}$$

$$C_{ox} E_o = en_s \tag{5.9}$$

The conductivity of graphene surface is of a scalar value in the presence of the DC electrostatic bias (E_o) and in the absence of DC magnetostatic bias as described in eq 5.8.[41] This concept derives a linear relation between the conductivity of graphene (σ_g) and the chemical potential (μ), a function of the external electric field (E_o). The DC gate voltage can be expressed methodically as $V_g = E_o d$, where d stands for the substrate thickness. Thus, the chemical potential (μ) of graphene can be nurtured by applying external electric field (E_o). The surface charge density (n_s) of two-dimensional (2D) graphene relating to gate capacitance (C_{ox}) is recalculated in eq 5.9 in terms of μ stated above, where e signifies the electronic charge, v_f is the Fermi velocity of electrons in graphene measured as 10^6 m/s and E_o is applied DC electrostatic bias. The quantum capacitance has a negligible importance here because the dielectric substrate is large in thickness.[42] E_o is solved under the condition of undoped and ungated case ($\mu = n_s = 0$); eq 5.12 is deduced from this fact where ε_b is the dielectric constant of the substrate in the equation Dn = $\varepsilon_b E_o = en_s/2$. D_n is the normalized component of the displacement vector on each side of a charged sheet in a homogeneous dielectric substrate with a dielectric constant of ε_b.[43] The maxima in chemical potential (μ_{max}) can be reached by using the eq 5.12, expressed in the eqs 5.13 and 5.14 where E_{bd} stands for the breakdown voltage of a dielectric substrate. The value of E_{bd} is ~ 1.5 V/nm for silicon dioxide (SiO_2) substrate having $\varepsilon_r = 3.9$.[44]

$$n_s = \frac{2}{\pi \hbar^2 v_f^2} \int_0^\infty \varepsilon \left[f_d(\varepsilon - \mu) - f_d(\varepsilon + \mu) \right] d\varepsilon \tag{5.10}$$

$$\frac{2\varepsilon_b E_o}{e} = \frac{2}{\pi\hbar^2 v_f^{\,2}} \int_0^\infty \varepsilon\left[f_d(\varepsilon-\mu) - f_d(\varepsilon+\mu)\right] d\varepsilon \tag{5.11}$$

$$E_o = \frac{e}{\pi\varepsilon_b b\hbar^2 v_f^{\,2}} \int_0^\infty \left[f_d(\varepsilon-\mu) - f_d(\varepsilon+\mu)\right] d\varepsilon \tag{5.12}$$

$$E_o \simeq \frac{e}{\varepsilon_o\varepsilon_r} \frac{1}{\pi} \left[\frac{\mu}{\hbar v_f}\right]^2 \tag{5.13}$$

$$\mu_{max} = h v_f \sqrt{\frac{\pi\varepsilon_o\varepsilon_r E_{bd}}{e}} \tag{5.14}$$

Graphene was historically first integrated with electronics as a transistor operating twice the speed of its silicon counterpart by IBM in 2011.[45] Later, integration of wafer-scale in the graphene-made electronic devices has been reported.[46] Flexible electronics find a way for the transistors' high performance by the synthesis of large-area graphene.[47] Hybrid graphene-based transistors came into the picture recently due to the least base transit time induced by the single atomic graphene layer.[48] Similarly, graphene-based RF field-effect transistors, 2D heterostructures, 2D-metal oxide semiconductor capacitor, and frequency multipliers have also been reported to date.[49–52]

5.4 EVOLUTION OF SPP: FROM DIELECTRIC-METAL INTERFACES TO DIELECTRIC-GRAPHENE INTERFACES

An electromagnetic surface wave is generally created at the interface between two different materials occupying two half-spaces. The surface wave will be absent if the two materials under consideration are identical. This fact confirms that a surface wave cannot be created at any interface of two different materials comprising distinct refractive indices. Surface plasmon polaritons (SPP), an essential aspect in plasmonics, is generated from the coupling between the electric fields and oscillations of the electron plasma in the material.[53] The wave confinement occurs at dielectric-metal interfaces due to the distinct refractive indices for using two different dielectric media when the EM wave falls upon the periodic metallic metasurface structures.[54] The schematic representations of the single-layer dielectric-metal interface and multilayer dielectric-metal interfaces have been demonstrated in Figure 5.2a and b, where q_{spp} is the wave vector of the SPP mode.

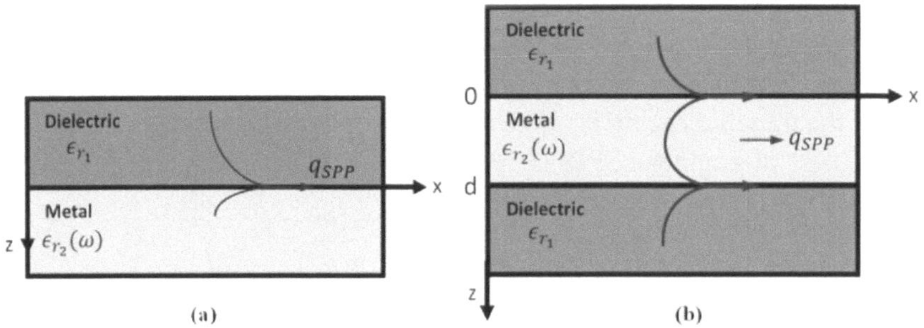

FIGURE 5.2 Schematic representation of the formation of SPP at (a) a dielectric-metal interface and (b) interfaces between dielectric-metal-dielectric structures.

In this section, the formation of surface plasmon polaritons (SPPs) has been illustrated in two different configurations, viz., a dielectric-metal interface and other interfaces between dielectric-metal-dielectric. In the first case, that is, dielectric-metal interface, both the media are assumed to be semi-infinite and isotropic with the dielectric occupying half-space (Z < 0), as shown in Figure 5.2a. The dielectric material can be defined using a positive real dielectric constant $\in_{r1}$, and the metal is defined by a frequency-dependent complex dielectric function of the following form:

$$\in_{r2}(\omega) = 1 - \frac{\omega_p^{\ 2}}{\omega^2 + i\omega\gamma} \tag{5.15}$$

The above equation follows the Drude model of the metal at higher frequencies in which it is regarded as a free electron gas.[55] The complex dielectric function for metal has been discussed in eq 5.15, where $\omega_p^{\ 2} = \frac{e^2 n_e}{m_e \in_0}$ denotes the square of the plasma frequency of the electron gas, γ is the scattering rate of the electrons and $\in_{r2}(\omega)$ is the complex dielectric function of the metal. The condition $Re\{\in_{r2}(\omega)\} < 0$ makes the diclcctric-metal interface support longitudinal waves that are tightly confined to the interface between the above said two media. These types of surface waves are called as SPP modes.

A three-layer heterostructure has been illustrated in Figure 5.2b. Generally, it can be of two combinations, dielectric-metal-dielectric (DMD) or metal-dielectric-metal (MDM) where SPP is formed at each of dielectric-metal interfaces and metal-dielectric junction. The communication between the electromagnetic fields of the SPP modes of the two opposite junctions is possible when the thickness of the metal layer is very small, and is known as a hybridized mode. The hybridized mode is demonstrated in Figure 5.2b by marking it using a red line.

If the metallic region is replaced with graphene, the dispersion relation in the case of graphene SPPs is defined by the relation provided in (5.16).[56]

$$\frac{\epsilon_{r1}}{k_1(q,\omega)}+\frac{\epsilon_{r2}}{k_2(q,\omega)}+i\frac{\sigma(\omega)}{\omega\epsilon_0}=0 \tag{5.16}$$

The notations denote their usual meanings where $\sigma(\omega)$ is frequency-dependent surface conductivity of graphene, k_j is the SPP wave number function with the variable q and frequency (ω). The above-said equation has the solution of real values if the imaginary portion of $\sigma(\omega)$ is positive and the real part of $\sigma(\omega)$ is zero. The solution of the above equation reduces to a complex one when the real part of $\sigma(\omega)$ is of finite value.

5.5 GRAPHENE-BASED METASURFACE: SUBSEQUENT DEVELOPMENTS

Surface plasmon polaritons (SPP) formed in graphene exhibits promising properties[57] to become one of the mainstream research topics in nanophotonics. Graphene surface plasmons (GSPs) are excited most efficiently in periodic graphene-based devices, including diffraction gratings, graphene plasmonic crystals, corrugated graphene, and conductivity gratings, etc.[58–61] The metasurface is a strong candidate for electromagnetic wave modulation in optical systems because EM wave can be manipulated in terms of amplitude and phase.[58] Metallic metasurfaces suffer from unwanted correlation between amplitude and phase control of the EM wave as one meta-atom has only one resonance condition. But graphene metasurface can achieve tunable conductivity under the exposure of external biasing, chemical doping or stretching the graphene surface mechanically.[62–65] Therefore, the graphene meta-atoms can achieve multiple resonance conditions within a single design, which further allows them to complete the wavefront's complete control. Therefore, periodic graphene ribbons have been used independently on top of the dielectric, working as tunable terahertz (THz) metasurfaces.[66] Graphene has been integrated with the metallic nanostructures as a hybrid combination to enhance the performance of existing metallic metasurface devices.[67, 68] Later, only graphene patterns have been employed to enhance surface plasmon polariton (SPP) of the electromagnetic wave at the graphene-dielectric interface for several applications like cross-polarization converters, absorbers, and spatial filters, etc.[69–72]

Terahertz gap (THz gap) is an emerging field of research. Scientific community prefers THz gap now a days because it is interference-free unlike the GHz domain apart from the availability of the wide bandwidth. A few literatures have discussed previously about the need of THz gap applications for modern communication.[73, 74] Some of the noble metals and 2D materials, viz., graphene is very useful for their durability in the THz domain. Silicon dioxide (SiO_2), zinc oxide (ZnO) dielectric materials have shown a very promising characteristics in the THz and optical domain.[75, 76] Over the last decade, research on graphene-based metasurface devices operating in the THz gap has gained attention among the research community globally. Periodic single-layer graphene patterns have been used in the metadevices for various applications in the THz domain.[70, 71] Recently, two separately patterned layers of graphene have been employed on the two opposite sides of a dielectric substrate for achieving maximum absorption possible of EM wave within the THz gap.[77] The tunable surface conductivity of graphene makes it possible to attain versatile characteristics by altering the Fermi level in graphene.[62–65]

5.6 TUNABLE PROPERTIES OF GRAPHENE

The 2D conductivity of the graphene surface plays an essential role in achieving the versatile properties of graphene-based devices. There are several ways to play with the properties of graphene conductivity.[72–76] Generally, the conductivity of graphene surface is very high due to the delocalized electrons on its π-orbital on the 2D graphene surface. The recent development in transistor technology is growing rapidly with graphene, as several techniques have been evolved.[78]

5.7 GRAPHENE-BASED METASURFACE ABSORBERS

EM wave absorbing structures have been evolved with the advent of time in terms of Salisbury screen, Jauman layer, and Dalenbach layer.[79–81] Several literatures have already discussed different theories about absorption where metal-dielectric-metal (MDM) layers have been used to absorb the incident EM wave.[82–84] The lossy-dielectric and magnetic materials[85,

[86] were implemented earlier in achieving EM wave absorbers; however, they are comparatively heavy. The absorbers used in practical applications like in anechoic chamber are also bulky. The considerably high thickness and bulkiness of the previously designed absorbers limit their applications in various fields, viz., RFID system, thermal detector system, stealth technology, and radar cross-section (RCS) reduction, etc.[81, 86–88] This leads to the research about the absorbers employing the metasurface concept where both sub-wavelength thickness and periodicity can be realized to achieve compactness. Currently, metasurface-based EM wave absorbers have been proposed to offer a single band as well as multiband absorbers within the GHz range.[89–92] Modern wireless communication needs wideband interference-free communication; so, the researchers are shifting toward the THz range, called THz gap (0.1–10 THz).[73, 74, 77] The tuning of the absorption band has also been realized by implementing reconfigurable elements.[93–97] The introduction of graphene plays an important role here because of its high electrical conductivity to build wideband EM wave absorbers to meet modern communication systems' demand.

When the electromagnetic wave is incident at an interface between two distinct media, a part of the wave is transmitted through the second medium while the other part of the wave is reflected back. When both the reflection and transmission of the wave from the interface become minimum, the wave is absorbed. The absorption of EM wave is measured in terms of absorbed power called absorptivity, $A(f)$ as provided in (5.17). The reflectivity and transmittance of the wave are termed as $R(f)$ and $T(f)$, respectively. The expressions of $R(f)$ and $T(f)$ contain both the co- and cross-polarized components of the wave as given in (5.18) and (5.19), respectively.

$$A(f) = 1 - |R(f)| - T(f)| \tag{5.17}$$

$$R(f) = |r_{co}(f)|^2 + |r_{cross}(f)|^2 \tag{5.18}$$

$$T(f) = |t_{co}(f)|^2 + |t_{cross}(f)|^2 \tag{5.19}$$

5.8 GRAPHENE METASURFACE-BASED TUNABLE TERAHERTZ ABSORBER

Graphene oxide was first proposed to make the ultraviolet absorbers in the year 2012.[98] Later, graphene oxide has also been experimentally found to offer absorption in the gigahertz region too.[98, 99] Graphene was introduced with the metasurface concept for the first time in the absorption of electromagnetic waves in 2013.[72] A graphene metasurface-based terahertz (THz) EM wave absorber has been proposed where two types of graphene patterns have been used as a top layer to check different absorption behavior of the device.[72] Multiband graphene-based absorber in the lower terahertz region has been reported.[101–103] But these absorbers are narrowband in nature.

5.9 SINGLE-LAYER TUNABLE AND POLARIZATION-INSENSITIVE GRAPHENE METASURFACE ABSORBER

A wideband tunable polarization-insensitive graphene metasurface absorber has been illustrated in Figure 5.3 to meet the criteria of modern electromagnetic measurement.[71] The device contains a graphene layer (thickness, $t_g = 0.001$ μm) on top of an amorphous silicon dioxide (SiO_2) substrate with an effective dielectric constant of 2.1 and a loss tangent (tan δ) of 0.0006. The thickness of the substrate (SiO_2) is 7 μm(t_{SiO2}). The backside of the substrate (SiO_2) is covered with a 0.1 μm thick gold layer with an electrical conductivity (σ) of 4.56×10^7 S/m. The device's top face is mainly a periodic distribution of the structural geometry defined in the unit cell. The single-layer graphene pattern is oriented along the horizontal plane. This has been implemented by providing periodic boundary conditions to the unit cell of the structure. The open boundary is defined along the vertical axis *w.r.t* the horizontal plane to allow the EM wave incidence to the device's top face. The different views of the unit cell geometry are shown in Figure 5.3a. The optimized dimensions of the unit cell geometry are provided as periodicity $a = 6$ μm, while the rest of the dimensions have been optimized as $q = 1$ μm, $d = 1$ μm, $n = 1$ μm, $b = 2.50$ μm, $c = 2.6$ μm.

The skin depth has the least value compared with the thickness of a conductor in the optical spectrum.[104] The incident electromagnetic wave experiences total reflection from the bottom gold layer with a thickness of 0.1 μm as it surpasses the skin depth value calculated from the operating frequency, thereby resulting in zero transmission of the wave. In this case, the absorption depends on the reflectivity totally, which can be enhanced with the reduction of the reflection from the top surface. The reflectivity, as well as the absorptivity curve are shown in Figure 5.3b. It can be experienced from Figure 5.3b that the level of reflectivity is less than -10dB within a spectrum ranging from 3.69 THz to 9.77 THz. This provides a rise to the fractional bandwidth calculated as ~90.34% *w.r.t.* the center point of the spectral band of interest for absorption above 90% absorptivity. Differently positioned three minima have been achieved at 4.21 THz, 6.78 THz, and 9.27 THz corresponding to the reflectivities of -37 dB, -18 dB, and -23 dB as well as absorptivities of 99.9%, 97%, and 99.4%, as observed from the reflectivity curve in Figure 5.3b.

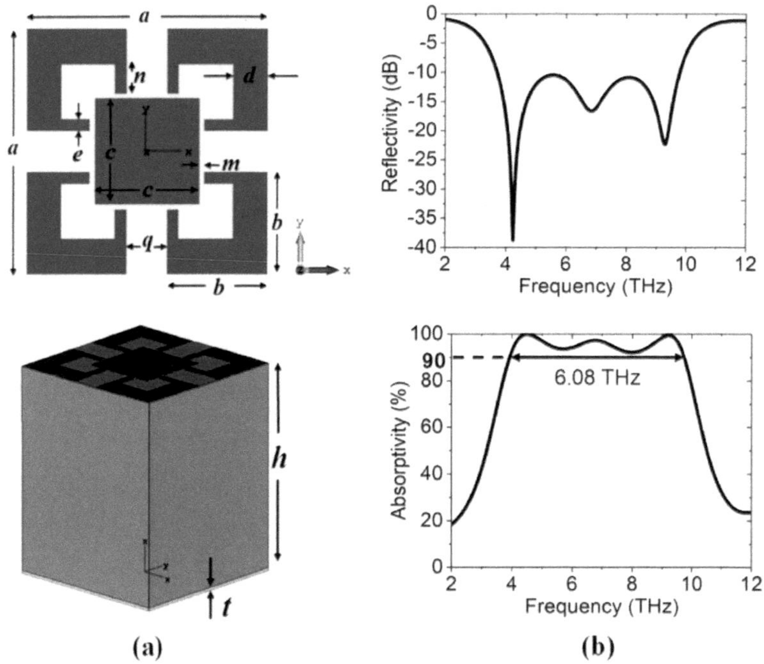

FIGURE 5.3 (a) Top view, 3-D perspective view and (b) reflectivity, absorptivity response of the wideband polarization-insensitive graphene metasurface absorber in lower MIR region. *Source*: Reprinted with permission from Ref. [71]. © 2020. IEEE.

The structure's performance has been observed under different polarization angles (ϕ) in Figure 5.4a, and the spectral response of the absorptivity is invariant with respect to ϕ, which further concludes that the device is polarization insensitive. Moreover, from the spectral responses of the absorptivities under different incident angles (θ) of the EM wave, the device shows stability in absorptivity response till θ = 40° for both transverse electric and transverse magnetic polarizations as depicted in Figure 5.4b and c, accordingly.

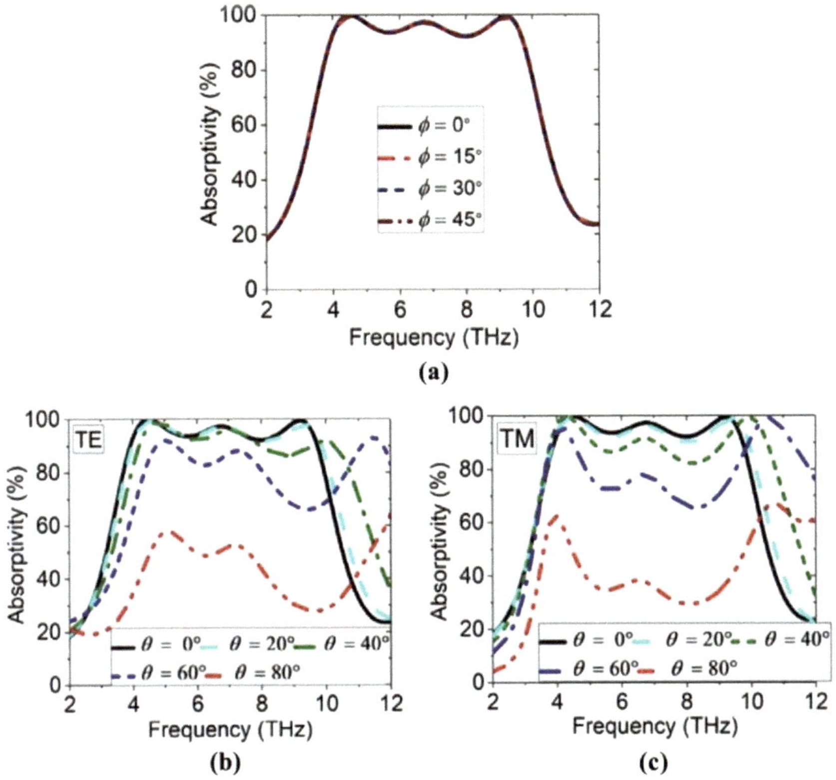

FIGURE 5.4 Response of absorptivity under (a) different angles of polarization (ϕ), and different angles of incidence (θ) for (b) TE polarization, and (c) TM polarization of the EM wave.

Source: Reprinted with permission from Ref. [71]. © 2020. IEEE.

The high absorptions at 4.21 THz, 6.78 THz, and 9.27 THz are explained in view of the derived electromagnetic parameters and distribution of the electric fields. The effective permittivity (ε_{eff}) and effective permeability (μ_{eff}) are derived by using eq 5.20 where d denotes the propagating path covered by the incident wave and k_0 stands for the propagation constant of the EM wave at free space. The derived values of ε_{eff} and μ_{eff} are included in Table 5.1 for the three optimized absorption peaks, viz., 4.21 THz, 6.78 THz, and 9.27 THz, respectively. The almost equal values of the real and imaginary valued ε_{eff} and μ_{eff} of the effective medium at these three absorption peak frequencies imply the impedance matching between the absorber and free space. The derived values of ε_{eff} and μ_{eff} are plotted in Figure 5.5a and b, respectively.

$$\chi_{es} = \frac{2j}{k_0}\frac{1-S_{11}}{S_{11}+1}$$

$$\chi_{ms} = \frac{2j}{k_0}\frac{S_{11}+1}{1-S_{11}}$$

$$\varepsilon_{eff} = 1+\frac{\chi_{es}}{d}$$

$$\mu_{eff} = 1+\frac{\chi_{ms}}{d} \tag{5.20}$$

TABLE 5.1 Calculation of the Electromagnetic Parameters of the Graphene Metasurface Absorber.

Frequency (THz)	Real part		Imaginary part	
	ε_{eff}	μ_{eff}	ε_{eff}	μ_{eff}
4.21	1.02	0.98	3.13	3.34
6.78	0.90	1.18	1.53	2.34
9.27	1.11	0.91	1.61	1.33

Source: Reprinted with permission from Ref. [71]. © 2020. IEEE.

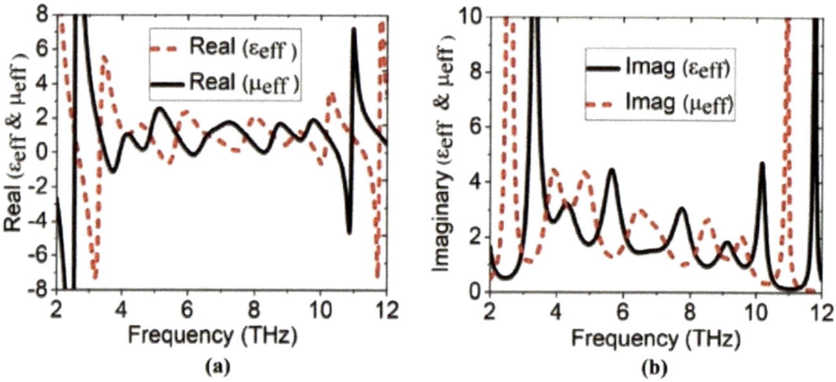

FIGURE 5.5 Spectral responses of (a) real-valued ε_{eff} and μ_{eff} and (b) imaginary valued ε_{eff} and μ_{eff}.

Source: Reprinted with permission from Ref. [71]. © 2020. IEEE.

The surface current vectors on the top and bottom surfaces of the device at 4.21 THz, 6.78 THz, and 9.27 THz, respectively are illustrated in Figure 5.6a. The top and bottom surface current vectors are antiparallel, indicating magnetic excitation at the first two frequency bands. In contrast, the same current vectors form parallel current loop resulting in the induced electric resonance at 9.27 THz. Therefore, both the electric and magnetic resonances are responsible for the high absorption at the above three frequencies.

The power loss density patterns at different resonant frequencies provide an idea about the coupling between the specifically patterned top graphene metasurface and the bottom gold plane. The power loss density patterns also mimic the electric field distributions, that is, field localizations at those resonant frequencies. At the lowest resonant frequency (4.21 THz), the power loss concentration is at the edges of the horizontal slot of the graphene metasurface, as it is known from Figure 5.6b. The power loss density is partially distributed on the edges of both the graphene metasurface orthogonal slots at 6.78 THz. At the highest resonant frequency (9.27 THz), the power loss density is concentrated on the edges of both the slots on graphene metasurface. It has been observed that the distributions of the power loss density are identical for both TE and TM polarization of the EM wave.

5.10 BILAYER GRAPHENE METASURFACE FOR WIDEBAND POLARIZATION-INSENSITIVE ABSORPTION

In this section, a wideband absorber within the THz gap has been discussed, incorporating the application of graphene layers on both surfaces of a silicon dioxide (SiO_2) substrate.[77] The top graphene pattern has been designed in such a way that it acts as a matching layer between air and SiO_2 interface under a certain condition of chemical potential value (μ). The bottom of the SiO_2 substrate is completely covered with a graphene layer with a high chemical potential value so that this layer can act as a metal sheet to bar the transmission of the EM wave. The unit cell's design details of the periodic graphene metasurface are illustrated in Figure 5.7a. A modified-fractal-shaped graphene pattern has been deposited on an amorphous silicon dioxide (SiO_2) substrate ($\varepsilon_r = 3.9$, $\tan \partial = 0.0006$) and the bottom side of the substrate is totally covered with graphene.

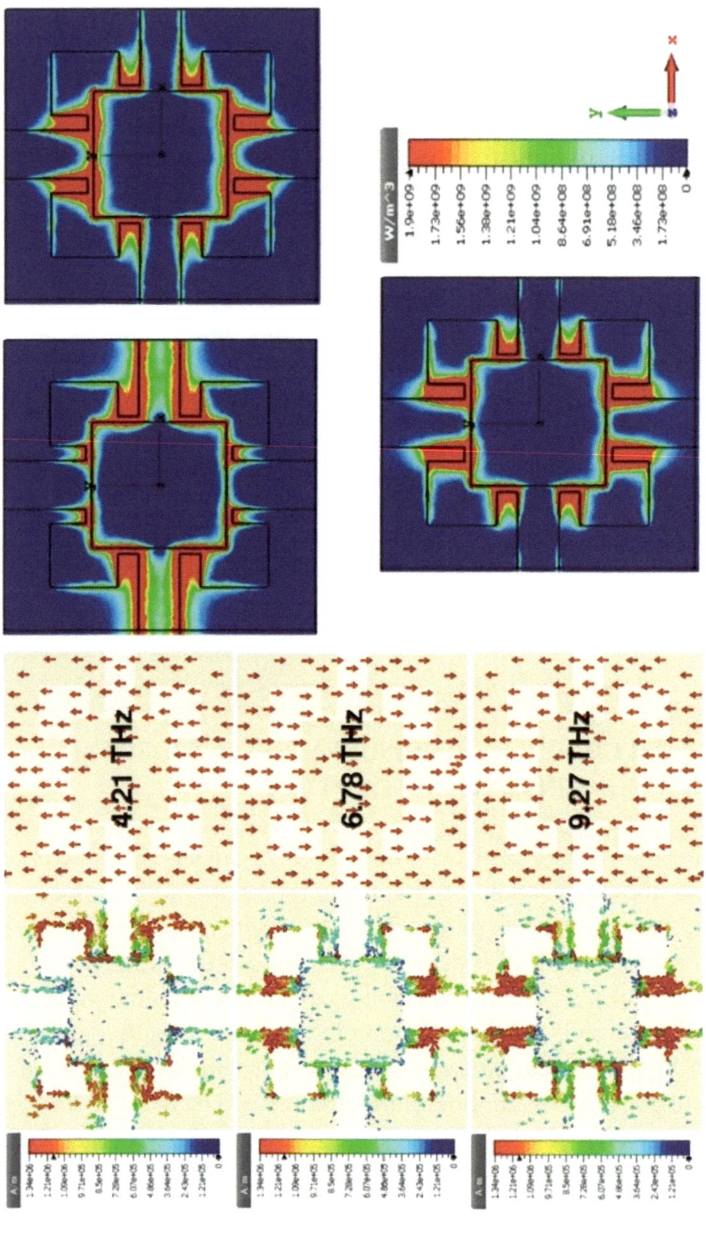

FIGURE 5.6 (a) Distribution of surface current vectors on top and bottom surfaces of the device and (b) pattern of power loss on the structured graphene metasurface at 4.21 THz, 6.78 THz and 9.27 THz.

Source: Reprinted with permission from Ref. [71]. © 2020. IEEE.

The optimized dimensions of the device geometry are described in Figure 5.7a. The main reason behind the absorption for this particular device is not only because of the multiple reflections and superposition of electromagnetic wave but also the coupling of multiple plasmonic resonances between the two differently patterned graphene surfaces separated by a very small distance. In the case of conventional absorbers, the absorption of EM wave occurs only because of the multiple reflections and superposition of the EM wave. Graphene can be designed as a very thin metal sheet with a plasma frequency, a function of the Fermi level, as given in eq 5.21.[72, 105, 106] Graphene exhibits "more-metallic"-type conductivity than the semiconductor-type counterpart as the resonant frequency increases with Fermi energy.[72] Therefore, a patterned graphene metasurface further achieves a wide spectral span in the absorptivity response by integrating plasmonic resonance.[72]

$$\omega_p = [\frac{2e^2 k_B T}{\pi \hbar^2 \varepsilon_o \Delta} ln\left\{ 2\cosh \frac{E_f}{2k_B T} \right\}]^{\frac{1}{2}} \tag{5.21}$$

The frequency response of the reflectivity is provided in Figure 5.7b while the responses of reflection (r) and transmission coefficient (t) are evident from Figure 5.7c. The increment in chemical potential (μ) of the bottom continuous graphene surface results in the rise in resonance frequency (ω_p) and the bottom graphene surface offers "more-metallic" behavior as seen from eq (5.5) which debars the further propagation of the wave under a certain value of chemical potential (μ).[72] The real component of graphene's surface conductivity is large enough and the imaginary component results in the oscillations due to its negative value.[72] It can be concluded from here that the bottom continuous graphene surface acts as a reflector withstanding multiple plasmonic resonances, and as a result of this, the transmission from the bottom graphene surface is zero as shown in Figure 5.7c.[77] The graphene metasurface discussed here, offers a wide spectral bandwidth of 9.74 THz in terms of 90% absorptivity in between 2.6 THz and 11.80 THz because the reflection level is under -10 dB within this whole range of frequencies as can be verified from Figure 5.7b. A fractional bandwidth of 140.86% calculated over 90% absorptivity *w.r.t.* the central spectral point has been realized. The reflection response achieves near unity absorptivities at the three resonance dips at 3.25 THz, 6.16 THz, and 9.83 THz, respectively. The absorptivity response becomes worse if the bottom graphene layer is replaced by a continuous metallic layer as evident from Figure 5.7e.

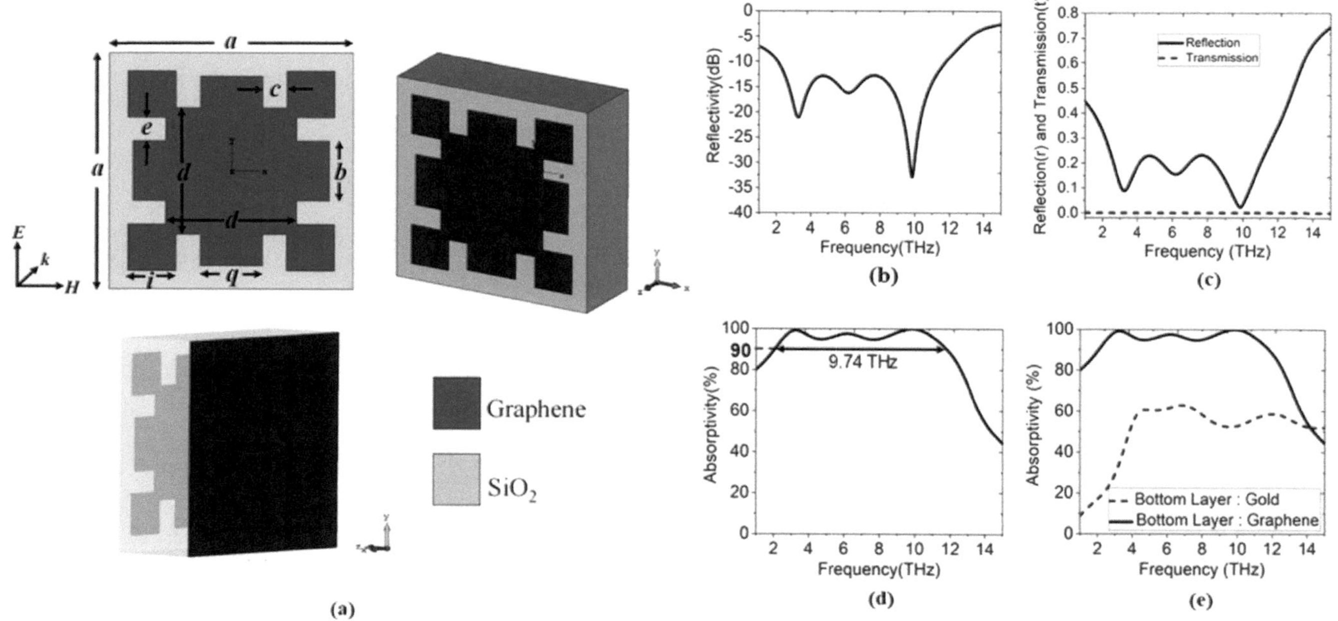

FIGURE 5.7 (a) Top view, perspective view and back view of the bilayer graphene metasurface for wideband polarization-insensitive absorption in lower MIR region ($a = 6.7$ μm, $c = 0.9$ μm, $e = 0.9$ μm, $b = 1.8$ μm, $d = 3.8$ μm, $i = 1.4$ μm, $q = 1.8$ μm); frequency responses of (b) reflectivity (dB), (c) transmission (t) coefficient, (d) absorptivity (%), and (e) comparison of absorptivity between bottom gold-backed metasurface and bottom graphene-backed metasurface.

Source: Reprinted with permission from Ref. [77]. © 2020 IIT Banaras Hindu Univ. Varanasi.

A diagrammatic setup to bias the two differently shaped graphene surfaces has been demonstrated in Figure 5.8a. External electric field can be imposed to these graphene patterns to control the μ values practically. A gold-backed silicon dioxide (SiO_2) substrate is placed just below the proposed device to provide the mechanical support as well as a common gate for biasing of the two separate graphene surfaces at the same time. The conductivity of the two separate graphene surfaces can be nurtured at the same time for several values of V_{g1} and V_{g2} exploiting T_μ and B_μ, respectively as illustrated in Figure 5.8a.[77] The additional two layers have no impact on the absorptivity response as there is zero transmission from the backside of the device, as confirmed from the Figure 5.8b. The bottom layer of the device is working as a reflector in the THz domain by changing the properties of the graphene surface chemically. The bottom graphene surface acts as a reflector when the chemical potential (μ) value is equal to 1.25 eV. The structure also can offer two-fold tunability of absorption band by the variations in T_μ and B_μ.

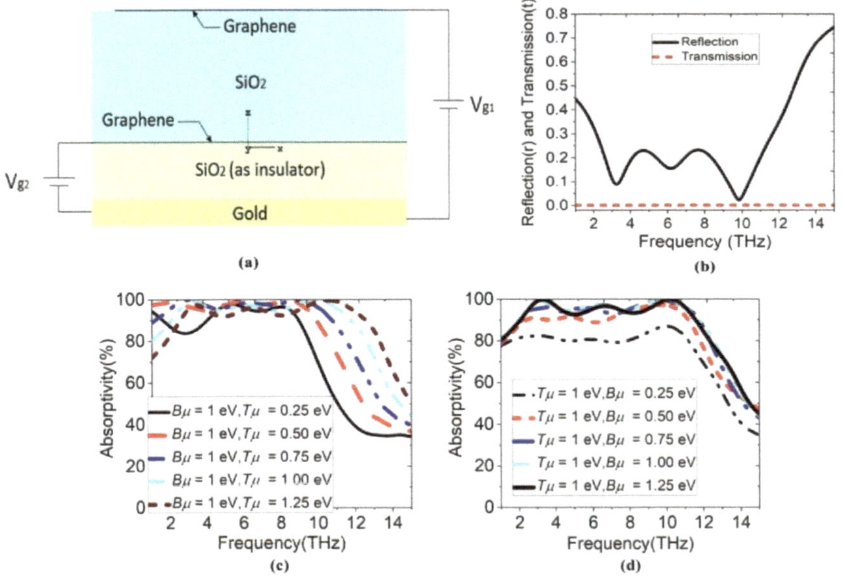

FIGURE 5.8 (a) The proposed device incorporating the biasing setup. (b) Transmission with reflection characteristics of the proposed design. (c) Absorptivity performance variations under different values of chemical potential ($T\mu$) of the top graphene metasurface, and (d) chemical potential ($B\mu$) of the bottom continuous graphene surface.

Source: Reprinted with permission from Ref. [77]. © 2020 IIT Banaras Hindu Univ. Varanasi.

5.11 POLARIZATION CONVERTERS USING GRAPHENE-BASED METASURFACE

Previously, the polarization state of the electromagnetic wave was manipulated for sensing, optical communication, and spectroscopies.[107] However, the bulky shape of the devices and narrow-band operation limit several applications. Metallic metasurfaces have been introduced earlier for polarization state manipulation of the EM wave within the GHz spectrum. Graphene-based THz dual band cross-polarization converters (CPC) have been first reported in 2014.[108]

5.12 WIDEBAND TUNABLE CROSS-POLARIZATION CONVERTER USING MONOLAYER GRAPHENE METASURFACE IN THE MID-INFRARED REGION

A graphene metasurface-based wideband cross-polarization converter (CPC) has been discussed here.[70] The top graphene surface of the device contains periodically slotted circular split rings placed on the SiO_2 substrate. The backside of the SiO_2 substrate is coated with gold layer to hinder the further propagation of wave from the backside of the device. This particular pattern of the graphene metasurface reflects back the x-polarized wave when a y-polarized wave is an incident on it.

The top face of the structure mainly contains an array of split ring slots on the graphene surface. The split ring slots are oriented in a 45° angled version *w.r.t.* the horizontal axis. The other face of the SiO_2 substrate (with an effective permittivity of 3.9) is covered with a 0.5 μm thick gold layer. The front and side views of the device are illustrated in Figure 5.9a. The dimensions of the device have been presented as: periodicity $p_x = p_y = p$ = 17.5 μm, r_1 = 8.5 μm, r_2 = 6.5 μm, w = 5 μm, dielectric thickness d_1 = 25 μm and thickness of backside gold layer d_2 = 0.5 μm. A y-polarized electromagnetic wave is incident on the top face of the device (as shown in Figure 5.9a) and the following formula has calculated the polarization conversion ratio (PCR):

$$PCR = \frac{|R_{xy}|^2}{|R_{yy}|^2 + |R_{xy}|^2} \tag{5.22}$$

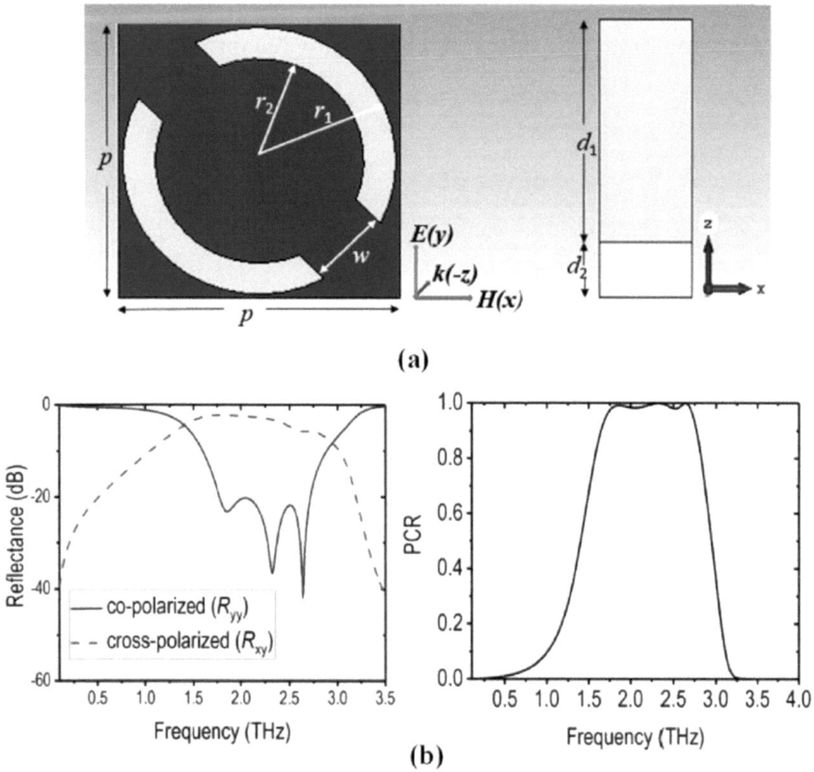

FIGURE 5.9 (a) Details of the unit cell of the graphene metasurface-based CPC and (b) spectral responses of the co-polarized, cross-polarized reflection coefficients, corresponding PCR of the device exhibiting cross-polarization conversion.

Source: Reprinted with permission from Ref. [70]. © 2019 IEEE.

The co-and cross-polarized (R_{yy} and R_{xy}) reflection coefficients are plotted in Figure 5.9b. It can be concluded that the minimum value of R_{yy} are –23.20 dB, –36.37 dB, and –41.64 dB at 1.86 THz, 2.33 THz, and 2.64 THz, respectively. The value of R_{yy} has been maintained under -10 dB from 1.60 THz to 2.89 THz and the value of R_{xy} has been observed more than -1 dB within the identical spectral span. The polarization conversion ratio (PCR) has been calculated from eq 5.22, where the notations are defined above. The value of the PCR has been assessed in Figure 5.9b. The device yields a fractional bandwidth about 51.92% *w.r.t.* the central spectral point of 2.16 THz by considering the PCR value above 0.9 from 1.62 THz to 2.75 THz. The device also achieves three near-unity PCR values at 1.86

THz, 2.33 THz, and 2.64 THz which further signifies the total conversion of the polarization state of the incident EM wave into its orthogonal form.

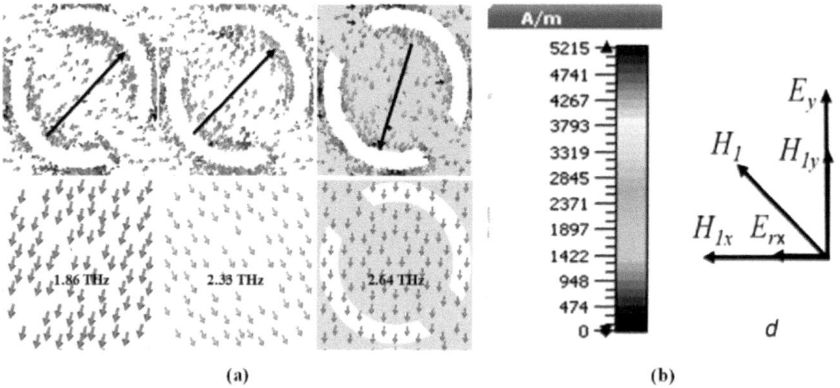

FIGURE 5.10 (a) Orientation of the surface current vectors on top and bottom faces of the CPC and (b) schematic of the field vectors to validate surface current distributions. *Source:* Reprinted with permission from Ref. [70]. © 2019 IEEE.

The surface current vectors on top graphene metasurface (first row) and bottom gold layer (second row) at the three near-unity PCR spectral points, viz., 1.86 THz, 2.33 THz, and 2.64 THz have been graphed together in Figure 5.10a. The surface currents on top and bottom surfaces are antiparallel, leading to magnetic resonance at 1.86 THz, while the current vectors on those surfaces are parallel at 2.64 THz resulting in the formation of electric resonance.[109] At 1.86 THz, the y-polarized incident wave generates antiparallel surface current, i.e., magnetic dipole excitation, which induces a magnetic field component H_1 as shown in Figure 5.10b. This component H_1 can be sorted out into another two components H_{1x} and H_{1y} along the x-axis and the y-axis, respectively. H_{1x} cannot produce any cross-polarized part while H_{1y} oriented along the same direction of the incoming electric field generates an induced electric field along the x-axis to form the cross-polarized part of the reflected wave.

The variation in PCR response of the device *w.r.t.* relaxation time (τ) is shown in Figure 5.11a. As per the quality and strain, the value of the patterned graphene surface's relaxation time varies in the range 0.02-2 ps.[110–112] The relaxation time is also called the scattering time because it is inversely proportional to the electrons' scattering rate. More cross-polarized reflection is caused when the free electron carrier loss decreases due to the increment in the scattering time of graphene. The best PCR performance (over 90%) can be seen in Figure 5.11a for the scattering time values in the range of 0.2–1 ps. The tunability in CPC performance has also been observed due to the conductivity modulation in graphene metasurface which is shown in Figure 5.11b for different chemical potential values over the range of 0.3–0.7 eV.[44, 113–115]

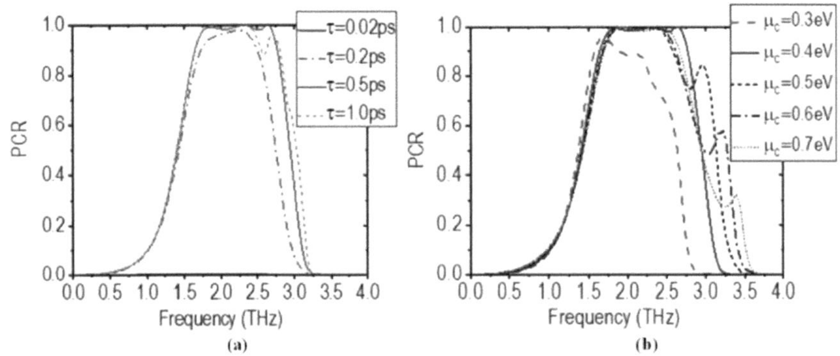

FIGURE 5.11 Spectral responses of the PCR under the variation of (a) relaxation time (τ) and (b) chemical potential (μ_c) of the specific graphene pattern used in the graphene-based CPC.

Source: Reprinted with permission from Ref. [70]. © 2019 IEEE.

5.13 TUNABLE BROADBAND TERAHERTZ CROSS-POLARIZATION CONVERTER OVER A WIDE ANGLE OF INCIDENCE

Another CPC with graphene metasurface has been designed here for the better angular stability in the PCR characteristics.[69] The design details of a unit cell of the CPC have been shown in Figure 5.12a. The front face of this particular design is a periodic pattern of ϕ-shaped graphene geometry printed on an insulating SiO_2 substrate backed by a 1-μm thick gold plate (with a conductivity of 4.56×10^7 S/m). The ϕ-shaped graphene patterns are tilted with an angle of 45° *w.r.t.* *x*- and *y*-axis. The thickness (d_1) and relative dielectric constant of the SiO_2 layer are taken as 8.5 μm and 2.1 within the operating frequency range (2–12 THz). The optimized dimensions are taken as periodicity $p_x = p_y = p = 2$ μm, $l = 2.5$ μm, $w = 0.2$ μm, $r_1 = 0.3$ μm, $r_2 = 0.8$ μm, $d_1 = 8.5$ μm, and $d_2 = 1$ μm.

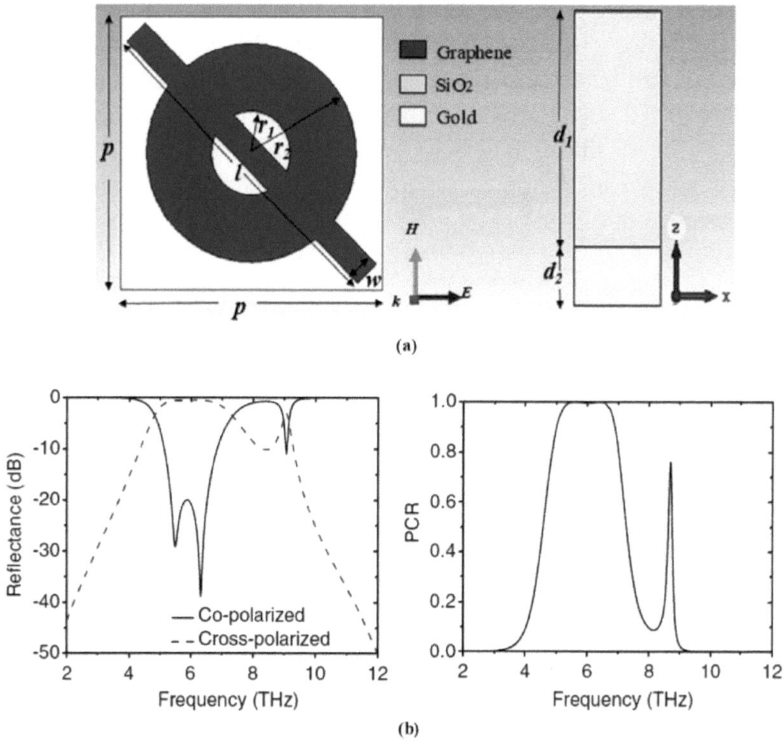

FIGURE 5.12 (a) Design details of the unit cell of the graphene metasurface exhibiting broadband CPC and (b) spectral responses of the co- and cross-polarized reflection coefficients and PCR of the graphene metasurface in the lower THz region.

Source: Reprinted with permission from Ref. [69]. © 2018 Optical Society of America.

The performance of this design has been discussed in this section. The PCR (polarization conversion ratio) of this particular graphene metasurface is above 0.9 over the spectral range 5.08–6.89 THz supporting a bandwidth of 1.81 THz (with a fractional bandwidth of 31.80% *w.r.t.* the mid-frequency of the above-said spectral range), as experienced in Figure 5.12b. Two near-unity PCR peaks have been achieved at 5.57 THz and 6.43 THz within the spectral range stated above, as confirmed in Figure 5.12b. The PCR performance has been checked under different incident angles of the EM wave for both transverse electric (TE) and transverse magnetic (TM) polarizations. It can be concluded from Figure 5.13a and b that the device offers similar PCR performances in both cases. Further observation confirms that the PCR is more than 0.9 over the spectral range stated above till 40° incident angles of the incoming EM wave to the top face of the device.

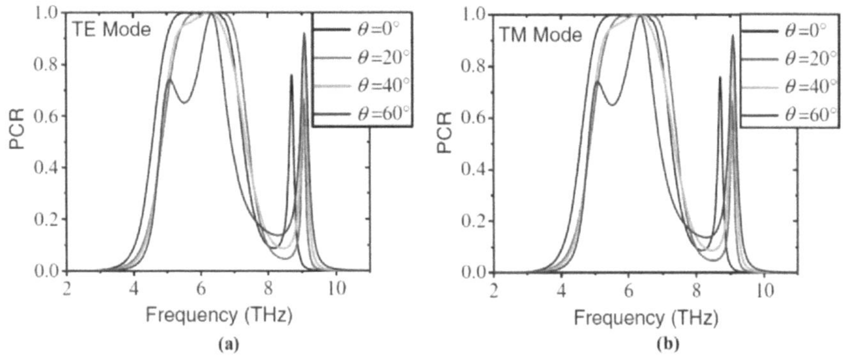

FIGURE 5.13 Responses of the PCR of the graphene-made CPC under (a) TE and (b) TM polarization of the incident EM wave.

Source: Reprinted with permission from Ref. [69]. © 2018 Optical Society of America.

Recently, graphene-based metasurface structures have also been proposed as linear to circular polarization converters.[116–118] Moreover, graphene has been proposed in the pixelated meta-atom approach for enhanced multifunctionality.[119]

5.14 SUMMARY

In summary, metasurface concept has been integrated with graphene to realize a few graphene metasurface-based devices for potential applications, viz., absorbers, polarization converters, etc. in the THz region. Two types of graphene metasurface-based ultrathin and compact wideband tunable absorbers are presented within the THz gap region. Both of them are polarization insensitive in nature and provide angular stability till 40° incident angles in better absorption bandwidth. Graphene-based metasurfaces for linear cross-polarization converters (CPC) have also been discussed in this chapter. Both multiband and wideband CPC incorporating graphene metasurface have been discussed in detail. These proposed graphene metasurface-based devices can find applications in detection, sensing, spectroscopic applications, polarization modulators, pocket cells, and Kerr cells, etc.

KEYWORDS

- **graphene-based metasurface**
- **sub-wavelength thickness**
- **graphene surface plasmon**
- **tunable surface conductivity**
- **surface plasmon polariton**
- **metadevice**

REFERENCES

1. Caloz, C.; Itoh, T. *Electromagnetic Metamaterials: Transmission Line Theory and Microwave Applications*; John Wiley & Sons, 2006.
2. Yu, N.; Genevet, P.; Kats, M. A.; Aieta, F.; Tetienne, J. P.; Capasso, F.; Gaburro, Z. Light Propagation with Phase Discontinuities: Generalized Laws of Reflection and Refraction. *Science* **2011**, *334*, 333–337. DOI: 10.1126/science.1210713.
3. Ni, X. J.; Emani, N. K.; Kildishev, A. V.; Boltasseva, A.; Shalaev, V. M. Broadband Light Bending with Plasmonic Nanoantenna. *Science* **2012**, *335*, 427. DOI: 10.1126/science.1214686.
4. Sun, S.; He, Q.; Xiao, S.; Xu, Q.; Li, X.; Zhou, L. Grandient-index Meta-surfaces as a Bridge Linking Propagating Waves and Surfaces Waves. *Nat. Mater.* **2012**, *11*, 426–431. DOI:10.1038/nmat3292.
5. Chen, H. T.; Taylor, A. J.; Yu, N. A Review of Metasurface: Physics and Applications. *Rep. Prog. Phys.* **2016**, *79*, 076401. DOI: 10.1088/0034-4885/79/7/076401.
6. Ding, F.; Pors, A.; Bozhevolnyi, S. I. Gradient Metasurface: A Review of Fundamentals and Applications. *Rep. Prog. Phys.* **2018**, *81*, 026401. DOI: 10.1088/1361-6633/aa8732.
7. Bhattacharyya, S.; Ghosh, S.; Srivastava, K. V. Triple Band Polarization-Independent Metamaterial Absorber with Bandwidth Enhancement at X-band. *J. Appl. Phys.* **2013**, *114* (9), 094514.
8. Aman, N.; Bhattacharyya, S.; Chakrabarti, P. Frequency and Time-domain Analyses of Multiple Reflections and Interference Phenomena in a Metamaterial Absorber. *J. Opt. Soc. Am. B* March **2020**, *37* (3), 586–592.
9. Munaga, P.; Bhattacharyya, S.; Ghosh, S.; Srivastava, K. V. An Ultra-thin Compact Polarization-Independent Hexa-Band Metamaterial Absorber. *Springer Appl. Phys. A* April **2018**, *124* (4), 331.
10. Bhattacharyya, S.; Ghosh, S.; Srivastava, K. V. A Wideband Cross Polarization Conversion Using Metasurface. *Radio Sci.* November **2017**, *52* (11), 1395–1404.
11. Nama, N. L.; Bhattacharyya, S.; Chakrabarti, P. A Metasurface-based Broadband Quasi Non-Dispersive Cross Polarization Converter for Far Infrared Region. *Wiley Int. J. RF and Microwave Computer-Aided Eng.* October **2019**, *29* (10), Article No. e21889.

12. Samantaray, D.; Bhattacharyya, S. A Gain-Enhanced Slotted Patch Antenna Using Metasurface as Superstrate Configuration. *IEEE Trans. Antennas Propag.* September **2020**, *68* (9), 6548–6556.
13. Ghosh, S.; Das, S.; Samantaray, D.; Bhattacharyya, S. Meander Line Based Defected Ground Microstrip Antenna Slotted with SRR for Terahertz Range. *Wiley Eng. Rep.* January **2020**, *2* (1), Article No. e12088.
14. Varuna, A. B.; Ghosh, S.; Srivastava, K. V. A Miniaturized-Element Bandpass Frequency Selective Surface using Meander Line Geometry. *Microwave Opt. Technol. Lett.* **2017**, *59* (10), 2484–2489.
15. Chakrabarti, T.; Sarkar, S.; Ghosh, S. K.; Bhattacharyya, S. In *An Ultra-thin FSS Bandpass Filter in Terahertz Region*, International Conference on Microwave Integrated Circuits, Photonics and Wireless Networks (IMICPW-2019), Trichy, India, May 22–24, 2019, pp. 454–456.
16. Novoselov, K. S.; Geim, A. K.; Morozov, S. V.; Jiang, D.; Zhang, Y.; Dubonos, S. V.; Grigorieva, I. V.; Firsov, A. A. Electric Field Effect in Atomically Thin Carbon Films. *Science* **2004**, *306*, 5696. DOI: 10.1126/science.1102896.
17. Novoselov, K. S.; Geim, A. K.; Morozov, S. V.; Jiang, D.; Katsnelson, M. I.; Grigorieva, I. V.; Dubonos, S. V.; Firsov, A. A. Two-dimensional Gas of Massless Dirac Fermions in Graphene. *Nature* **2005**, *438*, 197–200. DOI: 10.1038/nature04233.
18. Shi, G.; Wang, X. Introduction to the Chemistry of Graphene. *Phys. Chem. Chem. Phys.* **2015**. DOI: 10.1039/C5CP05212B.
19. Papageorgiou, G. D.; Kinloch, I. A.; Young, R. J. Mechanical Properties of Graphene and Graphene-based Nanocomposites. *Prog. Mater. Sci.* **2017**, *90*, 75–127. https://doi.org/10.1016/j.pmatsci.2017.07.004.
20. Gomez-Diaz, J. S.; Perruisseau, J. C. Graphene-based Plasmonic Switches at Near Infrared Frequencies. *Opt. Exp.* **2013**, *21*, 15490–15504. DOI: 10.1364/OE.21.015490.
21. Vakil, A.; Engheta, N. Transformation Optics Using Graphene. *Science* **2011**, *332*, 1291–1294. DOI: 10.1126/science.1202691.
22. Grigorenko, A. N.; Polini, M.; Novoselov, K. S. Graphene Plasmonics. *Nat. Photon.* **2012**, *6*, 749–758. DOI: 10.1038/nphoton.2012.262.
23. Su, Z.; Yin, J.; Zhao, X. Terahertz Dual-band Metamaterial Absorber Based on Graphene/MgF_2 Multilayer Structures. *Opt. Exp.* **2015**, *23*, 1679–1690. DOI: 10.1364/OE.23.001679.
24. Ning, R.; Bao, J.; Jiao, Z. X. Y. Omnidirectional Polarization-insensitive Tunable Absorption in Graphene Metamaterial of Nanodisk Structure. *J. Appl. Phys.* **2015**, *118*, 203101. DOI: 10.1063/1.4936222.
25. Zhang, Y.; Shi, Y.; Liang, C. H. Broadband Tunable Graphene-based Metamaterial Absorber. *Opt. Mater. Exp.* **2016**, *9*, 3036–3044. DOI: 10.1364/OME.6.003036.
26. Wang, Y.; Song, M.; Pu, M.; Gu, Y.; Hu, G.; Zhao, Z.; Wang, C.; Yu, H.; Luo, X. Stacked Graphene for Tunable Terahertz Absorber with Customized Bandwidth. *Plasmonics* **2016**, 1–6. DOI:10.1007/s11468-015-0162-5.
27. Ghosh, S. K.; Bhattacharyya, S.; Das, S. In *A Graphene Based Metasurface with Wideband Absorption in the Lower Mid Infrared Region*, 2018 IEEE MTT-S International Microwave and RF Conference (IMaRC), Kolkata, India, 2018, pp. 1–4. DOI: 10.1109/IMaRC.2018.8877373.

28. Ghosh, S. K.; Das, S.; Bhattacharyya, S. In *A Graphene Based Broadband Metasurface Absorber in the Terahertz Region*, 2019 URSI Asia-Pacific Radio Science Conference (AP-RASC), New Delhi, India, 2019, pp. 1–4. DOI: 10.23919/URSIAP-RASC.2019.8738642.
29. Ghosh, S. K.; Bhattacharyya, S.; Das, S. In *Broadband Graphene Based Reflective Cross Polarization Converter Metasurface Design with Unity Efficiency in the Lower Terahertz Gap*, 2019 IEEE MTT-S International Microwave and RF Conference (IMARC), Mumbai, India, 2019, pp. 1–4. DOI: 10.1109/IMaRC45935.2019.9118725.
30. Yadav, V. S.; Ghosh, S. K.; Bhattacharyya, S.; Das, S. Wideband Tunable Mid-infrared Cross-polarization Converter Using Monolayered Graphene-based Metasurface over a Wide Angle of Incidence. *IET Microwaves Antennas Propagat.* **2018**. DOI: 10.1049/iet-map.2018.5373.
31. Wei, Z.; Li, X.; Yin, J.; Huang, R.; Liu, Y.; Wang, W.; Liu, H.; Meng, H.; Liang, R. Active Plasmonic Band-stop Filters Based on Graphene Metamaterial at THz Wavelengths. *Opt. Exp.* **2016**, *24*, 14344–14351. DOI: 10.1364/OE.24.014344.
32. Farhat, M.; Rockstuhl, C.; Bagc, H. A 3D Tunable and Multi-frequency Graphene Plasmonic Cloak. *Opt. Exp.* **2013**, *21*, 12592–12603. DOI: 10.1364/OE.21.012592.
33. Shi, Y.; Zhang, Y. Generation of Wideband Tunable Orbital Angular Momentum Vortex Waves Using Graphene Metamaterial Reflectarray. *IEEE Access* **2018**, *6*, 5341–5347. DOI: 0.1109/ACCESS.2017.2740323.
34. Wei, Z.; Li, X.; Yin, J.; Huang, R.; Liu, Y.; Wang, W.; Liu, H.; Meng, H.; Lian, R. Active Plasmonic Band-stop Filters Based on Graphene Metamaterial at THz Wavelengths. *Opt. Exp.* **2016**, *24*, 14344–14351. DOI: 10.1364/OE.24.014344.
35. Carrasco, E.; Perruisseau, J. C. Reflectarray Antenna at Terahertz Using Graphene. *IEEE Antennas Wireless Propagat. Lett.* **2013**, *12*, 253–256. DOI: 10.1109/LAWP.2013.2247557.
36. Huang, H.; Xia, H.; Xie, W.; Guo, Z.; Li, H.; Xie, D. Design of Broadband Graphene Metamaterial Absorbers for Permittivity Sensing at Mid-infrared Regions. *Sci. Rep.* **2018**, *8*, 4183.
37. Yao, Y.; Kats, M. A.; Genevet, P.; Yu, N.; Song, Y.; Kong, J.; Kapaso, F. Broad Electrical Tuning of Graphene-loaded Plasmonic Antennas. *ACS Nano Lett.* **2013**, *13*, 1257–1264.
38. Shi, B.; Cai, W.; Zhang, X.; Xiang, Y.; Zhan, Y.; Geng, J.; Ren, M.; Xu, J. Tunable Band-stop Filters for Graphene Plasmons Based on Periodically Modulated Graphene. *Sci. Rep.* **2016**, *6*, 26796.
39. Hao, R.; Peng, X.; Li, E.; Xu, Y.; Jin, J.; Zhang, X.; Chen, H. Improved Slow Light Capacity in Graphene-based Waveguide. *Sci. Rep.* **2015**, *5*, 15335.
40. Zhang, F. M.; He, Y.; Chen, X. Guided Modes in Graphene Waveguides. *Appl. Phys. Lett.* **2009**, *94*, 212105.
41. Hanson, G. W. Dyadic Green's Functions and Guided Surface Waves for a Surface Conductivity Model of Graphene. *J. Appl. Phys.* **2008**, *103*, 064302(1-8).
42. Morote, M. E.; Gómez-Di; Diaz, J. S. G.; Carrier, J. P. Sinusoidally Modulated Graphene Leaky-Wave Antenna for Electronic Beam Scanning at THz. *IEEE Trans. Terahertz Sci. Technol.* **2014**, *4* (1), 116–122.
43. Hanson, G. W. Dyadic Green's Functions for an Anisotropic, Non-local Model of Biased Graphene. *IEEE Trans. Antennas Propag.* **2008**, *56*, 747–757.

44. Fuscaldo, W.; Burghignoli, P.; Baccarelli, P.; Galli, A. Graphene Fabry-Perot Cavity Leaky-wave Antennas: Plasmonic versus Nonplasmonic Solutions. *IEEE Trans. Antennas Propag.* **2017,** *65,* 1651–1660.
45. Wu, Y.; Lin, Y.; Bol, A. et al. High-frequency, Scaled Graphene Transistors on Diamond-like Carbon. *Nature* **2011**, *472*, 74–78. https://doi.org/10.1038/nature09979.
46. Tian, H.; Yang, Y.; Xie, D. et al. Wafer-Scale Integration of Graphene-based Electronic, Optoelectronic and Electroacoustic Devices. *Sci. Rep.* **2014**, *4*, 3598. https://doi.org/10.1038/srep03598.
47. Polat, E.; Balci, O.; Kakenov, N. et al. Synthesis of Large Area Graphene for High Performance in Flexible Optoelectronic Devices. *Sci. Rep.* **2015**, *5*, 16744. https://doi.org/10.1038/srep16744.
48. Liu, C.; Ma, W.; Chen, M. et al. A Vertical Silicon-graphene-germanium Transistor. *Nat. Commun.* **2019**, *10*, 4873. https://doi.org/10.1038/s41467-019-12814-1.
49. Moon, J. S. et al. Epitaxial-graphene RF Field-effect Transistors on Si-face 6H-SiC Substrates. *IEEE Electr. Device Lett. 30* (6), 650–652.
50. Kumar, C.; Das, S.; Jit, S. Chapter 7: Device Physics and Device Integration of Two-dimensional Heterostructures. In *2D Nanoscale Hetero-structured Materials*; Elsevier, 2020; pp. 195–214. https://doi.org/10.1016/B978-0-12-817678-8.00007-5.
51. Das, S.; Kumar, C.; Kumar, R.; Srivastava, A.; Jit, S. Two-Dimensional MoS2-Based Photosensitive Al/MoS_2/SiO_2/Si/Ag MOS Capacitor. *IEEE Photon. Technol. Lett.* Jan 1, **2020**, *32* (1), 67–70. DOI: 10.1109/LPT.2019.2957260.
52. Wang, H.; Nezich, D.; Kong, J.; Palacios, T. Graphene Frequency Multipliers. *IEEE Electr. Device Lett.* **2009**, *30* (5), 547–549.
53. Maier, S. A. *Chapter 5: Plasmonics: Fundamentals and Applications*; Springer; 2006, pp. 65–87.
54. Koppens, F. H. L.; Chang, D. E.; de Abajo, F. J. G. Graphene Plasmonics: A Strong Light-matter Interaction. *ACS Nano Lett.* **2011**, *11*, 3370–3377.
55. D'Souza, F.; Kadish, M. K. *Chapter 1: Handbook of Carbon Nanomaterials*; *5*; 2014.
56. Xiao, S.; Zhu, X.; Li, B. et al. Graphene-plasmon Polaritons: From Fundamental Properties to Potential Applications. *Front. Phys.* **2016**, *11*, 117801. https://doi.org/10.1007/s11467-016-0551-z.
57. Zheng, Z. B.; Li, J. T.; Ma, T.; Fang, H. L.; Ren, W. C.; Chen, J.; She, J. C.; Zhang, Y.; Liu, F.; Chen, H. J.; Deng, S. Z.; Xu, N. S. Tailoring of Electromagnetic Field Localizations by Two-dimensional Graphene Nanostructures. *Light: Sci. Appl. Nat.* **2017**, *6*, e17057.
58. Ju, L.; Geng, B.; Horng, J.; Girit, C.; Martin, M.; Hao, Z.; Bechtel, H. A.; Liang, X.; Zettl, A.; Shen, Y. R.; Wang, F. Graphene Plasmonics for Tunable Terahertz Metamaterials. *Nat. Nanotechnol.* **2011**, *6*, 630–634.
59. Yan, H.; Low, T.; Zhu, W.; Wu, Y.; Freitag, M.; Li, X.; Guinea, F.; Avouris, P.; Xia, F. Damping Pathways of Mid-infrared Plasmons in Graphene Nanostructures. *Nat. Photonics* **2013**, *7*, 394–399.
60. Bludov, Y. V.; Ferreira, A.; Peres, N. M. R.; Vasilevskiy, M. I. A Primer on Surface Plasmon-polaritons in Graphene. *Int. J. Mod. Phys. B* **2013**, *27*, 1341001.
61. Bludov, Y. V.; Ferreira, A.; Peres, N. M. R.; Vasilevskiy, M. I. Graphene-based Polaritonic Crystal. *Phys. Rev. B* **2012**, *85*, 245409.

62. Das, S.; Sudhagar, P.; Verma, V.; Song, D.; Ito, E.; Lee, S. Y.; Kang, Y. S.; Choi, W. Amplifying Charge-Transfer Characteristics of Graphene for Tri-iodide Reduction in Dye-Sensitized Solar Cells. *Adv. Funct. Mater.* **2011**, *21*, 3729–3736.
63. Das, S.; Sudhagar, P.; Ito, E.; Lee, D. Y.; Nagarajan, S.; Lee, S. Y.; Kang, Y. S.; Choi, W. Effect of HNO_3 Functionalization on Large Scale Graphene for Enhanced Tri-iodide Reduction in Dye-sensitized Solar Cells. *J. Mater. Chem.* **2012**, *22*, 20490–20497.
64. Das, S.; Sudhagar, P.; Ito, E.; Lee, D. Y.; Nagarajan, S.; Lee, S. Y.; Kang, Y. S.; Choi, W. Synthesis of Graphene-CoS Electro-catalytic Electrodes for Dye Sensitized Solar Cells. *Carbon* **2012**, *50*, 4815–4821.
65. Gomez-Diaz, J. S.; Moldovan, C.; Capdevila, S.; Romeu, J.; Bernard, L. S.; Magrez, A.; Lonescu, A. M.; Carrier, J. P. Self-biased Reconfigurable Graphene Stacks for Terahertz Plasmonics. *Nat. Commun.* **2015**, *6*, 1–8.
66. Liu, Y.; Sun, J.; Zhang, X.; Li, L.; Yin, H. In *Tunable Excitation of Surface Plasmon Polaritons on the Graphene Metasurfaces by Electron Beam*, 2018 11th UK-Europe-China Workshop on Millimeter Waves and Terahertz Technologies (UCMMT), HangZhou, China, 2018, pp. 1–3. DOI: 10.1109/UCMMT45316.2018.9015825.
67. Ghosh, S. K.; Yadav, V. S.; Bhattacharyya, S.; Das, S. In *A Graphene Based Bandwidth Enhanced Metamaterial Absorber using Circular Ring*, 2018 IEEE International Symposium on Antennas and Propagation & USNC/URSI National Radio Science Meeting, Boston, MA, 2018, pp. 1491–1492. DOI: 10.1109/APUSNCURSINRSM.2018.8608226.
68. Peng, L.; Li, X. M.; Liu, X.; Jiang, X.; Li, S. M. Metal and Graphene Hybrid Metasurface Designed Ultra-wideband Terahertz Absorbers with Polarization and Incident Angle Insensitivity. *Nanoscale Adv.* **2019**, *1*, 1452.
69. Yadav, V.; Ghosh, S.; Bhattacharyya, S.; Das, S. Graphene-based Metasurface for a Tunable Broadband Terahertz Cross-polarization Converter over a Wide Angle of Incidence. *Appl. Opt.* **2018**, *57*, 8720–8726.
70. Yadav, V. S.; Ghosh, S. K.; Das, S.; Bhattacharyya, S. Wideband Tunable Mid-infrared Cross-polarization Converter Using Monolayered Graphene-based Metasurface over a Wide Angle of Incidence. *IET Microw. Antennas Propag.* **2019**, *13*, 82–87. https://doi.org/10.1049/iet-map.2018.5373.
71. Ghosh, S. K.; Yadav, V. S.; Das, S.; Bhattacharyya, S. Tunable Graphene Based Metasurface for Polarization-independent Broadband Absorption in Lower Mid Infrared (MIR) Range. *IEEE Trans. Electromagn. Compat.* **2020**, *62* (2), 346–354.
72. Andryieuski, A.; Lavrinenko, A. V. Graphene Metamaterials Based Tunable Terahertz Absorber: Effective Surface Conductivity Approach. *Opt. Exp.* **2013**, *21*, 9144–9155.
73. Elayan, H.; Amin, O.; Subair, R. M.; Alouini, M. S. In *Terahertz Communication: The Opportunities of Wireless Technology beyond 5G*, International Conference on Advanced Communication Technologies and Networking (CommNet), IEEE. Retrieved from https://ieeexplore.ieee.org/document/8360286.
74. Sirtori, C. Bridge the Terahertz Gap. *Nature* **2002**, *417*, 132–133.
75. Carmen, R. T.; Nóvoa, J. A.; Martín, J.; Guitián, F.; Salgueiro, J. R.; Gil, A. Broadband Terahertz ZnO Photonic Crystals Fabricated by 3D Printing. *Ceram. Int.* **2019**, *45* (5), 6223–6227, ISSN 0272-8842.
76. Park, S. J.; Ahn, Y. H. Accurate Measurement of THz Dielectric Constant Using Metamaterials on a Quartz Substrate. *Curr. Optics Photon.* **2017,** *1* (6), 637–641.

77. Ghosh, S. K.; Das, S.; Bhattacharyya, S. Graphene Based Metasurface with near Unity Broadband Absorption in the Terahertz Gap. *Wiley Int. J. RF Microwave Comput. Aided Eng.* **2020,** e22436.
78. Schwierz, F. Graphene Transistors: Status, Prospects, and Problems. *Proc. IEEE 101* (7), 1567–1584.
79. Munk, B. A. *Frequency Selective Surfaces Theory and Design*; John Willey & Sons; 2000; pp. 315–335.
80. Chambers, B.; Tennant, A. In *Active Dallenbach Radar Absorber*, IEEE International Symposium on Antennas and Propagation, Albuquerque, New Mexico, USA, July 9–14, 2006, pp. 381–384.
81. Saville, P. Review of Radar Absorbing Materials. Defense R and D Canada-Atlantic, January 2005.
82. Dhillon, A. S.; Mittal, D.; Sidhu, E. THz Rectangular Microstrip Patch Antenna Employing Polyimide Substrate for Video Rate Imaging and Homeland Defence Applications. *Optik* **2017,** *144*, 634–641.
83. Bhattacharyya, S.; Srivastava, K. V. Triple Band Polarization-Independent Ultra-thin Metamaterial Absorber Using Electric Field-driven LC Resonator. *J. Appl. Phys.* **2014**, *115*, 064508.
84. Tao, H.; Bingham, C. M.; Pilon, D.; Fan, K.; Strikwerda, A. C.; Shrekenhamer, D.; Padilla, W. J. A Dual Band Terahertz Metamaterial Absorber. *J. Phys. D: Appl. Phys.* May **2010**, *43* (22), 225102.
85. Matous, K.; Dvorak, G. J. Optimization of Electromagnetic Absorption in Laminated Composite Plates. *IEEE Trans. Magnetics* May **2003**, *39* (3), 1827–1835.
86. Parida, R. C.; Singh, D.; Agarwal, N. K. Implementation of Multilayer Ferrite Radar Absorber Coating with Genetic Algorithm for Radar Cross-section Reduction at Xband. *Indian J. Radio Space Phys.* April **2007**, *36*, 145–152.
87. Bahret, W. F. The Beginning of Stealth Technology. *IEEE Trans. Aerospace Electron. Syst.* October **1993**, *29* (4), 1377–1385.
88. Costa, F.; Genovesi, S.; Monorchio, A. A Chipless RFID Based on Multiresonant High-impedance Surface. *IEEE Trans. Microwave Theory Techniques* **2013**, *61* (1), 146–153.
89. Hu, X.; Wang, J. High-speed Gate-tunable Terahertz Coherent Perfect Absorption Using a Split-ring Graphene. *Opt. Lett.* **2015**, *40*, 5538–5541.
90. Zhang, J.; Guo, C.; Liu, K.; Zhu, Z.; Ye, W.; Yuan, X.; Qin, S. Coherent Perfect Absorption and Transparency in a Nanostructured Graphene Film. *Opt. Exp.* **2014**, *22*, 12524–12532.
91. Ning, Y.; Dong, Z.; Si, J.; Deng, X. Tunable Polarization-independent Coherent Perfect Absorber Based on a Metal-graphene Nanostructure. *Opt. Express* **2017**, *25*, 32467–32474.
92. Masyukov, M.; Vozianova, A.; Grebenchukov A. et al. Optically Tunable Terahertz Chiral Metasurface Based on Multi-layered Graphen. *Sci. Rep.* **2020**, *10*, 3157. https://doi.org/10.1038/s41598-020-60097-0.
93. Ghosh, S.; Srivastava, K. V. Polarization-Insensitive Switchable Absorber/ Reflector and Its Realization Using a Novel Biasing Technique. *IEEE Trans. Antennas Propag.* **2016**, *64* (8), 3665–3670.

94. Ghosh, S.; Srivastava, K. V. A Polarization-Independent Broadband Multi-Layer Switchable Absorber Using Active Frequency Selective Surface. *IEEE Antennas Wireless Propag. Lett.* **2017**, *16*, 3147–3150.
95. Ghosh, S.; Srivastava, K. V. Polarisation-Independent Tunable Absorber with Embedded Biasing Network. *IET Electron. Lett.* **2017**, *53* (17), 1176–1178.
96. Ghosh, S.; Lim, S. Fluidically Switchable Metasurface for Wide Spectrum Absorption. *Nat. Sci. Rep.* **2018**, *8*, 10169.
97. Ghosh, S.; Lim, S. A Multifunctional Reconfigurable Frequency Selective Surface Using Liquid Metal Alloy. *IEEE Trans. Antennas Propag.* **2018**, *66* (9), 4953–4957.
98. He, W.; Lu, L. Revisiting the Structure of Graphene Oxide for Preparing New-Style Graphene-Based Ultraviolet Absorbers. *Adv. Funct. Mater.* **2012**. DOI: 10.1002/adfm.201102998.
99. Bhattacharyya, R.; Prakash, O.; Roy, S.; Singh, A. P.; Bhattacharyya, T. K.; Maiti, P.; Bhattacharyya, S.; Das, S. Graphene Oxide-ferrite Hybrid Framework as Enhanced Broadband Absorption in Gigahertz Frequencies. *Nat. Sci. Rep.* August **2019**, *9*, Article No. 12111.
100. Bhattacharyya, R.; Singh, V. K.; Bhattacharyya, S.; Maiti, P.; Das, S. Defect reconstruction in graphene for excellent broadband absorption properties with enhanced bandwidth. Accepted for publication in Elsevier Applied Surface Science.
101. Bhattacharyya, S.; Baradiya, H.; Chaudhary, R. K.; Srivastava, K. V. In *An Electric Field Driven LC Resonator Structure as Ultra-Thin Metamaterial Absorber*, 5th Annual Conference, Antenna Test and Measurement Society, Mumbai, India, February 2–3, 2012.
102. Bhattacharyya, S.; Srivastava, K. V. In *An Ultra-Thin Electric Field Driven LC Resonator Structure as Meta-material Absorber for Dual Band Applications*, 2013 URSI International Symposium on Electromagnetic Theory (EMTS), Hiroshima, Japan, May 20–24, 2013.
103. Xu, B.; Gu, C. Q.; Zhuo, L.; Zhen, N. Y. A Novel Structure for Tunable Terahertz Absorber Based on Graphene. *Opt. Exp.* **2013**, 21 (20), 23803–23811.
104. Paquin, R. A. *Properties of Metals*, Chapter 35; University of Arizona: Tucson.
105. Su, Z.; Chen, X.; Yin, J.; Zhao, X. Graphene-based Terahertz Metasurface with Tunable Spectrum Splitting. *Opt. Lett.* **2016**, *41*, 3799–3802.
106. Zhihua, P.; Jingcui, P.; Yangfeng, P.; Yangyu, O.; Yantao, N. Complex Permittivity and Microwave Absorption Properties of Carbon Nanotubes/Polymer Composites: A Numerical Study. *Phys. Lett. A* **2008**, *372* (20), 3714–3718.
107. Wiesauer, K.; Jordens, C. Recent Advances in Birefringence Studies at Terahertz Frequencies. *J. Infrared Millim. Te*. **2013**, *34*, 456–467.
108. Ding, J.; Arigong, B.; Ren, H.; Shao, J.; Zhou, M.; Lin, Y.; Zhang, H. Mid-Infrared Tunable Dual-Frequency Cross Polarization Converters Using Graphene-Based L-Shaped Nanoslot Array. *Plasmonics, Springer* **2014**, *10*, 351–356.
109. Mei, Z. L.; Ma, X.M.; Lu, C. et al. High-efficiency and Wide-bandwidth Linear Polarization Converter Based on Double U-shaped Metasurface. *AIP Adv*. **2017**, *7*.
110. Yao, G.; Ling, F.; Yue, J. et al. Dynamically Tunable Terahertz Cross Polarization Amplitude Based on Graphene Metamaterials. *Opt. Dev. Int.* **2015**. DOI: 10.1364/OEDI.2015.JW3A.38.

111. Maeng, I.; Lim, S.; Chae, S. J. et al. Gate-controlled Nonlinear Conductivity of Dirac Fermion in Graphene Field-effect Transistors Measured by Terahertz Time-domain Spectroscopy. *Nano Lett.* **2012**, *12*, 551–555.
112. Neugebauer, P.; Orlita, M.; Faugeras, C. et al. How Perfect Can Graphene Be? *Phys. Rev. Lett.* **2009**, *103*, 1–4.
113. Chen, M.; Chang, L.; Gao, X. et al. Wideband Tunable Cross Polarization Converter Based on a Graphene Metasurface with a Hollow-carved 'H' Array. *IEEE Photonics J.* **2017**, *9* (5), 1–12.
114. Hanson, G. W. Dyadic Greens Functions and Guided Surface Waves for a Surface Conductivity Model of Graphene. *J. Appl. Phys.* **2008**, *103*, 064302.
115. Figotin, A.; Vitebskiy, I. Frozen Light in Photonic Crystals with Degenerate Band Edge. *Phys. Rev. E* **2006**, *74*, Article no. 066613.
116. Wu, S.; Zha, D.; Yun, H.; Miao, L.; Jiang, J. Design of Stable and High-efficiency Graphene-based Polarizers for Oblique Angle of Incidence in Terahertz Frequency. *Appl. Opt.* **2019**, *58*, 492–497.
117. Guo, T.; Argyropoulos, C. Broadband Polarizers Based on Graphene Metasurfaces. *Opt. Lett.* **2016**, *41*, 5592–5595.
118. Ghosh, S. K.; Das, S.; Bhattacharyya, S. Transmittive-type Triple-band Linear to Circular Polarization Conversion in THz Region Using Graphene-based Metasurface. *Opt. Commun.* **2021**, *480*, 126480. https://doi.org/10.1016/j.optcom.2020.126480.
119. Kumar, P.; Lakhtakia, A.; Jain, P. K. Tri-controllable Pixelated Metasurface for Absorbing Terahertz Radiation. *Appl. Opt.* **2019**, *58*, 9614–9623.

CHAPTER 6

Silicon Nanotube FETs: From Device Concept to Analytical Model Development

ARUN KUMAR*[1,2] and PRAMOD KUMAR TIWARI[1]

[1]*Department of Electrical Engineering, Indian Institute of Technology, Patna, Bihar, India*

[2]*Department of Electronics and Communication Engineering, SRM University-AP, India*

**Corresponding author. E-mail: arun.pee16@iitp.ac.in*

ABSTRACT

Multigate MOSFET structures are likely to take up the baton for future applications. Among different multigate structures, silicon nanotube (Si-NT) FETs exhibited promising electrical characteristics and have been hailed as the most promising device structure for the mainstream CMOS technology owing to their geometrical structure that provides excellent electrostatics control as well as better drive current capability. Thus, this chapter highlights the detailed Si-NT FETs device concept, their structure, and also presents an insight into their operation. Efforts in the fabrication of better-scaled devices like Si-NT FETs must be supported by proper circuit design to get the best performance figures. In order to facilitate circuit simulation with such advanced technologies, the analytical model formulation has also been presented in this chapter. The main focus has been the development of a unified drain current model of Si-NT FETs incorporating short-channel effects (SCEs) and quantum effects. The model fits TCAD data very well across a wide range of voltage biases and physical device parameters. The model has been developed using a surface-potential approach, which is in line with the current commercial industry standard BSIM-CMG framework.

6.1 INTRODUCTION

Semiconductor technology has been evolving day by day, from SONY's production of the world's first transistor radio in the year 1957 to almost every switching device in 2020.[1] CMOS devices have greatly influenced the digital transition over the last five decades and still continue to be a motivating force for the invention of new semiconductor device technology.[2] CMOS technology evolution trend over the last 50 years has been to increase the speed, density, and reduce power consumption through the downscaling of the device dimension complemented by the rapid advancement of the lithographic techniques.[3] Although the device downscaling has been continued for the last 50 years, it is essential to point out that the scaling is finally facing challenges owing to severe short-channel effects (SCEs) and other bottleneck poised due to gate oxide tunneling, dielectric breakdown, etc.[4] The pace of MOSFET downscaling has forced technologists to come up with noble solutions. CMOS technology boosters such as continued oxide scaling using High-K dielectric as gate oxide materials, improving channel mobility by using III–V group materials, and nonconventional multigate MOSFET structures for enhanced electrostatic control have emerged as a potential solution to extend the CMOS scaling down to sub-22 nm and beyond.

Different multigate FETs have been actively investigated in order to extend further scaling. These devices have the potential for extending the scaling of MOSFET structures beyond the limitations of planar MOSFET structures. Nonconventional multigate MOSFET devices are less susceptible to SCEs like threshold voltage roll-off, subthreshold leakage current, DIBL, etc. as well as have increased drive current capability. Increasing the number of the gate around the channel region effectively suppresses the SCEs, offers stronger electrostatics control of the inversion channel region, making multigate MOSFET highly scalable as compared with the classical bulk MOSFETs. Moreover, good electrostatics control due to multigates also allows silicon channel to be lightly doped, which translates into an improvement in carrier mobility (less scattering), reduced random dopant fluctuation (RDF), and decreased variability in device operation.

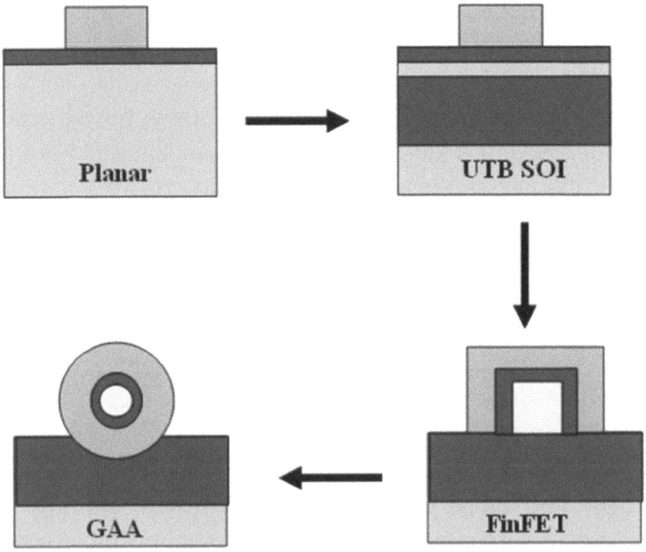

FIGURE 6.1 Evolution of multigate MOSFET structures.

Evolution of the multigate FETs structures are shown in Figure 6.1. Among all the multigate structures, in particular, the nanowire gate-all-around (GAA) structure provides many possibilities and hence drawn much fanfare owing to its geometrical structure that provides excellent control over other nonplanar MOS devices. Although GAA FETs display superior electrostatics control and appears to be a probable candidate for future CMOS devices. GAA FETs suffer from low drive current capability in their operation at ultrascaled device dimension owing to their inherent small effective channel width. Stacking of nanowires can enhance the drive current capability but at the cost of reduced device density over the given chip area and enhanced source/drain series resistance.[5] Moreover, heat dissipation can also be a severe issue, especially for the topmost nanowire in the stack.

6.2 SILICON NANOTUBE FETS

The incessant device dimension downscaling of MOSFETs could be continued only if somehow large drive current could be obtained while maintaining the SCEs suppression as of GAA FETs. This is where a new device structure called silicon nanotube FETs steps in. The structure of Si-NT FETs offers a unique characteristic of high-drive current while maintaining the SCEs immunity.[6] This innovative concept employs a unique core-shell gate stacking in order to achieve full volume inversion in the channel region. A surge in the carrier concentration at the center of the channel region complemented with a rapid roll-off in energy bands at source-channel junction constituted a high-drive current in Si-NT FETs' structure. The researchers also made an observation that the core-shell gate architecture not only enhances the drive current capability but also provides superior SCEs immunity than its counterpart, particularly nanowire GAA FETs with efficient chip area consumption.[6] This extraordinary enhancement of drive current in Si-NT FETs comes from the inherent volume inversion phenomenon. Here, the channel charge centroid shifts away from the surface toward the center of the channel region, reducing carrier scattering, thereby improving the carriers' mobility. Moreover, volume inversion also allows for strong carrier energy quantization leading to improved on-state drain current coupled with robust electrostatics minimizing off-state leakage and SCE performance degradation.

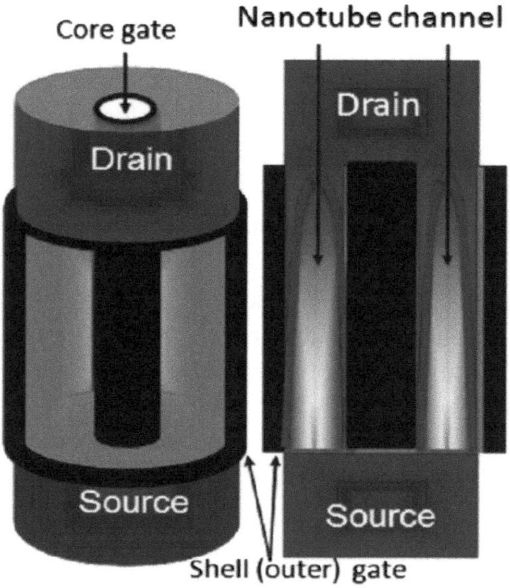

FIGURE 6.2 3-D simulated structure of Si-NT FET depicting volume inversion phenomenon.
Source: Reprinted with permission from H. M. Fahad et al., "Silicon nanotube field effect transistor with core-shell gate stacks for enhanced high-performance operation and area scaling benefits," *Nano Lett.*, 2011, Copyright (2011) American Chemical Society.[6]

The 3-D Si-NT FET structure is shown in Figure 6.2. It is with a tubular channel operated by the two gates, that is, inner (core) and outer (shell) gate, respectively. Similar to GAA FETs, Si-NT FETs are also composed of source, drain, gate oxide, and a tubular channel region, wrapped by thin inner and outer oxide layers. The outer and inner gates can each be independently biased or can be tied together. In the channel region, the inner gate enclosed by the oxide layer supports increased charge control, while the outer gate is identical to the GAA FETs. The device is symmetrically biased, with the same voltage biasing for both inner and outer gates.

6.3 ANALYTICAL MODEL DEVELOPMENT FOR SILICON NANOTUBE FETS

Considering the introduction of Si-NT FETs in CMOS integrated circuits, it is important to have good analytical models available that can easily be used for the possible exploration of any circuit design. Every IC design, whether analog, digital, or mixed signal, is designed and verified beforehand using a circuit simulator before going for physical production. However, in order to perform the circuit simulation, device characteristics in terms of analytical mathematical equations are required to be used in the back-end of the circuit simulator.

The analytical model could be seen as a bridge between the circuit design and IC manufacturers with which the circuit designers can design circuits without thinking about the intricate know-how of device manufacturing. Developing analytical models that are computationally robust and efficient for Si-NT FETs not only helps in analyzing the design space of individual device structure but also supports exploring the performance of the entire system.

Considering the cylindrical structure of Si-NT FETs, the 1-D Poisson equation is solved along the radial direction combined with Gauss's law to formulate a relationship between applied gate voltage and surface potentials. The second step involved in the development of the core I–V model is to formulate an output current equation which relates the output current to surface potential with input terminal voltages as the parameter, that is, $I_{ds} = f(V_{gs}, V_{ds}, \varphi_s)$. This approach gives an inherently explicit and physics-based model with an accurate description in all operating regions of device operation. The key to this approach is the development of unified surface-potential expression, which is valid and continuous in all operating regions. A good I–V model is not only designed for accuracy but also for the ability to catch hidden underlying physics such as quantum confinement effects (QCEs). The QCEs, which are highly prominent in the ultrathin channel region and lead to the discretization of the energy levels, have also been incorporated in the developed analytical drain current model.

6.3.1 MODELING OF UNIFIED SURFACE POTENTIAL

The 2-D cross-sectional view of the Si-NT FETs is shown in Figure 6.3, representing various regions of the device structure. The inner gate radius, gate oxide thickness, and channel thickness are noted as t_c, t_{ox} and t_{si}, respectively.

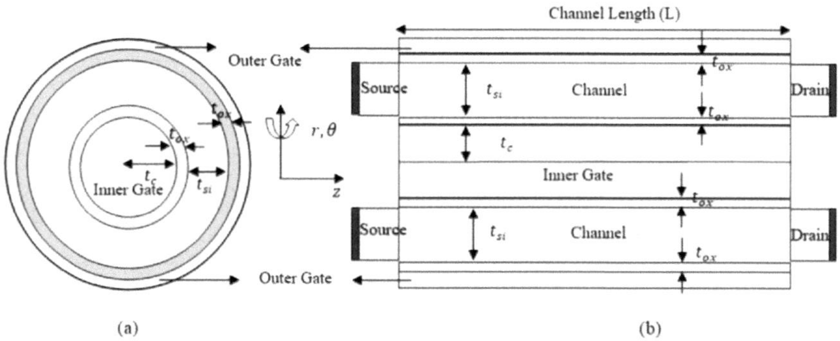

FIGURE 6.3 Schematic of Si-NT FET. (a) 2-D cross-sectional view along the radial direction, (b) 2-D cross-sectional view along the channel.

As the cross-section channel thickness is decreased, it results in increased quantum containment, affecting subband energies.[7] The charge density is calculated in a 2-D confined system of Si-NT FET device with the ultrathin channel area by taking the summation over the entire energy subband using the Fermi-Dirac function and 1-D density of state (DOS) as follows[8]:

$$Q_i = \frac{2q}{\hbar}\sqrt{\frac{2m_e KT}{\pi}} \sum_i \mathcal{F}_{-\frac{1}{2}}\left(\frac{E_F - E_i}{KT}\right) \tag{6.1}$$

where q, $\hbar$, m_e represents the electronic charge, modified Planck's constant, the effective mass of the electron, respectively. K refers to Boltzmann's constant. The temperature is denoted by. The quasi-Fermi energy level is represented by E_F. $\mathcal{F}_{-1/2}$ denotes the (−1/2) order Fermi-Dirac integral, E_i refers to the ith energy levels referenced with the conduction band minima, and is given as[8]

$$E_i = \frac{E_g}{2} - q\varphi_s + E_{l,n} \tag{6.2}$$

Here $E_{l,n}$ is the discretized energy levels obtained by solving the Schrödinger equation in the channel region of Si-NT device structure using proper boundary conditions.[8] φ_s is the potential at the channel–oxide interface (surface potential). E_g is silicon band-gap. The semiconductor charge density can also be written as a function of the applied gate bias (V) at the inner surface using Gauss law as follows:

$$Q_{i1} = C_{ox1}(V_{gs} - V_{fb} - \varphi_{s1} - V(z)) \tag{6.3}$$

where V_{gs} is the applied gate to source voltage, V_{fb} is the flat band voltage, and φ_{s1} is the inner surface potential. C_{ox1} is the capacitance due to inner gate oxide and can be written as[9] follows:

$$C_{ox1} = \frac{\in_{ox}}{r_1 ln\left(\frac{r_1}{r_1 - t_{ox}}\right)} \tag{6.4}$$

On equating eqs 6.1 and 6.3 for obtaining the expression for inner surface potential, the following expression is obtained

$$\varphi_{s1} = V_{gs} - V_{fb} - V(z) - \frac{1}{C_{ox1}}\frac{2q}{\hbar}\sqrt{\frac{2m_e KT}{\pi}}\sum_{l,n}\mathcal{F}_{-\frac{1}{2}}\left(\frac{E_F + q\varphi_{s1} - E_{l,n} - \frac{E_g}{2})}{KT}\right) \tag{6.5}$$

The above equation is implicit, and a computational iteration method would be needed to determine the surface potential that is usually not preferred for applications like circuit simulation. The surface potential equation is therefore calculated by deriving separately for both the subthreshold region of operation and above the threshold region and subsequently stitching those expressions using an interpolation function.[8]

The number of charge carriers available in the higher subbands is negligible in the subthreshold regime of device operation with no drain voltage applied. Thus, ignoring the higher subband, the expression for surface potential under the subthreshold region of operation (φ_a) is given employing the analytical solution of the Fermi-Dirac integral[8] as follows:

$$\varphi_a = V_{gs} - V_{fb} - V(z) - \frac{C}{C_{ox1}}\left(D + \frac{exp\left(-\left(\frac{E_F + q\varphi_a - E_g/2)}{KT}\right)\right)}{\sqrt{\pi}}\right)^{-1} \tag{6.6}$$

where

$$C = \frac{2q}{\hbar}\sqrt{\frac{2m_e KT}{\pi}} \tag{6.7}$$

$$D = \left(2.99 + 2\left(\frac{E_F + q(V_{gs} - V_{fb}) - E_g/2}{KT}\right) + \left(2\left(\frac{E_F + q(V_{gs} - V_{fb}) - E_g/2}{KT} - 1.495\right)^{2.828} + 4.46\right)^{0.353}\right)^{-0.5} \tag{6.8}$$

Using the Binomial expansion[10] and ignoring the higher order terms, Eq. (6) can be rewritten as follows:

$$\varphi_a = V_{gs} - V_{fb} - V(z) - \frac{C}{C_{ox1}D} + \frac{C}{2\sqrt{\pi}C_{ox1}D^2}exp\left(-\left(\frac{E_F + q(V_{gs} - V_{fb}) - E_g/2}{KT}\right)\right) \tag{6.9}$$

The availability of the carriers in the conduction band would be sufficient above the threshold region of operation. The exponential term will be dominant for any applied gate voltage above the threshold. The expression for surface potential φ_b above threshold can be given using the *Lambert-W* mathematical function[11] and algebraic calculation as follows:

$$\varphi_b = V_{gs} - V_{fb} - V(z) - V_T \cdot \mathcal{L}\left(\frac{C}{V_T C_{ox1}}exp\left(\frac{\left(V_{gs} - V_{fb} - \frac{E_g}{2q} - \frac{E_{l,n}}{q} - V(z)\right)}{V_T}\right)\right) \tag{6.10}$$

By stitching the evaluated surface-potential expressions of eqs 6.9 and 6.10, a unified expression for the inner surface potential that is valid for the entire region of operation can be formulated using the interpolation function as follows[12]:

$$\varphi_{s1} = \frac{1}{2}\left(\varphi_a + \varphi_b - \sqrt{(\varphi_a - \varphi_b)^2 + \alpha\varphi_a\varphi_b}\right) \tag{6.11}$$

Here α is a smoothing parameter.

Similarly, the same set of expressions like eqs 6.1, 6.3, and 6.4 could be written for the outer surface potential (φ_{s2}) of Si-NT FET following the same approach as discussed above while calculating the inner surface-potential expression. Thus,

$$Q_{i2} = \frac{2q}{\hbar}\sqrt{\frac{2m_e KT}{\pi}}\sum_{l,n}\mathcal{F}_{-\frac{1}{2}}\left(\frac{E_F - E_{l,n} - \frac{E_g}{2} + q\varphi_{s2}}{KT}\right) \tag{6.12}$$

and

$$Q_{i2} = C_{ox2}(V_{gs} - V_{fb} - \varphi_{s2} - V(z)) \tag{6.13}$$

where φ_{s2} is the outer surface potential. C_{ox2} is the capacitance due to outer gate oxide and can be written as[9] follows:

$$C_{ox2} = \frac{\epsilon_{ox}}{r_2 ln\left(\frac{r_2 + t_{ox}}{r_2}\right)} \tag{6.14}$$

On equating eqs 6.12 and 6.13 in order to obtain an expression for outer surface potential, the following expression is obtained.

$$\varphi_{s2} = V_{gs} - V_{fb} - V(z) - \frac{1}{C_{ox2}}\frac{2q}{\hbar}\sqrt{\frac{2m_e KT}{\pi}}\sum_{l,n}\mathcal{F}_{-\frac{1}{2}}\left(\frac{E_F + q\varphi_{s2} - E_{l,n} - \frac{E_g}{2})}{KT}\right) \tag{6.15}$$

The above expression for the outer surface potential is also transcendental in nature. By adopting the same approach as done for the calculation of inner surface potential as above, the expressions for the outer surface potential under subthreshold region and above-threshold region can also be written as follows:

$$\varphi_c = V_{gs} - V_{fb} - V(z) - \frac{C}{C_{ox2}D} + \frac{C}{2\sqrt{\pi}C_{ox2}D^2} exp\left(-\left(\frac{E_F + q(V_{gs} - V_{fb}) - E_g/2}{KT}\right)\right) \tag{6.16}$$

$$\varphi_d = V_{gs} - V_{fb} - V(z) - V_T \cdot \mathcal{L}\left(\frac{C}{V_T C_{ox2}} exp\left(\frac{\left(V_{gs} - V_{fb} - \frac{E_g}{2q} - \frac{E_{l,n}}{q} - V(z)\right)}{V_T}\right)\right) \tag{6.17}$$

By stitching the evaluated surface-potential expressions of eqs 6.16 and 6.17, a unified expression for the outer surface potential that is valid for the entire region of operation can be formulated using the interpolation function as follows[12]:

$$\varphi_{s2} = \frac{1}{2}\left(\varphi_c + \varphi_d - \sqrt{(\varphi_c - \varphi_d)^2 + \alpha\varphi_c\varphi_d}\right) \tag{6.18}$$

6.3.2 MODELING OF UNIFIED DRAIN CURRENT

For obtaining the complete I–V characteristics in terms of applied external voltage, explicit surface potential, and charge density expressions are required. Under a 2-D confined system, the charge density contribution (Q_{i1}, Q_{i2}) and surface potentials (φ_{s1}, φ_{s2}) separately for both the inner and outer surface are already evaluated in the last section, and thus, the surface-potential expressions can be readily utilized here for the formulation of the short-channel drain current model.

By applying the different voltages at the drain and the source ends, respectively, would transfer the carriers from the source to the drain region through a combination of both diffusion and drift transport mechanisms leading to the flow of the drain current. As a result, the drain current is obtained by integrating carriers over the channel region from the source to the drain end. In view of the current continuity principle, the normalized value of drain current flowing through the device can be given as[13]follows:

$$I_{ds}(z) = Q_n \cdot v(E_z) \tag{6.19}$$

where the carriers' velocity is calculated utilizing the following model

$$v(E_z) = \begin{cases} \mu_n \dfrac{dV(z)}{dz} & E_z < E_{sat} \\ v_{sat}, & E_z >> E_{sat} \end{cases} \tag{6.20}$$

where μ_n refers to the low-field electron mobility model that incorporates the carriers scattering under the influence of gate and drain bias. $E_{sat} = \frac{2v_{sat}}{\mu_n}$ is the saturation electric field with $v_{sat} = 2.5\times10^7\,\text{cm/s}$ being the saturation velocity. The average charge density incorporating both drift and diffusion components in the channel region can also be written as[14]

$$Q_n = \left(\frac{(Q_{S1}+Q_{D1})}{2} + 2V_T C_{ox1}\right) + \left(\frac{(Q_{S2}+Q_{D2})}{2} + 2V_T C_{ox2}\right) \tag{6.21}$$

where $Q_{S1,2} = C_{ox1,2}(V_{gs} - V_{fb} - \varphi_{s1,2})$ and $Q_{D1,2} = C_{ox1,2}(V_{gs} - V_{fb} - \varphi_{s1,2} - V(z))$ are charge densities present at the source side and drain side, respectively, and can be derived using the eqs 6.11 and 6.18.

For a short-channel length device structure, the carriers in the channel region undergo enhanced scattering when they move under the influence of both lateral and transverse electric fields due to both gates and drain bias, respectively. Moreover, the high lateral electric field also breaks the gradual channel approximation. The carrier velocity is no longer proportional to the lateral field rather tends to saturate after achieving the saturation limit. The dependence of mobility on the lateral electric field, that is, the velocity–electric field relation that takes into account the saturation velocity is given by the following relation[15]:

$$\mu_{eo} = \frac{\mu_n}{\sqrt{1+\left(\frac{\mu_n V_{ds}}{v_{sat} L}\right)^2}} \tag{6.22}$$

Drain current models valid into linear and saturation regions are developed separately and stitched together to obtain a unified model. Initially, as the drain biasing voltage $V(z) = V_{ds}$ increases, the current also increases linearly and tend to saturate with the carrier reaching its saturation velocity v_{sat} at $V(z) = V_{dsat}$. Under the linear region, eq 6.19 can be integrated in terms of charge density, which leads to[16]

$$I_{ds}(z) = Q_n \cdot \mu_{eo} \frac{dV(z)}{dz} \tag{6.23}$$

Considering the relation of $dV(z) = dQ_n/C_{ox}$ in the linear region of device operation, eq. 6.23 can be integrated in terms of charge densities at source and drain ends respectively as follows:

$$I_{ds} = \frac{\mu_{eo}}{L}\left(2V_T\left(Q_{S1,2} - Q_{D1,2}\right) + \frac{Q_{S1,2}^2 - Q_{D1,2}^2}{2C_{ox1,2}}\right) \tag{6.24}$$

For a conventional long channel device structure, under the saturation region, the drain current does not increase with further increase in drain voltage otherwise saturates at some maximum value. However, this definition is not valid for the case of short-channel devices because of the fact that the carriers' velocity saturation occurs much before the drain end at some particular drain voltage called saturation voltage. In other words, the saturation voltage is the intercept between linear and saturation regions, which is determined by the saturation current at which the carrier velocity reaches the saturation velocity.[16]

In order to obtain the saturation voltage, using the theory of velocity saturation effect, the drain current can be given as[16] follows:

$$I_{dsat} = v_{sat} \cdot C_{ox1,2}(V_{gs} - V_{fb} - \varphi_{s1,2} - V_{dsat}) \tag{6.25}$$

Here, V_{dsat} is the saturation voltage at which current saturates.

Replacing V_{ds} and I_{ds} with V_{dsat} and I_{dsat} in eq 6.24 and equating with eq 6.25, V_{dsat} has been obtained ignoring diffusion components as follows:

$$V_{dsat} = V_{gs} - V_{fb} + V_T \cdot ln\left(\frac{C_1}{v_{sat}LC_{ox1,2}D\mu_{eo} - D\sqrt{\left(v_{sat}LC_{ox1,2}\mu_{eo}\right)^2 + Q_{S1,2}^2}}\right) \tag{6.26}$$

The major distinction between the two piece-wise drain current expressions of eqs 6.24 and 6.25 is only the drain-source bias voltage value. Thus, an effective drain biasing voltage $V_{dseff} = f(V_{ds}, V_{dsat})$ continuous from linear region to saturation region is introduced using an interpolation function as follows:

$$V_{dseff} = \frac{1}{2}\left(V_{ds} + V_{dsat} - \sqrt{\left(V_{ds} - V_{dsat}\right)^2 + \alpha V_{ds}V_{dsat}}\right) \tag{6.27}$$

where α is a smoothing parameter

Therefore, a unified drain current model is given by integrating charge density at both the inner surface and the outer surface of Si-NT FET, which leads to[9, 16]

$$I_{ds} = \frac{2\pi r_1\mu_e}{L}\left(2V_T\left(Q_{S1} - Q_{DS1}\right) + \frac{Q_{S1}^2 - Q_{DS1}^2}{2C_{ox1}}\right) + \frac{2\pi r_2\mu_e}{L}\left(2V_T\left(Q_{S2} - Q_{DS2}\right) + \frac{Q_{S2}^2 - Q_{DS2}^2}{2C_{ox2}}\right) \tag{6.28}$$

where

$$Q_{S1,2} = C_{ox1,2}(V_{gs} - V_{fb} - \varphi_{s1,2}) \tag{6.29}$$

$$Q_{DS1,2} = C_{ox1,2}(V_{gs} - V_{fb} - \varphi_{s1,2} - V_{dseff}) \tag{6.30}$$

and the effective mobility used in Eq. (22) is also modified incorporating the effective drain bias voltage as follows:

$$\mu_e = \frac{\mu_n}{\sqrt{1 + \left(\frac{\mu_n V_{dseff}}{v_{sat} L}\right)^2}} \tag{6.31}$$

6.4 SHORT-CHANNEL EFFECTS CORRECTIONS

Once the drain current of Si-NT FET incorporating the velocity saturation effects using current continuity principle and accounting the mobility degradation owing to low-field and high-field is obtained, the applicability of the developed model for short-channel device by integrating the multiple short-channel effects (SCEs) can be done. The following section discusses the modeling of different SCEs, which will be incorporated in the drain current expression of eq 6.28.

6.4.1 MOBILITY MODEL CONSIDERING THE VELOCITY OVERSHOOTS EFFECTS

The classical drift-diffusion carrier transport tends to underestimate the output drain current as it does not take into account of nonlocal fields generated with high-field gradients in nanoscale devices, and velocity overshoot is one of them which is inherent in the short-channel device structure. In such a scenario, the scattering events are no longer localized, and some portions of the carriers have a finite probability of acquiring higher thermal energy, especially near the drain end.[18] Under this situation, it is highly possible for the carriers to exceed their saturation velocity. In order to incorporate the velocity overshoot effects, the energy balance equations are required to be solved, as proposed by *Stratton* and *Bloteckjaer.*[19] However, the effect of velocity overshoot of the carrier has been incorporated by modifying the mobility model by following empirical relation as proposed by Roldan et al.[20]

$$\mu_e = \frac{\mu_n}{\sqrt{1 + \left(\frac{\mu_n V_{dseff}}{v_{sat} L}\right)^2}} + \frac{\lambda_a}{L} \tag{6.32}$$

Here, $\lambda_a = 40 \times 10^5 \, \text{cm}^3 / \text{V} - \text{s}$ is the velocity overshoot parameter, and the same value is used as reported in.[20]

6.4.2 CHANNEL LENGTH MODULATION EFFECTS

The drain current expression is developed using two piece-wise drain current models for linear region and saturation region, which is later stitched using an interpolation function. For long channel devices, it assumes that as the lateral electric field reaches saturation condition, the carriers drift velocity will be saturated as v_{sat}. However, this is not true for short-channel devices. For $V_{ds} > v_{dsat}$, the channel is partially saturated, and the effect of channel length modulation (CLM) is significant. The CLM effect makes the channel length shorter by $L_{eff} = L - \Delta L$. ΔL is referred to as the amount of channel length modulation by the drain voltage, and this effect is incorporated through the following empirical equation[20]

$$\Delta L = \delta \lambda_C \times \sinh^{-1}\left(\frac{V_{ds} - V_{dsat}}{E_{sat} \lambda_C}\right) \tag{6.33}$$

where, λ_C is the characteristics length, δ is the fitting parameter, and ΔL is the distance from drain end and where pinch-off occurs.

6.4.3 THRESHOLD VOLTAGE ROLL-OFF AND DIBL EFFECTS

The threshold voltage roll-off, subthreshold slope degradation, and DIBL effects are also some other important SCEs that are related to the reduction of the source-channel energy barrier. The threshold voltage roll-off and DIBL cause the reduction of the threshold voltage in short-channel devices by lowering the source barrier that is governed by the 2-D Poisson equation. The voltage applied at the drain end generates an electric field that perturbs the channel electrostatics and may lower the potential separating the source and drain regions, thereby allowing more injection of carriers at the source-channel junction.

In the proposed model, the threshold voltage roll-off and DIBL is included by the method of threshold voltage shift and is done by replacing the gate voltage V_{gs} with an effective gate voltage.[21] Therefore, the effective gate voltage is given as follows:

$$V_{gseff} = V_{gs} + \phi_{min} \tag{6.34}$$

where ϕ_{min} is the minimum surface potential as derived in Ref.[22]

It should be mentioned that in order to obtain a drain current, the surface potential is first estimated, and the minimum electrostatic potential ϕ_{min} is then evaluated and added to the gate voltage afterwards. The surface potential is evaluated again for effective (modified) V_{gseff}, and subsequently, the charge density and drain current for the device is determined later.

6.5 RESULTS AND DISCUSSION

The current–voltage characteristics obtained from the proposed model are compared with the Sentauras TCAD[23] results to assess the accuracy of the proposed model. The TCAD results are calibrated using data measured from the results reported in Ref. [5] By reproducing the result of Si-NT FET with a channel thickness of 10 nm and a gate length of 20 nm, the simulation method was optimized to simulate Si-NT FETs, as shown in Figure 6.4. The simulation results are in close agreement with the reported data of Ref. [5], which means that the desired results can be reproduced effectively by our simulations.

The Si-NT FET output characteristic at various gate bias voltages is illustrated in Figure 6.5. As can be seen from the figure, the proposed model results are in good agreement with the TCAD results at 20 nm channel length, thus accurately capturing the behavior of the short-channel device structure including carrier velocity saturation.

The Si-NT FET transfer characteristic for various channel thicknesses is shown in Figure 6.6. With an increase in drain voltage biasing, the drain current increases. Increased channel thickness effectively allows more carriers to sweep from the source to the drain end resulting in an improvement in the drain current value. The proposed model outcomes are in good agreement with the TCAD results and correctly reproduce the current behavior of the short-channel device structure for the complete region of operation.

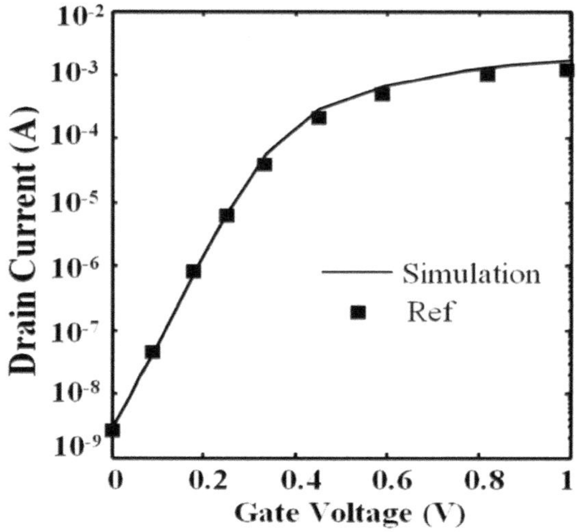

FIGURE 6.4 Comparison of transfer characteristics obtained from TCAD simulation with measured data for Si-NT FET. 3-D TCAD simulations results are calibrated with results reported in Ref. [5] for the same device dimension.[13]

Source: Reprinted with permission from A. Kumar et al., in Proc. 14th *Nanotechnology Materials and Devices Conference*, 2019 (©: 2019 IEEE).

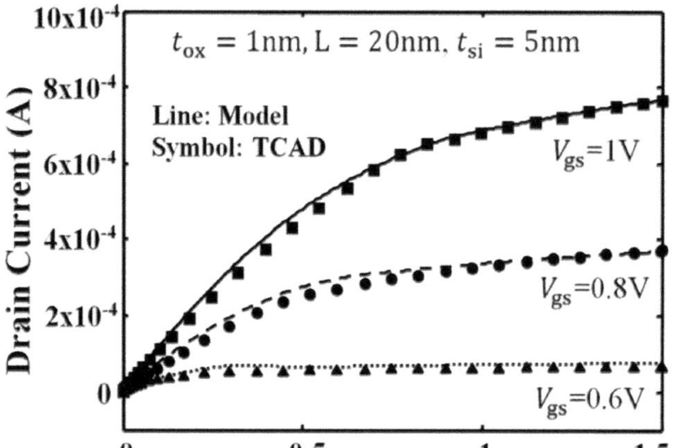

FIGURE 6.5 Plot of variation of drain current (A) against variable drain voltage (V) for different gate voltage values[13]

Source: Reprinted with permission from A. Kumar et al., in Proc. 14th *Nanotechnology Materials and Devices Conference*, 2019 (©: 2019 IEEE).

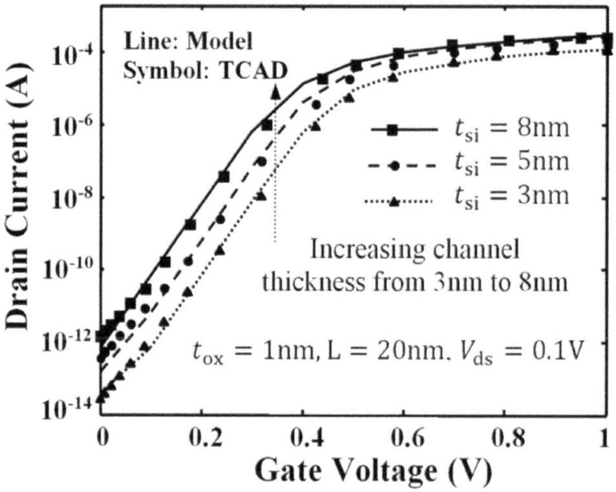

FIGURE 6.6 Plot of variation of drain current (A) against varying gate voltage (V) for different channel thickness values.[13]

Source: Reprinted with permission from A. Kumar et al., in Proc. 14th *Nanotechnology Materials and Devices Conference*, 2019 (©: 2019 IEEE).

6.6 SUMMARY AND CONCLUSION

This chapter has been committed to the introduction of Si-NT FET as the next-generation transistor architecture for future CMOS applications. Moreover, in this chapter, an explicit analytical model for unified drain current of Si-NT FET incorporating short-channel effects has also been presented. A drain voltage expression which is continuous for both the linear and saturation case is evaluated using two piece-wise drain current models. In order to accurately predict the transfer and output characteristics, the short-channel effects are also well integrated in the derived drain current model. The results obtained from the presented drain current model have been verified using the results of the Sentauras TCAD simulator. Since the Si-NT FET application extension is critically dependent on the availability of an analytical model that can be conveniently used to efficiently explore any circuit design application, this chapter has been devoted to providing a reliable model of the Si-NT FET drain current.

KEYWORDS

- **silicon nanotube (Si-NT) FET**
- **electrostatics control**
- **unified drain current model**
- **surface-potential approach**
- **BSIM-CMG framework**

REFERENCES

1. Jung, E. 4th Industrial Revolution and Boundry: Challenges and Opportunities, *2018 IEEE International Electron Devices Meeting (IEDM)*, San Francisco, CA, 2018, pp. 1.1.1–1.1.10.
2. Kuzmicz, W. The Future of CMOS: More Moore or the Next Big Thing?, *2017 MIXDES* - 24th *International Conference Mixed Design of Integrated Circuits and Systems*, Bydgoszcz, 2017, pp. 21–26.
3. Dennard, R. H.; Gaensslen, F. H.; Yu, H. N.; Rideout, V. L.; Bassous, E.; Blanc, A. L. Design of Ion-implanted MOSFET's with Very Small Physical Dimensions. *IEEE J. Solid-State Circuits* **1974**, *SC-9*, 256–268.
4. Radamson, H. H. et al. Miniaturization of CMOS. *Micromachines* **2019**, *10* (5), 293.
5. Fahad, H. M.; Hussain, M. M. Are Nanotube Architectures More Advantageous than Nanowire Architectures for Field Effect Transistors? *Nat. Sci. Rep.* June **2012**, *2* (475), 1–7.
6. Fahad, H. M.; Smith, C. E.; Rojas, J. P.; Hussain, M. M. Silicon Nanotube Field Effect Transistor with Core-shell Gate Stacks for Enhanced High-performance Operation and Area Scaling Benefits. *Nano Lett.* Sep. **2011**, *11* (10), 4393–4399.
7. Dasgupta, A.; Rastogi, P.; Agarwal, A.; Hu, C.; Chauhan, Y. S. Compact Modeling of Cross-Sectional Scaling in Gate-All-Around FETs: 3-D to 1-D Transition. *IEEE Trans. Electron Devices* March **2018**, *65* (3), 1094–1100.
8. Kumar, A.; Tiwari, P. K. An Explicit Unified Drain Current Model for Silicon-Nanotube-Based Ultrathin Double Gate-All-Around MOSFETs. *IEEE Trans. Nanotech.* Nov. **2018**, *17* (6), 1224–1234.
9. Hong, C.; Zhou, J.; Huang, J.; Wang, R.; Bai, W.; Kuo, J. B.; Chen, Y. A General and Transformable Model Platform for Emerging Multi-gate MOSFETs. *IEEE Electron Device Lett.* Aug. **2017**, *38* (8), 1015–1018.
10. Hilder, D. W.; Sweetenham, J. G. Binomial Expansions. *Model Answers Ordinary Natl. Certif. Math. Eng.* **1966**, 20–27.
11. Dence, T. P. A Brief Look into the Lambert W Function. *Appl. Math.* **2013**, 04 (06), 887–892.

12. He, J.; Tao, Y.; Liu, F.; Feng, J.; Yang, S. Analytic Channel Potential Solution to the Undoped Surrounding-gate MOSFETs. *Solid State Electron.* May **2007**, *51*, 802–805.
13. Kumar, A.; Srinivas, P. S. T. N.; Kumar Tiwari, P. Compact Drain Current Model of Silicon-Nanotube based Double Gate-All-Around (DGAA) MOSFETs Incorporating Short Channel Effects, *2019 IEEE 14th Nanotechnology Materials and Devices Conference (NMDC)*, Stockholm, Sweden, 2019, pp. 1–4,
14. Khandelwal, S. et al. Unified Compact Model Covering Drift-diffusion to Ballistic Carrier Transport. *IEEE Electr. Device Lett.* Feb. **2016**, *37* (2), 134–137.
15. Caughey, D. M.; Thomas, R. E. Carrier Mobilities in Silicon Empirically Related to Doping and Field. *Proc. IEEE* **1967**, *55*, 2192–2193.
16. Trevisoli, R. D.; Doria, R. T.; de Souza, M.; Das, S.; Ferain, I.; Pavanello, M. A. Surface-potential-based Drain Current Analytical Model for Triple-gate Junctionless Nanowire Transistors. *IEEE Trans. Electr. Devices* Dec. **2012**, *59* (12), 3510–3518.
17. Smaani, B.; Latreche, S.; Iñíguez, B. Compact Drain-current Model for Undoped Cylindrical Surrounding-gate Metal-oxide Semiconductor Field Effect Transistors Including Short Channel Effects. *J. Appl. Phys.* **2013**, *114* (22), 224507.
18. Roldan, J. B.; Gámiz, F.; Lopez-Villanueva, J. A.; Carceller, J. E. Modeling Effects of Electron-velocity Overshoot in a MOSFET. *IEEE Trans. Electr. Devices* May **1997**, *44*, 841–846.
19. Blotekjaer, K. Transport Equations for Electrons in Two-valley Semiconductors. *IEEE Trans. Electr. Devices* Jan. **1970**, *17* (1), 38–47.
20. Roldan, J. B. et al. An Analytical I–V Model for Surrounding-gate Transistors that Includes Quantum and Velocity Overshoot Effects. *IEEE Trans. Electr. Devices* Nov. **2011**, *57* (11), 2925–2933.
21. Trevisoli, R. D.; Doria, R. T.; de Souza, M.; Das, S.; Ferain, I.; Pavanello, M. A. Surface-potential-based Drain Current Analytical Model for Triple-gate Junctionless Nanowire Transistors. *IEEE Trans. Electr. Devices* Dec. **2012**, *59* (12), 3510–3518.
22. Kumar, A.; Bhushan, S.; Tiwari, P. K. A Threshold Voltage Model of Silicon-Nanotube-Based Ultrathin Double Gate-All-Around (DGAA) MOSFETs Incorporating Quantum Confinement Effects. *IEEE Trans. Nanotech.* Sept. **2017**, *16* (5), 868–875.
23. Sentaurus Device User Guide, Version N-2017.09, Synopsys, Mountain View, CA, USA, 2016.

CHAPTER 7

Techniques for Deposition of Diamond-Like Carbon and Their Potential Applications in Teaching and Training Materials Engineers

RAJNISH SHARMA*

Chitkara University Institute of Engineering and Technology, Chitkara University, Punjab, India

**E-mail: rajnish.sharma@chitkarauniversity.edu.in*

ABSTRACT

Diamond considered to be the hardest known material on our planet has a variety of applications other than those usually found in ornaments. Owing to its unique characteristics like extreme hardness, high reflection coefficient, low friction, capability to act as both conductor and insulator and its noble character make it a potential candidate for applications in fields of ophthalmic glasses, sharp blades with long lives, field emission display devices and tribology, etc.

Diamond, mostly available only as a natural material and rare in supplies remains an extremely expensive item. Researchers have been able to develop diamond-like carbon (DLC), a material akin with properties of diamond. DLC boasts of different characteristics, when deposited under different conditions and different techniques.

Originally developed in 1971 by Aisenberg and Chabot,[1] research on DLC has matured quite a lot and researchers understand quite well the procedures and techniques required to be deployed to develop DLC best suiting a particular kind of application. The most important of these techniques are plasma-enhanced chemical vapor deposition, ion beam method, and saddle field fast atom beam source techniques. All these techniques come with a variety of internal modifications, their strengths, and limitations.

This chapter discusses the typical characteristics of diamond as the material, a detailed description of techniques to deposit and characterize these films. Taking care of a typical academician's angle, chapter also dwells on the kind of laboratory experiments which can be easily setup in an academic institute to further appreciate the concepts covered in the chapter.

7.1 INTRODUCTION

Hydrogenated amorphous carbon (a-C:H) or so-called diamond-like carbon (DLC) films are of considerable interest due to their numerous useful properties like high hardness, high electrical resistivity, excellent immunity to corrosive chemicals and wide range of optical transparency.[2,3] These films can be deposited by a variety of techniques including radio frequency (RF) and direct current (DC) plasma decomposition of hydrocarbon gases or by RF sputtering using low energy carbon ion/atom beams and ion plating techniques. Among these different deposition techniques, the saddle field fast atom beam (FAB) source method has emerged as a competitive candidate for a variety of applications.[4] The main advantage of the saddle - field FAB source lies in the possibility of getting a 100% neutral beam, as claimed by its manufacturers.[5]

Therefore, the problem of charging of the substrates should not pose a problem. The other advantages of the FAB source method, as compared to the more conventional RF self-bias technique are: (1) an uncomplicated DC power supply with no RF matching problems, (2) ease of scale up (modular design), and (3) no need to design the system geometry to realize the needed asymmetry as is required in the RF self-bias deposition of DLC films.

Section below describes techniques available for deposition of DLC films, while mainly focusing upon FAB which basically forms the basic crux of this chapter.

7.2 TECHNIQUES FOR DEPOSITION OF DLC FILMS

The most common technique to grow good DLC films is the RF self-bias method. Here, the two most important process parameters are the self-bias voltage and the pressure of the hydrocarbon gases. The advantages of RF self-bias technique are the following:

1. Uniform coatings on different substrates.
2. Insulating substrate can be coated without difficulty.
3. Higher deposition rates as compared to DC glow discharge technique.
4. The process parameters can be tailored to achieve desired deposition.
5. The interaction between the plasma and its surroundings is "*focussed*" to the small powered electrode, i.e., deposition and sputtering is almost totally restricted to the electrode with the substrate on top. Contamination from the reactor walls as well as undesirable coating of the vacuum system is minimized.
6. Due to the high negative self-bias voltage, the acceleration potential, which determines the average kinetic energy per positive ion impinging on the electrode, is at least one order of magnitude lower for the electrons in comparison to the positive ions. Therefore, the electrons contribution to the power dissipated on the substrate is negligible.
7. Larger particles (e.g., dust), like any electrically floating probe in contact with the plasma, charge up negatively. They are thus repelled from the negatively biased substrates, one prerequisite for pinhole-free coatings.

RF PECVD technique has got its own limitations in terms of huge cost of the equipment and a specialized skill to operate such equipment.

The present chapter focuses on relatively much economical technique, which can be very easily set up in academic institute. Technique is named as saddle field FAB source.[4] Structure of a typical FAB source is shown in the Figure 7.1 below.

It is a cold cathode device and consists of two anode rods (graphite) surrounded by a rectangular cathode also made of carbon (graphite). The source used is of twin (double) cell construction but designed as a modular unit for multiple combinations in a linear or matrix array. When a DC positive voltage is applied to the source (anode) by a power supply, the electrons originate from a sector of the cathode and travel through the anode region toward the opposite cathode sector. While travelling, they are retarded, return, and continue to oscillate about a central saddle field under the action of a DC electrostatic field. The electrons, therefore, travel long oscillatory paths before being captured by the anode. This increases the efficiency of ionization. When a hydrocarbon gas or vapor is introduced, the probability of ionization is high because of the long electron paths and this unique feature of the source provides the ability that the discharge can be maintained at a pressure approaching two orders of magnitude lower than that normally associated with cold cathode sources.

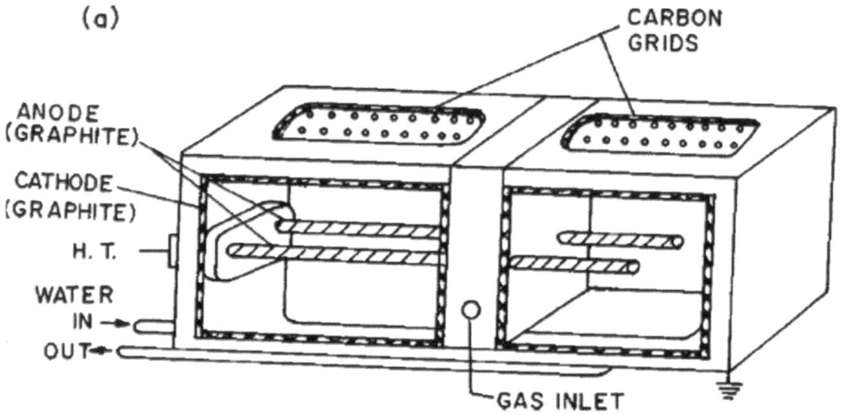

FIGURE 7.1 Structure of a typical FAB source.

Typical operating pressure can be from 10^{-4} to 10^{-3} mbar. The positive ions formed in the discharge travel readily toward the cathode and emerge along straight paths when an aperture is introduced. Another unique feature of the source is the neutral nature of the particle beam produced. The energetic atoms of the supply gas or the fast atoms originate as energetic ions (energy nearly corresponding to the applied voltage) which are positive within the body of the plasma but while coming out of the source they are neutralized. The two proposed mechanisms of neutralizations are: (1) the ions are neutralized by colliding with residual gas molecules before passing through the aperture of the source and (2) the electrostatic field close to the output grid helps electrons to recombine with ions with little loss of energy.

The plasma chamber of the source and electrodes are made of carbon and coupled with the cold cathode operation, this construction allows the source to be used with gases other than inert gases. For the deposition of DLC films, one needs to aspirate some hydrocarbon gas or vapor through the source. The operation of the source is dependent only on two parameters. These are the regulated current output from the power supply and the discharge voltage, which is determined by the gas flow (pressure) and consequently by the setting of the needle valve/mass flow controller used, etc.

A schematic view of a typical experimental setup for the deposition of DLC thin films alone is shown in Figure 7.2.

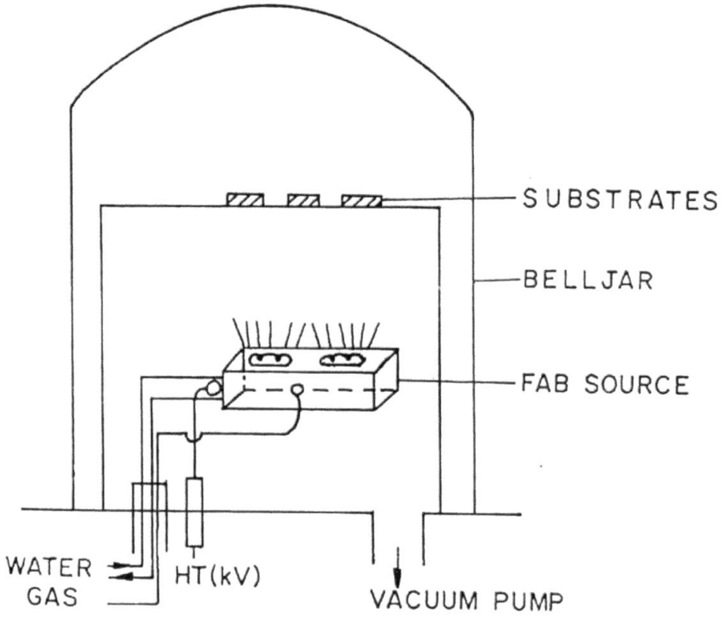

FIGURE 7.2 Typical setup for a saddle field fast atom beam source equipment to deposit DLC films.

It must be noted that the above-experimental setup has got the ability to deposit only thin DLC films. In the later part of the chapter, another setup is proposed which can be called as a more enhanced version of the above setup and can be a good model for use in academic setups.

7.3 PREPARATION OF THE SUBSTRATES TO DEPOSIT DLC FILMS

This is a very important step and proper cleaning of the substrates before being introduced into the deposition chamber should be paid particular attention in order to deposit high quality DLC films. Depending upon the requirement, substrates like Corning 7059 glass, single crystalline silicon wafer (5–10 Ω cm, p-type), well-polished germanium, optically polished stainless steel pieces and quartz plates may be used for the deposition of DLC films. The procedure for cleaning the glass, quartz, and stainless steel substrates comprises of an initial wash with the soap solution and double-distilled water followed finally by an ultrasonic cleaning for 15 min in pure alcohol. The Si wafers, however, are required to be given to etch by hydrofluoric acid to remove the native oxide layer. Overall cleaning toward the end of the process using distilled water is mandatory. The substrates are required to be dried by blowing with dry nitrogen gas.

7.4 CHARACTERIZATION OF DIAMOND-LIKE CARBON FILMS

Once deposited, DLC films are required to be characterized for knowing about the exact composition of the films or other bonding patterns, etc.

Following techniques and parameters are normally used for the same:

7.4.1 THICKNESS MEASUREMENT

The thickness of the thin films is one of the first important parameters to be evaluated. Initial estimation of the thickness can be done using any kind of Talystep profilometer.[6] A typical profilometer is equipped with a diamond stylus and measures the height of the step, which is amplified and recorded on a magnetic chart paper. It is very difficult to form a step after depositing DLC films on the substrates. Therefore, before loading the substrates in the deposition chamber, a dot mark is made on the substrate using a marker pen that acts as a step after the dot is erased, using acetone for instance. The thicknesses of the films can also be verified by ellipsometry, a technique which has been explained later.

7.4.2 HARDNESS MEASUREMENT

To qualify the DLC films as a protective coating it is very important to evaluate the hardness of the films. Hardness is also expected to reflect the other related properties of thin films, like porosity, density or surface quality.[7] Again, the hardness of a material is a parameter describing its ability to withstand penetration by a surface contact. The determination of the hardness of a material is normally made by driving a sharp indenter of a defined geometry into the surface under specified conditions for a fixed time. Usually a pressure, termed the hardness, H, is calculated from the applied load P and the contact area. A of the loaded indenter: $H = P/A$.

A Knoop hardness (HK) indenter equipped with a suitable equipment like Zwick 3212 instrument (Germany) can be used to evaluate the hardness of the DLC films.[8]

The HK is defined as the applied load P divided by the projected area of the impression, and can be calculated by the formula HK = 14,229 P/l^2, where l is the length of the diagonal of the indenter.

DLC films deposited onto Si substrates are usually used for this purpose.

7.4.2.1 ELLIPSOMETRY

Ellipsometry is a convenient and accurate technique for the measurement of thickness and refractive indices of very thin films on solid surfaces and for the measurement of optical constants on reflecting surfaces.[9] Ellipsometry technique can also be used for quantitative microstructural information of the film by analyzing ε_1 vs. hv and ε_2 vs. hv trace. In ellipsometry, the change in the state of polarization of elliptically polarized light due to Δ, reflection is measured and interpreted in terms of properties of the reflecting surface. Two characteristic parameters measured in ellipsometry are the change in the relative amplitude t and ψ and phase difference Δ of two orthogonal components of the incident light due to reflection. From these two measured quantities, two parameters of the reflecting surface can be determined. For a surface covered with an absorbing film, thickness and refractive index of the film can be determined, if the optical constants of the substrate are known.

In X-ray photoelectron spectroscopy (XPS) and X-ray Auger electron spectroscopy (XAES) studies, XPS is a chemical sensitive technique and the XPS chemical shift has proven to be a valuable method for understanding the local environment of an atom in a solid and to investigate whether a carbon film is sp^3 or sp^2 bonded.[10] Apart from the chemical shift, the XPS technique can provide other useful information such as film composition, plasmon loss for conducting materials and valence band density of states.

A relatively sensitive analysis of the sp^3 and sp^2 bonding in the DLC films is provided by AES or XAES. XAES has further advantage of not damaging the samples under investigation.

7.4.2.2 SCANNING ELECTRON MICROSCOPY STUDIES

Scanning electron microscopy (SEM) is a widely used technique for studying surface topography as well as cross-sectional features of thin films.[11] In SEM, a finely focused electron beam is scanned over the surface of the specimen. The secondary electrons emanating from about 10Å from film surface are extracted by a high potential applied to a scintillator-detector system. Secondary electron images are generated by synchronizing the optical output of the detector with raster of the electron probe across the sample.

7.4.2.3 IR STUDIES

IR spectroscopy is a non-destructive method capable of providing a lot of information about the material without having to form contacts. Difference in the structure and the composition show up in the IR absorption spectra. The absorption in the IR below the optical gap is characterized by an absorption tail extending toward lower energy, as well as C–H and C–C vibrational bands, the latter becoming allowed due to the lack of translational symmetry in the amorphous materials.

7.5 TYPICAL EXPERIMENTAL SETUP THAT CAN BE SETUP IN ACADEMIC INSTITUTES FOR DEPOSITION AND STUDYING THE APPLICATIONS OF DIAMOND-LIKE CARBON THIN FILMS

As mentioned above, out of many techniques for the deposition of DLC films, saddle field fast atom beam source technique remain relatively inexpensive and easy to work with method, which can be very easily setup in academic institutes.

Figure 7.3 shows a typical line diagram of such a setup.

As can be seen very clearly, equipment suggested in Figure 7.3 is a combination of three main sections of the experimental setup

a. Metallization unit
b. DC saddle field unit
c. VI characterization unit, computer controller cryostat system

A typical diffusion vacuum pump/rotary pump combination will be costing somewhere about US $ 15000 only and there are many leading suppliers who can create and customize this kind of experimental setup for academic institutes.

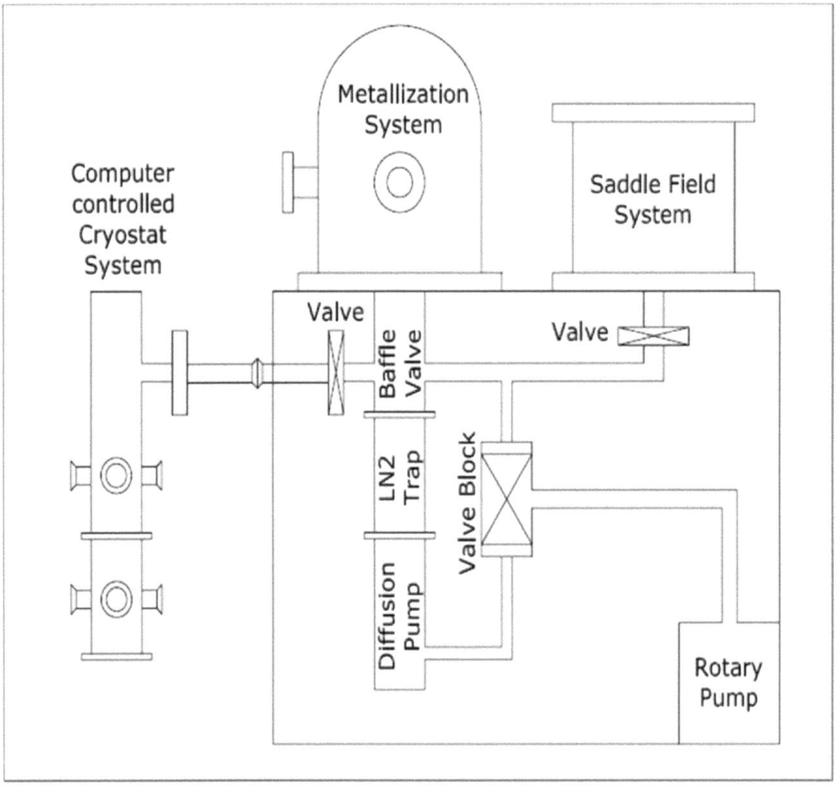

FIGURE 7.3 Saddle field fast beam source equipment along with a metallization unit.

Once put in operation, following simple experiments can be conducted at both under graduate and post graduate level like

1. Learn and understand the need of creating vacuum during deposition of thin films, where a student appreciates and gets a first-hand

experience of the field. A Roth's vacuum technology has worked as very good resource in this direction for last several years.[12]

2. Learn and understand the various parts of a vacuum system likewise appreciating the need of basic vacuum creation phenomenon like backing and roughing.
3. Student may get his hands laid on his first experiment to deposit thin mental films of aluminum/silver, etc., on the basic glass substrates. Before doing so they can be taught about the basic cleanroom ethics and requirement of a creation of a cleanroom culture.
4. With the basic understanding about the vacuum systems, their need and hands-on the experience of depositing thin metal films, student may be given an opportunity to make an attempt to deposit thin DLC films on glass/silicon substrates in the specific chamber meant for the same. Because of usage of high values of DC voltages and involvement of hydrocarbon gases, it is always suggested that these experiments should be conducted under the guidance and supervision of some trained lab staff.
5. Suggested lab setup is quite capable of even providing the basic characterization of thus deposited metal or DLC films as well. Learners can use that very much to understand the basic need of characterization techniques and lays their hands-on usage of the same.

A detailed lab manual can be made available on request through the editor. Suggested list of lab experiments is tentative only and users can devise many such basic lab experiments on the proposed lab setup.

7.6 CONCLUSION

The present chapter has tried to touch upon the subject of Material Science in more of a practical approach, where specific application a well-known material DLC has been provided.

Right from making the readers aware about the DLC as a material to commonly available techniques for the deposition of these films to suggesting an economical lab setup, this chapter has all for a potential material Science Engineer.

This chapter does not obviously claim to be a well detailed and complete resource on such a vast subject matter, but it would surely act as a good guideline if somebody wants to get a real feel of this exciting and amazing field. All kinds of support can be provided in terms of assistance of setup of the suggested lab setup.

ACKNOWLEDGMENT

The author would like to thank his student Preeti for updating the manuscript with some of the latest references.

KEYWORDS

- **DLC**
- **PECVD**
- **roughing**
- **backing**
- **deposition**

REFERENCES

1. Aisenberg, S.; Chabot, R. Ion-beam Deposition of Thin Films of Diamondlike Carbon. *J. Appl. Phys.* **1971**, *42*, 2953–2958.
2. Grill, A. Diamond-like Carbon: State of the Art. *Diamond Related Mater.* **1999**, *8*, 428–434.
3. Kim, D.-O.; Hong, H.-K.; Seo, D.-B.; Trung, T. N.; Hwang, C.-C.; Lee, Z. et al. Novel high-k gate dielectric properties of ultrathin hydrocarbon films for next-generation metal-insulator-semiconductor devices. *Carbon* **2020**, *158*, 513–518.
4. Panwar, O.; Sarangi, D.; Kumar, S.; Dixit, P.; Bhattacharyya, R. Diamond-like carbon films grown using a saddle field source. *J. Vacuum Sci. Technol. A: Vacuum Surfaces Films* **1995**, *13*, 2519–2524.
5. In Atom Tech Ltd. Island Form Avenue: West Molexy, Surrey KT82UZ, England.
6. Bhowmik, R.; Mitra, P.; Choudhury, R.; Reddy, V. Substrate Effect on the Structural Phase Formation and Magnetic Properties of α-Fe_2O_3 and Ti Doped α-Fe_2O_3 Thin Films. *Appl. Surface Sci.* **2020**, *501*, 144224.
7. Scaglione, S.; Emiliani, G. Mechanical Properties (Hardness and Adhesion) of a-C: H Thin Films Produced by Dual Ion Beam Sputtering. *J. Vacuum Sci. Technol. A: Vacuum Surfaces Film* **1989**, *7*, 2303–2306.

8. Kumar, S.; Dwivedi, N.; Rauthan, C.; Panwar, O. Properties of Nitrogen Diluted Hydrogenated Amorphous Carbon (n-type aC: H) Films and Their Realization in n-type aC: H/p-type Crystalline Silicon Heterojunction Diodes. *Vacuum* **2010**, *84*, 882–889.
9. Airaksinen, V.-M. Chapter 15 – Silicon Wafer and Thin Film Measurements. In *Handbook of Silicon Based MEMS Materials and Technologies*, Second Edition; Tilli, M., Motooka, T., Airaksinen, V.-M., Franssila, S., Paulasto-Kröckel, M., Lindroos, V., Eds.; William Andrew Publishing: Boston, 2015; pp. 381–390.
10. Lascovich, J.; Scaglione, S. Comparison among XAES, PELS and XPS Techniques for Evaluation of sp2 Percentage in aC: H. *Appl. Surface Sci.* **1994**, *78*, 17–23.
11. Pang, H.; Wang, X.; Zhang, G.; Chen, H.; Lv, G.; Yang, S. Characterization of Diamond-like Carbon Films by SEM, XRD and Raman Spectroscopy. *Appl. Surface Sci.* **2010**, *256*, 6403–6407.
12. Sharma, R.; Panwar, O.; Kumar, S.; Sarangi, D.; Goullet, A.; Dixit, P. et al. Effect of Substrate Bias on SE, XPS and XAES Studies of Diamond-like Carbon Films Deposited by Saddle Field Fast Atom Beam Source. *Appl. Surface Sci.* **2003**, *220*, 313–320.

CHAPTER 8

Modeling Approach of Work-Function Engineered Tunnel Field Effect Transistors: A Physical Insight

NAVJEET BAGGA[1*] and SAHELI SARKHEL[2]

[1]*Department of Electronics & Communication Engineering, PDPM-IIITDM-Jabalpur, India*

[2]*Department of Electronics and Communication Engineering, Netaji Subhash Engineering College, Kolkata, India*

[*]*Corresponding author. E-mail: navjeet.bagga9@gmail.com, navjeet@iiitdmj.ac.in*

ABSTRACT

Over the past six decades, the semiconductor research community has been primarily governed by the ongoing trend of device dimension miniaturization to explore various non-conventional devices with reduced power consumption and enhanced speed to achieve high package density state-of-the-art integrated circuits having superior performance metrics. Tunnel FETs, with their fundamental characteristics of achieving steep sub-threshold slope, have become one of the most promising alternatives for ultra-low power applications. In this chapter, we shall present an in-depth insight on the device physics and analytical modeling of work function engineered gate novel Tunnel FET structures where modulation of gate metal work-function improves device performance in terms of higher ON current to OFF current ratio (I_{ON}/I_{OFF}) and steeper sub-threshold slope.

8.1 INTRODUCTION

Over the last few decades, tremendous escalation of present-day nanoelectronics has been continuously dominated by the untiring efforts of the semiconductor research community to design and implement several nonconventional high-performance, high-speed devices with ultra-nano dimensions. These emerging devices can operate at low voltages and having low power consumption and in turn became more impressive as an alternative to the conventional MOSFET (Metal Oxide Semiconductor Field Effect Transistor).[1–5] Reduced dimension results in higher package density as well as reduced gate capacitances, thereby increasing the operating speed of the devices. In spite of several improvisations on the basic MOSFET suggested by the ongoing research, the crucial deterrent factor in the path of unhindered progress in MOS technology lies in the limitation of minimum achievable subthreshold slope (SS) of 60 mV/decade.[6, 7] If the channel length of a conventional MOSFET is scaled down to sub-50 nm node with a proportional reduction of supply voltage down to 0.5 V, it becomes almost a daunting challenge to overcome thermionic emission induced OFF state leakage current. As per the device physics of MOSFETs, gradual increment in the gate terminal voltage lowers the channel potential barrier enabling more carriers to move from the source to the drain resulting in enhanced drain current. Thus, dependence of the drain current on thermionic emission of the carriers from source to drain is the primary cause for higher OFF state current due to subthreshold conduction resulting in a higher value of subthreshold slope. Subthreshold slope of a MOSFET is defined to be the change in gate voltage required to increase the drain current by a factor of 10.[6] Mathematically, it can be expressed as

$$SS = \left(\frac{d(\log_{10} I_D)}{dV_G} \right)^{-1} = 2.3 \frac{kT}{q} \left(1 + \frac{C_{depletion}}{C_{oxide}} \right), \tag{8.1}$$

where $C_{\text{depletion}}$ is the capacitance of the depletion region, and C_{oxide} is the capacitance of the gate oxide. It can be well understood from the above equation that the lowest value of subthreshold slope possible for a MOSFET at room temperature is 2.3 kT/q, which amounts to be 60 mV/decade[7], i.e., for a decade change in current, the required gate voltage is 60 mV. This fundamental limitation poses the greatest difficulty for supply voltage down scaling as predicted by contemporary scaling trend, threatening the viability of MOSFETs in near future. Among many possible alternatives, tunnel field effect transistors (TFETs) have the potential to replace MOSFETs for low power consuming low voltage applications by their intrinsic virtues of reliability, low leakage current in OFF state, ability to achieve subthreshold slope value below the limiting value of 60 mV/decade, and significantly reduced SCEs.

8.2 TUNNEL FIELD EFFECT TRANSISTORS (TFETS)

8.2.1 DEVICE STRUCTURE AND OPERATION PRINCIPLE

Tunnel field effect transistor is basically a simple gated reverse biased p–i–n diode where the device source and drain regions are heavily doped with opposite dopants, and the channel region is ideally intrinsic as shown in Figure 8.1. Unlike conventional MOSFET, the device operates on the principle of band-to-band tunneling of carriers, which is not rely on the idea of barrier lowering.[8] This operating principle eliminates the limitation of 60 mV/decade on SS. In spite of having several aforementioned operational advantages, TFETs also have several limitations that obstruct its usages in industrial platform. The operating principle of TFETs indicates that the device current originates principally due to tunneling of carriers from the source valence band to the channel conduction band that results in low OFF state current on one hand, whereas reducing the ON state current on the other hand.[9–11] In addition, polarity reversal of the gate bias does not shut the transistor off, instead allows to conduct in opposite direction and results in ambipolar current. This necessitates the search for several improvisations on the basic TFET structure in order to deal with this limitation of the ON current and ambipolar behavior. Studies by different groups of researchers are available in the contemporary literature on TFETs with different modifications on gate work-function, modulation of material bandgap, and various structural changes.[12–16] Though the adoption of III-V material could be a good choice to improve the ON current, it is not compatible with CMOS fabrication setup. In a similar way, point TFET (tunneling occur perpendicular to the gate field) does not receive much attention due to its smaller current as compared to line TFET (tunneling occur parallel to the gate field). Therefore, a different strategical viewpoint is needed for elevating the performance of silicon (Si) based point TFET. In this chapter, we deal with few Si-TFET structures, their physical insight, and modeling approach to grab the clear understanding of the device characteristics.

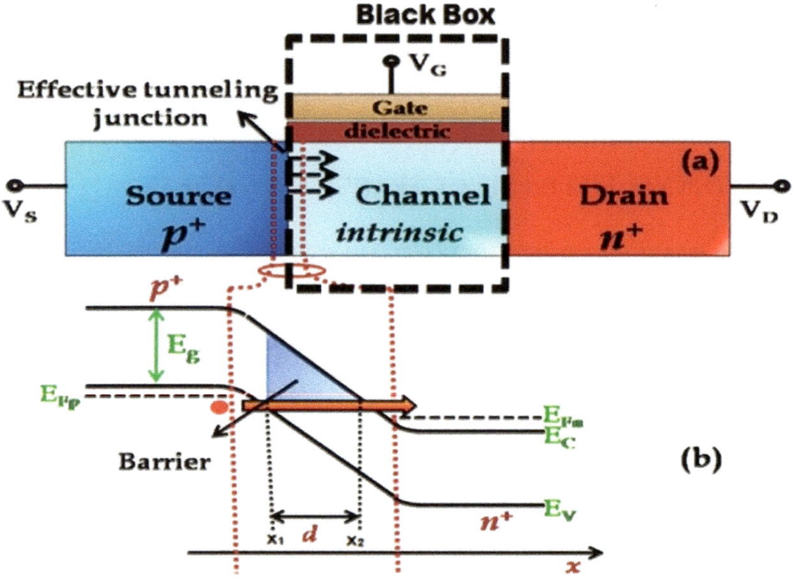

FIGURE 8.1 Schematic of a basic TFET along with its band diagram.

8.2.2 DIFFERENT WORK-FUNCTION MODULATED TFET STRUCTURES

In this chapter, we basically deal about two gate work-function modulated structures that show a significant improvement in ON current and reduction in OFF current. In addition, we discuss the modeling approach of these structures to get an analytical model of the surface potential and electric field profile using conventional Poisson's equation by considering Young's parabolic approximation.[17] The transfer characteristics of TFET is derived using a well-known Kane's model.[18, 19] In both the structures, we have not considered the source and drain resistances, and ideally, the model is governed by the channel electrostatics. For the time being, we are only considered two-dimensional (2D) TFET structures in this chapter, whereas with the suitable modifications and consideration of boundary conditions the discussed modeling approach is simply extended for three-dimensional (3D) TFETs like nanowires or gate all around (GAA) structures. Throughout this chapter, we use acronym S-A for structure-A and S-B for structure-B that are "Double Gate Triple Metal Tunnel FET" and "Double Gate Linearly Graded Gate Tunnel FET," respectively.

8.2.2.1 DOUBLE GATE TRIPLE METAL TUNNEL FET (S-A)

A 2D schematic view of S-A is shown in Figure 8.2. S-A comprises three different metals of unlike work-function values connected in cascade to create a gate electrode. The choice of adopting three different work-functions in the gate electrode supports the novelty in the structure S-A as compared to the conventional TFET in which the gate electrode is made up of single metal electrode. The work-function of metal M_1 is kept small as the effective tunneling takes place at the source channel junction, thus the smaller work-function engendered more band bending and helps to get smaller tunneling path.[20] The length of M_1 also plays an important role in effective tunneling. The purpose of metal M_2 is catchy in this S-A structure, as the larger work-function of M_2 helps to form a potential barrier in the channel region. The modulation of this potential barrier with suitable gate voltage improves the device performance in terms of ON current and reduced ambipolar current. The objective of metal M_3 is to create a band pass filter action[21] at the drain side. This ultimately widens the depletion region at the drain-channel junction and in turn reduces the reverse tunneling. The peculiarity of different work-functions justifies the reason of incorporating different lengths L_1, L_2, and L_3 of M_1, M_2, and M_3, respectively. The channel length in S-A is L which is the sum of all the three gate lengths, i.e., $L = L_1 + L_2 + L_3$. The device is simulated using Silvaco ATLAS *TCAD* tool[22] and also well modeled using MATLAB.[23] All the default device parameters are same for structures S-A and S-B (discussed in next sub-section) as enlisted in Table 8.1, unless stated otherwise.

8.2.2.2 DOUBLE GATE LINEARLY GRADED GATE TUNNEL FET (S-B)

A 2D schematic of S-B is shown in Figure 8.3. In this double gate structure, the gate electrode is linearly graded from the source to the drain end with two different metal (M_a and M_b) work-functions. The effective work-function $\Phi_{meff}(x)$ that specifies the work-function of the constituent gate materials results a binary metal alloy system is defined by[24, 25]:

$$\phi_{meff} = (x / L)\phi_{Mb} + (1 - x / L)\phi_{Ma}. \quad (8.2)$$

Here, Φ_{Ma} and Φ_{Mb} are the work-function of metals M_a and M_b, respectively, and L is the channel length.

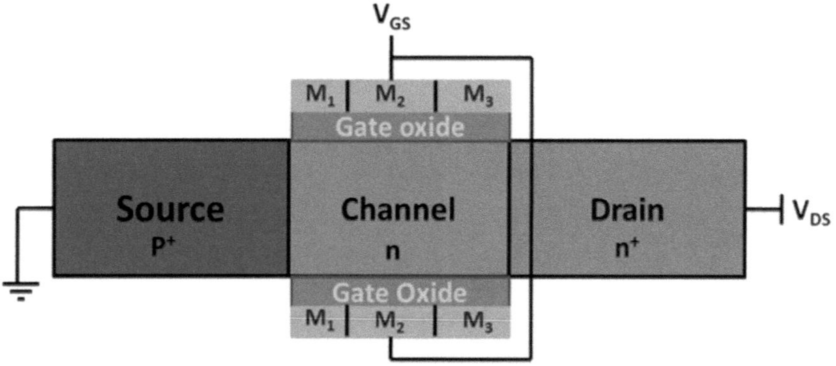

FIGURE 8.2 A 2D schematic view of "Double Gate Triple Metal Tunnel FET" (S-A).

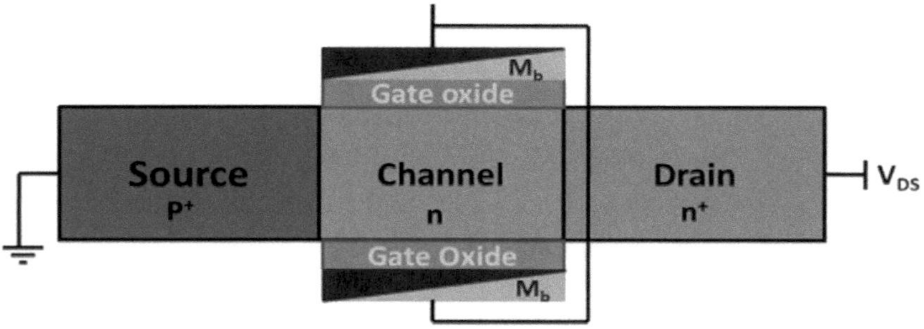

FIGURE 8.3 A 2D schematic view of "Double Gate Linearly Graded Gate Tunnel FET" (S-B).

The purpose of employing binary graded metal is to reduce short channel effects (SCEs) in nanodevices. This grading of work-function ultimately modulates the potential of the channel and in turn the vertical field that is responsible for drain-induced barrier lowering (DIBL) in conventional MOSFET and drain-induced barrier thinning (DIBT) in tunnel FETs. The structure S-B helps to reduce the impact of drain field effectively by proper modulation of gate work-function. In 2D devices, field at any point is a combination of vertical and horizontal fields. In this binary metal system $[M_{a,\alpha} - M_{b,(1-\alpha)}]$, 100% M_a is at source end and 100% M_b is at drain end while in the channel a continuous lateral variation of work-function value is taken.

8.2.3 ANALYTICAL MODELING APPROACH

In order to understand the modeling approach, we consider a basic "black-box" structure as shown in Figure 8.1 with no work-function engineering in its gate electrode. At this primary stage, we are considering that the gate electrode is having a simple metal layer with work-function φ. In general, the device characteristics is governed by the channel characteristics (ignoring the resistance effect of source and drain regions). Therefore, we apply a 2D Poisson's equation in the channel region only with consideration of Young's parabolic approximation of the potential variation.[17] This modeling approach is applicable for any geometrical structure and therefore at later part of this section, we apply the suitable boundary condition of S-A and S-B to obtain the results.

8.2.3.1 GENERAL APPROACH OF SURFACE POTENTIAL MODELING

In general, the surface potential profile, electric field, and finally the drain current define the characteristics of any device (we are not considering the capacitance–voltage [C/V] characteristics in this chapter). To start with, consider a 2D Poisson's equation applied in the channel region of Figure 8.1.

A 2D Poisson's equation employed in the channel region to govern the relationship between potential and charge:

$$\frac{\partial^2\varphi(x,y)}{\partial x^2}+\frac{\partial^2\varphi(x,y)}{\partial y^2}=\frac{qN_a}{\varepsilon_{Si}} \quad \text{for} \quad \left(0\le x\le L, 0\le y\le t_{Si}\right). \tag{8.3}$$

Here, $\varphi(x, y)$ is the potential in the channel as a function of both x and y direction, N_a is the uniform channel doping concentration, and ε_{Si} is silicon permittivity. The thickness and length of the channel is t_{si} and L, respectively. In general, Young's parabolic potential approximation (PPA) is chosen, and then it is fitted in the channel as per the geometrical structure and acquired boundary conditions of a particular device. Therefore, the considered PPA in the channel region is given as:

$$\varphi(x,y) = a_1(x) + a_2(x)y + a_3(x)y^2 \quad \text{for} \quad (0 \le x \le L, 0 \le y \le t_{Si}), \quad (8.4)$$

where a_1, a_2, and a_3 are arbitrary constants that need to be calculated as per the boundary conditions. These boundary conditions may vary as per the device geometry. In this section, we are dealing with a boundary condition of conventional signal gate TFET and this can be replicate in next sub-sections for structures S-A and S-B. A continuity in electric potential and electric flux inside the channel region as well as at the $Si\text{-}SiO_2$ interface can be ensured by considering the following set of boundary conditions:

The surface potential at the front gate side is given as:

$$\varphi(x,0) = \varphi_f(x).$$

i. The surface potential at the back gate side is given as:

$$\varphi(x,t_{Si}) = \varphi_b(x).$$

ii. Continuity of the electric field at the $Si\text{-}SiO_2$ interface toward front gate is:

$$\varepsilon_{Si} \left. \frac{d\varphi(x,y)}{dy} \right|_{y=0} = \varepsilon_f \frac{\varphi_f(x) - V_{gs} + V_{ffb}(x)}{t_f}$$

iii. Continuity of the electric field at the $Si\text{-}SiO_2$ interface toward back gate is:

$$\varepsilon_{Si} \left. \frac{d\varphi(x,y)}{dy} \right|_{y=t_{Si}} = \varepsilon_b \frac{V_{gs} - \varphi_f(x) - V_{bfb}(x)}{t_b}$$

iv. At the source–channel junction, the potential is given as:

$$\varphi(0,y) = V_{bi}.$$

v. At the drain–channel junction, the potential is given as:

$$\varphi(L,y) = V_{bi} + V_{DS}.$$

These six boundary conditions cover the entire boundary of the channel in which the drain current electrostatics takes place. Here, V_{ffb} (t_f) and V_{bfb} (t_b) are flat band voltage (oxide thickness) of the front and back gates, respectively. V_{bi} is the built-in volage of p–n junction and is given by $V_{bi} = \frac{kT}{q}\ln\left(\frac{N_A N_D}{n_i^2}\right)$. The boundary conditions are used to find the arbitrary constants of eq 8.4. Further, to solve the Poisson's equation as given in eq 8.3, we replace $\dfrac{\partial^2\varphi(x,y)}{\partial y^2}$ as a function of "x" only and solve a 1D equation for acquiring surface potential profile as:

$$\frac{\partial^2 \varphi(x,y)}{\partial x^2} + 2a_3(x) = \frac{qN_a}{\varepsilon_{Si}}. \tag{8.5}$$

This equation provides an exponential expression of surface potential profile that needs to be fitted according to the boundary conditions of a particular structure. In this chapter, we deal surface potential modeling of two separate structures S-A and S-B, respectively.

8.2.3.1.1 Modeling Approach for S-A (Double Gate Triple Metal Tunnel FET)

It is mentioned above that S-A is a Triple Metal Double Gate Tunnel FET in which the gate electrode consists of three metals of having unlike work-functions. The modeling approach will be same as we have explained above except few of the boundary conditions will get changed accordingly. As we have three different gate metals, therefore, we have considered three separate potential profiles and finally we merge with suitable boundary conditions. Thus, for S-A, eq 8.4 can be reframed as:

$$\phi_1(x,y) = a_{11}(x) + a_{12}(x)y + a_{13}(x)y^2 \tag{8.6a}$$

$$\phi_2(x,y) = a_{21}(x) + a_{22}(x)y + a_{23}(x)y^2 \tag{8.6b}$$

$$\phi_3(x,y) = a_{31}(x) + a_{32}(x)y + a_{33}(x)y^2 \tag{8.6c}$$

Due to the presence of three different metal work-functions, the boundary conditions (B.C.) (iii) and (iv) will be changed to maintain a proper continuity of electric flux and potential. Therefore,

B.C. (iii) for S-A:

$$\left.\frac{d\phi_1(x,y)}{dy}\right|_{y=0} = \frac{\varepsilon_{ox}}{\varepsilon_{si}} \frac{\phi_{s1}(x) - V'_{GS,1}}{t_f} \qquad \text{under Metal } M_1$$

$$\left.\frac{d\phi_2(x,y)}{dy}\right|_{y=0} = \frac{\varepsilon_{ox}}{\varepsilon_{si}} \frac{\phi_{s2}(x) - V'_{GS,2}}{t_f} \qquad \text{under Metal } M_2$$

$$\left.\frac{d\phi_3(x,y)}{dy}\right|_{y=0} = \frac{\varepsilon_{ox}}{\varepsilon_{si}} \frac{\phi_{s3}(x) - V'_{GS,3}}{t_f} \qquad \text{under Metal } M_3$$

B.C. (iv) for S-A:

$$\left.\frac{d\phi_1(x,y)}{dy}\right|_{y=t_{si}} = \frac{\varepsilon_{ox}}{\varepsilon_{si}}\frac{V'_{GS,1} - \phi_{B1}(x)}{t_b} \quad \text{under Metal } M_1$$

$$\left.\frac{d\phi_2(x,y)}{dy}\right|_{y=t_{si}} = \frac{\varepsilon_{ox}}{\varepsilon_{si}}\frac{V'_{GS,2} - \phi_{B2}(x)}{t_b} \quad \text{under Metal } M_2$$

$$\left.\frac{d\phi_3(x,y)}{dy}\right|_{y=t_{si}} = \frac{\varepsilon_{ox}}{\varepsilon_{si}}\frac{V'_{GS,3} - \phi_{B3}(x)}{t_b} \quad \text{under Metal } M_3$$

Here $V'_{GS,bj} = V_{GS} - V_{FB,j}$ is the effective gate voltage for $j = 1, 2$, and 3. t_f and t_b are thickness of front and back gate oxide, respectively. In addition to the general analysis, we have to employ the extra boundary condition as per structural requirement. Therefore,

- At the interface of two dissimilar metals, the surface potential is continuous:

$$[\phi_1(L_1,0) = \phi_2(L_2,0)] \text{and} [\phi_2(L_2,0) = \phi_3(L_3,0)]$$

- Electric flux at the interface of two dissimilar metals is continuous:

$$\left(\left.\frac{d\phi_1(x,y)}{dx}\right|_{x=L1} = \left.\frac{d\phi_2(x,y)}{dx}\right|_{x=L1}\right) \text{and} \left(\left.\frac{d\phi_2(x,y)}{dx}\right|_{x=L2} = \left.\frac{d\phi_3(x,y)}{dx}\right|_{x=L2}\right)$$

Using all these essential boundary conditions, the acquired arbitrary constants are:

$$a_{11}(x) = \phi_{s1}$$

$$a_{12}(x) = \frac{\varepsilon_{ox}}{\varepsilon_{si}}\frac{\phi_{s1}(x) - V_{GS1}}{t_f}$$

$$a_{13}(x) = \frac{V'_{GS,b1}}{t_{si}^2(1+2C_{si}/C_b)} + \frac{V_{GS1}(C_f/C_{si} + C_f/C_b)}{t_{si}^2(1+2C_{si}/C_b)} - \frac{\phi_{s1}(x)[1 + C_f/C_{si} + C_f/C_b]}{t_{si}^2(1+2C_{si}/C_b)}$$

$$a_{21}(x) = \phi_{s2}$$

$$a_{22}(x) = \frac{\varepsilon_{ox}}{\varepsilon_{si}}\frac{\phi_{s2}(x) - V_{GS2}}{t_f}$$

$$a_{23}(x)=\frac{V'_{GS,b2}}{t_{si}^2(1+2C_{si}/C_b)}+\frac{V_{GS2}(C_f/C_{si}+C_f/C_b)}{t_{si}^2(1+2C_{si}/C_b)}-\frac{\phi_{s2}(x)[1+C_f/C_{si}+C_f/C_b]}{t_{si}^2(1+2C_{si}/C_b)}$$

$$a_{31}(x)=\phi_{s3}$$

$$a_{32}(x)=\frac{\varepsilon_{ox}}{\varepsilon_{si}}\frac{\phi_{s3}(x)-V_{GS3}}{t_f}$$

$$a_{33}(x)=\frac{V'_{GS,b3}}{t_{si}^2(1+2C_{si}/C_b)}+\frac{V_{GS3}(C_f/C_{si}+C_f/C_b)}{t_{si}^2(1+2C_{si}/C_b)}-\frac{\phi_{s3}(x)[1+C_f/C_{si}+C_f/C_b]}{t_{si}^2(1+2C_{si}/C_b)}$$

Following the procedure adopted in eq 8.5, finally, we have derived the surface potential profile for S-A:

$$\phi_{s1}(x)=A\exp(\eta x)+B\exp(-\eta x)-\beta_1/\alpha \quad 0\leq x\leq L_1 \qquad \text{under Metal } M_1$$

$$\phi_{s2}(x)=C\exp(\eta(x-L1))+D\exp(-\eta(x-L1))-\beta_2/\alpha \qquad \text{under Metal } M_2$$
$$\text{for} \quad L_1\leq x\leq (L_1+L_2)$$

$$\phi_{s3}(x)=E\exp(\eta(x-L1-L2))+F\exp(-\eta(x-L1-L2))$$
$$-\beta_3/\alpha \text{ for } (L1+L2)\geq x\leq L \qquad \text{under Metal } M_3$$

where $\eta=\sqrt{\alpha}$, $\beta_j=\frac{-qN_{cn}}{\varepsilon_{si}}-\frac{2V'_{GS,bj}}{t_{si}^2(1+2C_{si}/C_b)}-\frac{2V_{GSj}(C_f/C_b+C_f/C_{si})}{t_{si}^2(1+2C_{si}/C_b)}$, $C_f=\varepsilon_f/t_f$ and $C_b=\varepsilon_b/t_b$.

The constants A, B, C, D, E, and F can be obtained by using above defined B.C. of S-A:

$$A=V_{bi}-B-\sigma_1$$

$$B=\frac{(V_{bi}+V_{DS})\exp(-\eta(L_1+L_3))(1-P)}{Q}+\frac{P(\sigma_3-\sigma_2)}{Q}+$$
$$+\frac{[\sigma_3\exp(-\eta(L_1+L_3))(P-1)]+[V_{bi}\exp(\eta L_2)(1-P)]+[\sigma_1\exp(\eta L_2)(1-P)]}{Q}$$
$$+\frac{(P(\sigma_1-\sigma_2)\cosh(\eta(L_2-L_1)))-\sinh(\eta(L_2-L_1))}{Q}$$

where $P = Z_1/Z_2$; $Z_1 = \exp(-\eta(L_1 + 2L_3)) + \exp(\eta L_1)$, $Z_2 = \exp$; $Z_1 = \exp(-\eta(L_1 + 2L_3))Z_2 = \exp(-\eta(L_1 + 2L_3)) - \exp(\eta L_1)$ and

$$Q = (2P\cosh(\eta L_2)) - 2\sinh(\eta L_2)$$

$$C = A\exp(\eta L_1) + \frac{\sigma_1 - \sigma_2}{2}$$

$$D = B\exp(-\eta L_1) + \frac{\sigma_1 - \sigma_2}{2}$$

$$E = \exp(-\eta L_3)((V_{bi} + V_{DS}) - \sigma_3 - F\exp(-\eta L_3))$$

$$F = \frac{2B\sinh(\eta L_2)}{Z_2} + \frac{\sigma_1\exp(\eta L_2)}{Z_2} + \frac{(V_{bi} + V_{DS})\exp(-\eta(L_1 + L_3))}{Z_2}$$

$$+ \frac{(\sigma_2 - \sigma_3)}{Z_2} + + \frac{\left[V_{bip}\exp(\eta L2)\right] - \left[(\sigma_1 - \sigma_2)\cosh(\eta(L2 - L1))\right] - \left[\sigma_3\exp(-\eta(L1 + L3))\right]}{2}$$

where $\sigma_1 = -\beta_1/\alpha, \sigma_2 = -\beta_2/\alpha$ and $\sigma_3 = -\beta_3/\alpha$.

- ***Electric Field Profile of S-A:***
  - The electric field, along the channel length, i.e., the lateral electric field can be obtained by differentiating the surface potential with respect to x.

$$E_{1x}(x) = -\left.\frac{d\phi_1(x,y)}{dx}\right|_{y=0}$$

$$= -A\eta\exp(\eta x) + B\eta\exp(-\eta x) \text{ for } 0 \le x \le L_1 \text{ under Metal } M_1$$

$$E_{2x}(x) = -\left.\frac{d\phi_2(x,y)}{dx}\right|_{y=0} = -C\eta\exp(\eta(x - L1))$$

$$+D\eta\exp(-\eta(x - L1)) \text{ for } L_1 \le x \le (L_1 + L_2) \text{ under Metal } M_2$$

$$E_{3x}(x) = -\left.\frac{d\phi_3(x,y)}{dx}\right|_{y=0} = -E\eta\exp(\eta(x - L1 - L2))$$

$$+F\eta\exp(-\eta(x - L1 - L2)) \text{ for } (L_1 + L_2) \le x \le L \text{ under Metal } M_3$$

  - The vertical electric field can be obtained as:

$$E_{1y}(x) = -\frac{d\phi_1(x,y)}{dy} = -C_{11}(x) - 2yC_{12}(x) \text{ for}$$

$$0 \le x \le L_1 \text{ under Metal } M_1$$

$$E_{2y}(x) = -\frac{d\phi_2(x,y)}{dy} = -C_{21}(x) - 2yC_{22}(x) \quad \text{for}$$
$$L_1 \le x \le (L_1 + L_2) \text{ under Metal } M_2$$
$$E_{3y}(x) = -\frac{d\phi_3(x,y)}{dy} = -C_{31}(x) - 2yC_{32}(x) \quad \text{for}$$
$$(L_1 + L_2) \le x \le L \text{ under Metal } M_3$$

These equations are plotted in MATLAB and same is found to be best fitted with simulation results. The results obtained with these equations are discussed in next section.

8.2.3.1.2 Modeling Approach for S-B (Double Gate Linearly Graded Gate TFET)

It is mentioned above that S-B is a linearly graded Gate Double Gate Tunnel FET in which the gate electrode consists of two metals of having different work-functions that linearly vary from source to drain as given in eq 8.2.

The boundary conditions for S-B are same as defined in general approach section except the flat band voltages. Now, by using the boundary conditions, we found the arbitrary constants as:

$$a_1(x) = \varphi_f(x)$$
$$a_2(x) = \frac{\varepsilon_f}{\varepsilon_{Si}} \frac{\varphi_f(x) - V'_{fg}}{t_f}$$
$$a_3(x) = \frac{\left[\left(\frac{\varepsilon_b}{\varepsilon_{Si}} \frac{V'_{bg} - \varphi_f(x)}{t_b}\right) - \left(\frac{\varepsilon_f}{\varepsilon_{Si}} \frac{\varphi_f(x) - V'_{fg}}{t_f}\right)\left(1 + \frac{\varepsilon_b t_{Si}}{\varepsilon_{Si} t_b}\right)\right]}{\left(2 + \frac{\varepsilon_b t_{SI}}{\varepsilon_{Si} t_b}\right) t_{Si}}$$

Here, $V'_{fg} = V_{gsf} - V_{ffb}(x)$ and $V'_{bg} = V_{gsb} - V_{bfb}(x)$. V_{gsf} and V_{gsb} are the gate voltages applied to the front and back gates, respectively. For a tied structure, $V_{gsf} = V_{gsb}$. By substituting the values of $a_1(x)$, $a_2(x)$ and $a_3(x)$ in eq 8.4 and replace $\frac{\partial^2 \varphi(x,y)}{\partial y^2}$ on eq 8.3 as a function of "x" only, the surface potential can be written as:

$$\frac{d^2\varphi_f(x)}{dx^2}+\alpha V'_{bg}-\alpha\varphi_f(x)-\beta\varphi_f(x)+\beta V'_{fg}=\frac{qN_a}{\varepsilon_{Si}}$$

where $\alpha=\dfrac{2\left(\varepsilon_b/\varepsilon_{Si}t_b\right)}{\left(2+\left(\varepsilon_b t_{Si}/\varepsilon_{Si}t_b\right)\right)t_{Si}}$ and $\beta=\dfrac{2\left(\varepsilon_f/\varepsilon_{Si}t_f\right)\left(1+\left(\varepsilon_b t_{Si}/\varepsilon_{Si}t_b\right)\right)}{\left(2+\left(\varepsilon_b t_{Si}/\varepsilon_{Si}t_b\right)\right)t_{Si}}$

$$\frac{d^2\varphi_f(x)}{dx^2}-\eta^2\varphi_f(x)-\eta^2\varphi_{meff}(x)=\frac{qN_a}{\varepsilon_{Si}}-\alpha\left(V_{bg}+\phi_S\right)-\beta\left(V_{fg}+\phi_S\right)$$

where $\eta^2=(\alpha+\beta)$.

The above differential equation can be solved by variable separation method to achieve the resultant expression of the front surface potential as:

$$\varphi_f(x)=K_1e^{\eta x}+K_2e^{-\eta x}+(\phi_{ma}-\phi_{mb})\left(x/L\right)+\left[\frac{\alpha}{\eta^2}\left(V_{bg}+\phi_S\right)+\frac{\beta}{\eta^2}\left(V_{fg}+\phi_S\right)-\phi_{ma}-\frac{qN_a}{\eta^2\varepsilon_{Si}}\right]$$

Now, the two coefficients K_1 and K_2 can be calculated as:

$$K_1=\frac{\phi_{bi,D}+V_D-\phi_{bi,S}e^{-\eta L}-(\phi_{ma}-\phi_{mb})-\left[\frac{\alpha}{\eta^2}\left(V_{bg}+\phi_S\right)+\frac{\beta}{\eta^2}\left(V_{fg}+\phi_S\right)-\phi_{ma}-\frac{qN_a}{\eta^2\varepsilon_{Si}}\right]\left(1-e^{-\eta L}\right)}{\left(e^{\eta L}-e^{-\eta L}\right)}$$

$$K_2=\frac{\phi_{bi,S}e^{\eta L}+(\phi_{ma}-\phi_{mb})-\phi_{bi,D}-V_D+\left[\frac{\alpha}{\eta^2}\left(V_{bg}+\phi_S\right)+\frac{\beta}{\eta^2}\left(V_{fg}+\phi_S\right)-\phi_{ma}-\frac{qN_a}{\eta^2\varepsilon_{Si}}\right]\left(1-e^{\eta L}\right)}{\left(e^{\eta L}-e^{-\eta L}\right)}$$

- *Electric Field Profile of S-B:*
  - The electric field can be obtained by differentiation of the potential profile with keeping a negative sign. Thus, the 2D transverse electric field is given as[20]:

$$E_y=-a_2(x)-2ya_3(x).$$

- The lateral surface electric field can be expressed as

$$E_x = -\left[\eta K_1 e^{\eta x} - \eta K_2 e^{-\eta x} + \frac{(\phi_{ma} - \phi_{mb})}{L}\right].$$

Thus, the overall effective electric field can be given as $E = \sqrt{E_x^{\ 2} + E_y^{\ 2}}$.

These resulting equations of S-B are plotted in MATLAB and same is found to be best fitted with simulation results. The results obtained with these equations are discussed in the next section.

8.2.3.2 GENERAL APPROACH OF DRAIN CURRENT MODELING

The drain current of a TFET is based on band-to-band tunneling mechanism which is governed by well-known Kane's model.[18, 19] Basically, this Kane's model states that the current of the device is an integral function of the generation rate of carriers, as given below:

$$I_D = q\int G_{BTB}\,dxdy, \tag{8.7}$$

where GBTB is the generation rate of the carrier which is given as:

$$G_{BTB}(E) = A_k E^{D_k} \exp\left[-\frac{B_k}{E}\right]. \tag{8.8}$$

Electric field (E) is bifurcated into local and average field[26], which is responsible for effective tunneling at source–channel junction. The default Kane's parameters are dependent on effective masses of the carrier at conduction and valance bands, respectively. The parameter D_k is taken as 2 and 2.5 for direct and indirect tunneling, respectively. Therefore, the current expression can be written as:

$$I_D = q\int A_k E_L E_{avg}^{D_k - 1} \exp(-B_k / E_{avg})dv. \tag{8.9}$$

Here, the local electric field can be obtained by $E_L = \sqrt{E_x^2 + E_y^2}$.

In general, all type of tunneling except along the channel is considered to be submissive in this modeling approach. In addition, the average electric field in eq 8.9 is a function of tunneling path, i.e., the distance "*d*" as shown in Figure 8.1. The points x_1 and x_2 are called turning points between which we need to integrate the generation rate to obtain the drain current,

$$E_{avg} = Eg / qd \tag{8.10}$$

where $d = (x_2 - x_1)$ represents the tunneling path.

The starting point of the tunneling is considered at the position where the difference of source and channel conduction band reaches to $E_g + \Delta\psi_1$. Therefore,

$$\phi_s(x)\Big|_{x=x_1} = \frac{(E_g + \Delta\Sigma_1)}{q}, \tag{8.11}$$

where $\Delta\Sigma_1$ can be calculated by Joyce and Dixon approximation,[27] which is a difference between valance band edge (E_{vs}) and Fermi level of the source (E_{Fs}). Similarly, the end point of the tunneling window can be obtained at the position where the difference of source and channel conduction band edge reaches to $E_g + \Delta\Sigma_2$:

$$\phi_s(z)\Big|_{x=x_2} = \frac{(E_g + \Delta\Sigma_2)}{q}. \tag{8.12}$$

Here, $\Delta\Sigma_2$ is nothing but the built-in potential as $\Delta\Sigma_2 = E_{vs}-E_{vch}$. Therefore, the drain current is modeled using Kane's model[18]:

$$I_D = 2\pi q A_k \int_{y=0}^{y=t_{si}} \int_{x=x_1}^{x=x_2} \frac{E_{avg}^{D_k-1} E_L}{q^{D_k-1} x^{D_k-1}} \exp\left(\frac{-B_k q}{Eg} x\right) dx dy. \tag{8.13}$$

The acquired results of the modeling are discussed in the next section in addition with the simulation results that justify the validity of our manual calculations.

8.3 RESULTS AND DISCUSSION

In this chapter, we have explored the general idea of getting an analytical model of 2D TFET structures. The surface potential model has been obtained by solving a 2D Poisson's equation in the channel region of the device, and then the electric field profile has been derived subsequently. The acquired results of analytical model have been matched with the simulation results obtained by Silvaco ATLAS TCAD tool. We will be explaining the results of structures S-A and S-B in separate sub-sections. The parameters used in the analysis are mentioned in Table 8.1 and considered as default, unless stated otherwise (data are adopted from[20] and[29]).

TABLE 8.1 Parameter Table.

Parameters	Value
Source doping (p+)	10^{20} cm^{-3}
Drain doping (n+)	10^{19} cm^{-3}
Channel doping (n)	10^{15} cm^{-3}
Channel length	60 nm
Front oxide thickness (t_f)	2 nm
Back oxide thickness (t_b)	2 nm
Channel thickness (t_{Si})	10 nm
For S-A (φ_{M1}, φ_{M2}, and φ_{M3})	(4.4 eV, 5.3 eV, and 5.0 eV)
For S-B (φ_{Ma}, φ_{Mb})	(4.4 eV and 5.3 eV)
Kane's parameter	
A	3.5×10^{21} m$^{-1/2}$V$^{-5/2}$s^{-1}
B	22.5×10^{6} V/cm
D	2.5 for indirect tunneling
	2 for direct tunneling

8.3.1 RESULTS OBTAINED FOR A DOUBLE GATE TRIPLE METAL TUNNEL FET (S-A)

In structure S-A, three metals of different work-functions are connected in cascade to form a gate electrode with work-functions, φ_{M1}, φ_{M2}, and $\varphi_{M3,}$ respectively. In the device, φ_{M1} and φ_{M3} are intentionally kept of lower valued as compared to φ_{M2} that helps to create a suitable barrier in the channel and results in an improved ON to OFF current ratio. To increase the effective tunneling at the source–channel junction, we require a very high field at that junction that can be achieved by using a lower value of φ_{M1}. We have kept the value of φ_{M3} to be a little higher than φ_{M1} to create a bandpass filter action at the drain–channel junction. This helps to reduce the effective tunneling at that junction and in turn reduces the ambipolar current. The modeled surface potential profile of structure S-A is shown in Figure 8.4. It is clear from the figure that the barrier formed in the channel can be modulated by employing proper gate bias.

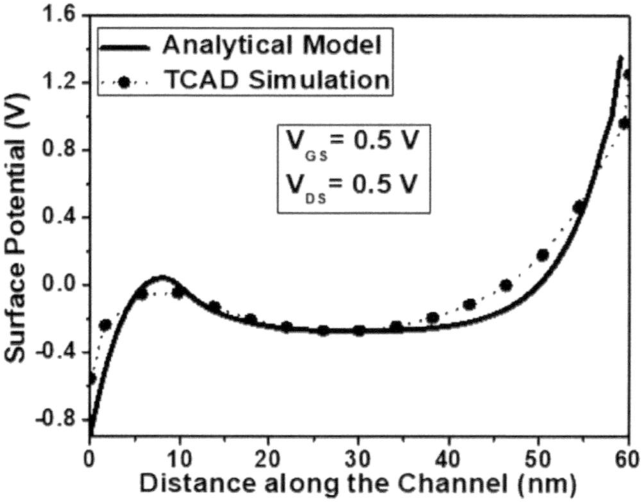

FIGURE 8.4 The surface potential profile of structure S-A plotted for the channel length $L = 60$ nm (with $L_1 = 10$ nm, $L_2 = 30$ nm, and $L_3 = 20$ nm, respectively).

In the OFF state of the device, the filled valance band of the source must not align to the vacant conduction band of the channel that ultimately stops the tunneling and results in very small OFF current as compared to conventional MOSFET. In contrast, the applied gate voltage modulates the tunneling path that results in ON current of the device.[20] The impact of applied gate voltage (V_{GS}) on the surface potential is depicted in Figure 8.5 that shows that the potential barrier present in the channel gets modulated with applied V_{GS}. The potential of using a higher work-function of metal M_2 can be justified using Figure 8.6 as the potential barrier formed under the M_2 restricts the impact of drain voltage at the source–channel junction that reduces the DIBT (drain-induced barrier thinning) in TFET.[28] It is evident from Figure 8.6 that the impact of drain voltage is not affecting the tunneling path at the source–channel junction.

The lateral electric field profile is plotted in Figure 8.7 for applied gate and drain voltages. It is clear that the work-function of metal M_1 helps to achieve higher field at the source–channel junction that significantly reduces the tunnel path.[20] The cascade connection of metal M_1 and metal M_2 creates a quantum well for electrons and holes under M_1 and M_2, respectively. Due to this, some resonant energy is observed; however, at this energy the transmission coefficient is very small. Therefore, the tunnel current is obstructed due to the combination of three unlike metals.

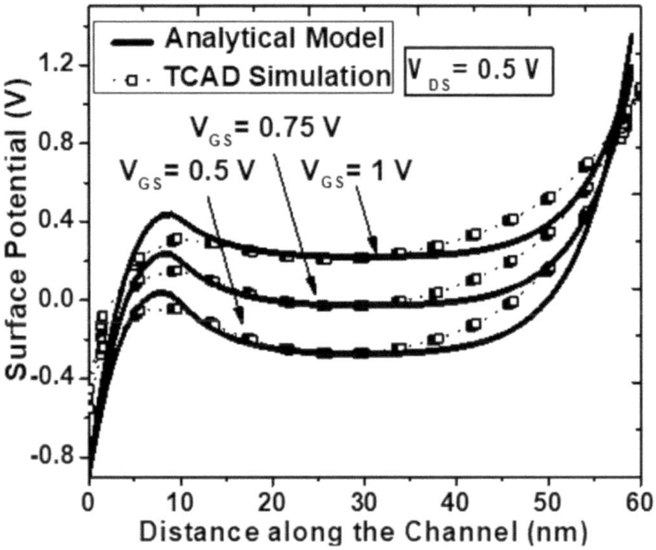

FIGURE 8.5 The surface potential profile of structure S-A for channel length $L = 60$ nm and fixed $V_{DS} = 0.5$ V with different varying gate voltages. Symbol represents the simulated data.

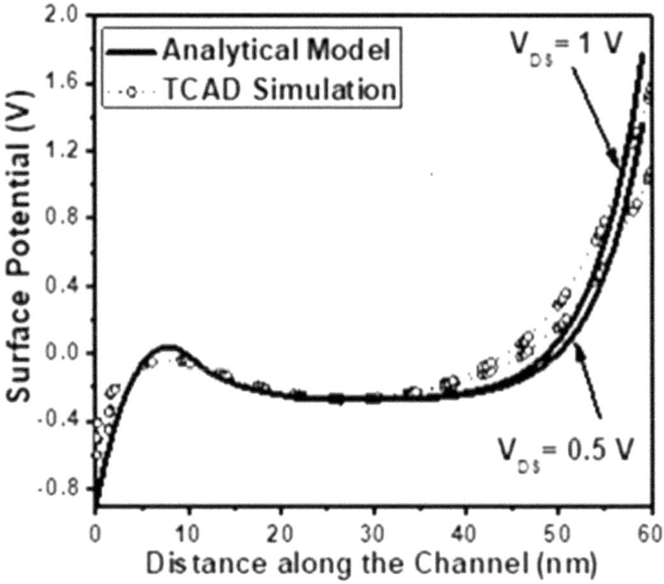

FIGURE 8.6 The surface potential profile of structure S-A for channel length $L = 60$ nm and fixed $V_{GS} = 0.5$ V with different varying drain voltages.

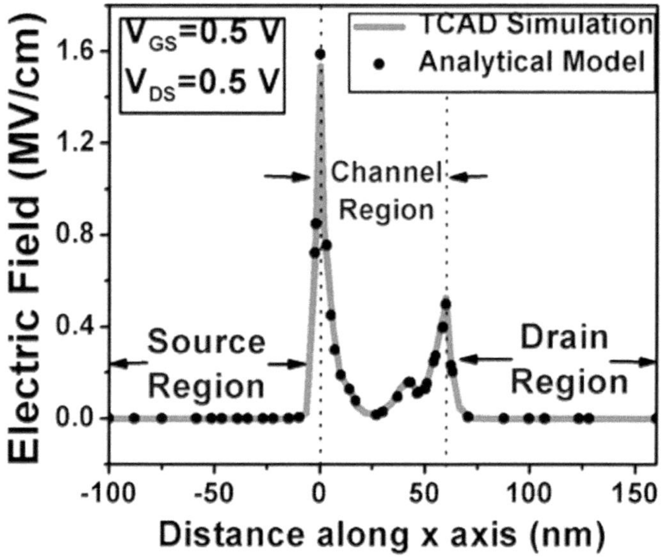

FIGURE 8.7 The electric field profile along the x axis for a fixed V_{GS} = 0.5 V and V_{DS} = 0.5 V.

To obtain the transfer characteristics of the structure S-A, we have employed the acquired electric field on the Kane's model. Figure 8.8 shows the I_D–V_{gs} characteristics for a fixed V_{DS} value. It is clear that the choice of smaller work-function at the source–channel junction provides a much improved tunneling current. However, still the ON current of the TFET is not up to the mark of conventional MOSFET. Much research has been done and still going to improve the ON current of the device so that we can easily replace MOSFET by TFET devices in low power applications.

Figure 8.9 depicts the transfer characteristics for different values of channel thicknesses that reveal that thinner the channel better the gate control. The decrease in channel thickness reduces the bulk capacitive effect and in turn it increases the current. The tunneling probability could be exponentially increased if the tunnel path gets reduced forming in quantum well; therefore, we choose the minimum tunnel path. In structure S-A, this said tunnel path is appeared under the region of metal M_1. The voltage applied to gate has less sensitivity to this tunnel path, which is mainly due to the work-function difference. In our analysis of analytical modeling, we have not incorporated the quantum effects in view of the agreement of the reference,[20] which clearly justifies that the accuracy does not affect much at this channel length and thicknesses (the calculated error with and without incorporating the quantum effect is less than 1%). This avoids the complexity in the analytical modeling to the great extent. In addition, the variation in different metal lengths is also been studied in Figure 8.10. It is clear that if we choose larger length of metal M_1 (L_1) then the current is increased as it increases the effective gate control on the tunneling junction.

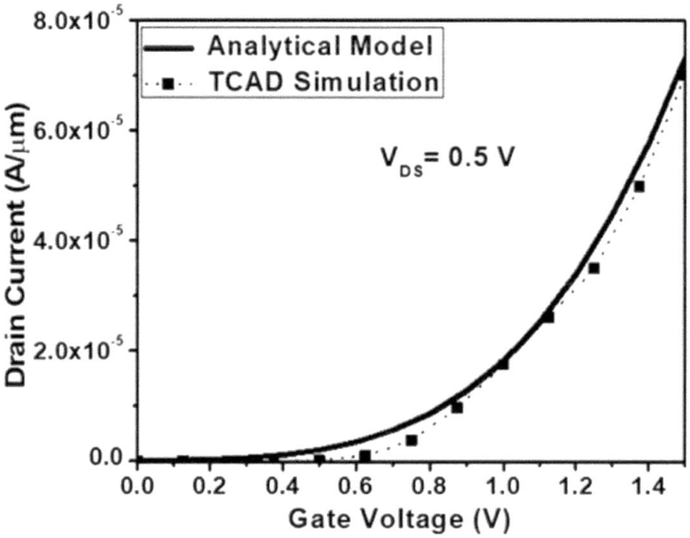

FIGURE 8.8 I_D–V_{GS} characteristics for a fixed V_{DS} = 0.5 V.

8.3.2 RESULTS OBTAINED FOR A DOUBLE GATE LINEARLY GRADED GATE TUNNEL FET (S-B)

In structure S-B, two metals of different work-functions are connected in linearly graded fashion to form a gate electrode with work-functions φ_{Ma} and φ_{Mb}, respectively. To prove the significance of opting linearly graded gate (LGG) electrode is justified with a suitable comparison of the acquired results with single metal gate (SMG) double gate TFET structure.[29] Figure 8.11 shows the surface potential profile along the channel length of structure S-B. It is clear that LGG structure is having higher surface potential as compared to SMG that ultimately helps to improve the tunneling at a given gate bias.

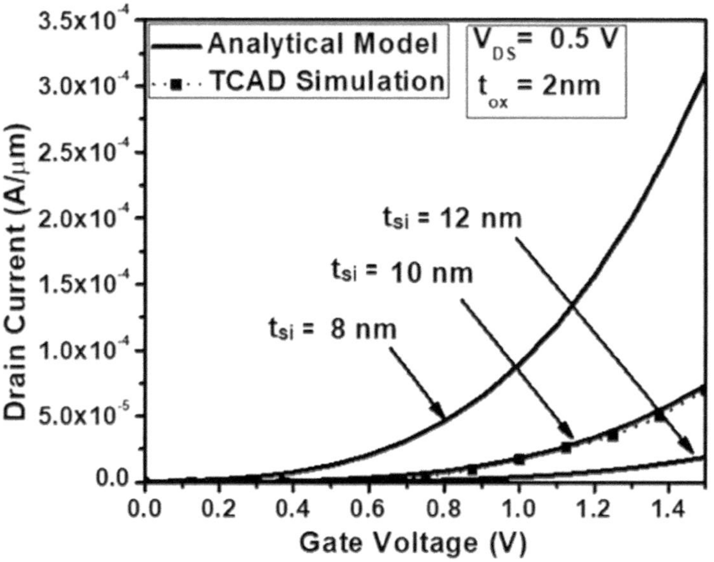

FIGURE 8.9 I_D–V_{GS} characteristics for a fixed V_{DS} = 0.5 V and varying channel thicknesses.

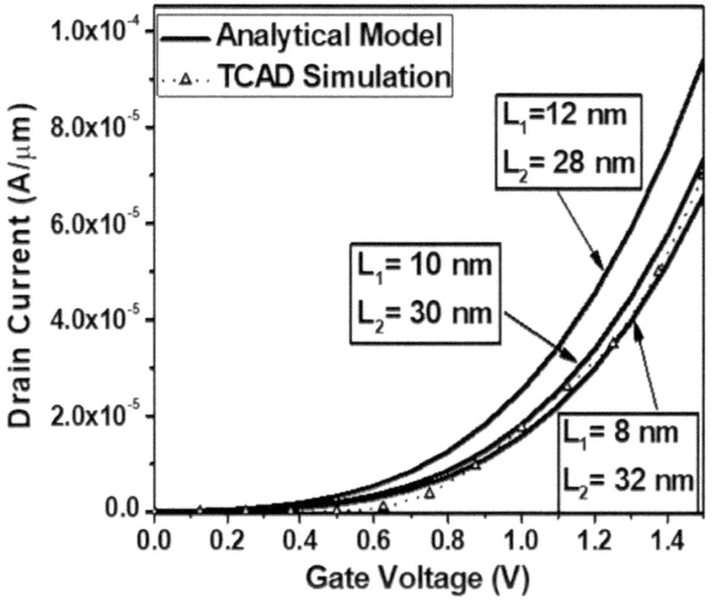

FIGURE 8.10 I_D–Vgs characteristics for a fixed V_{DS} = 0.5 V with varying the lengths of metals M_1 and M_2, respectively.

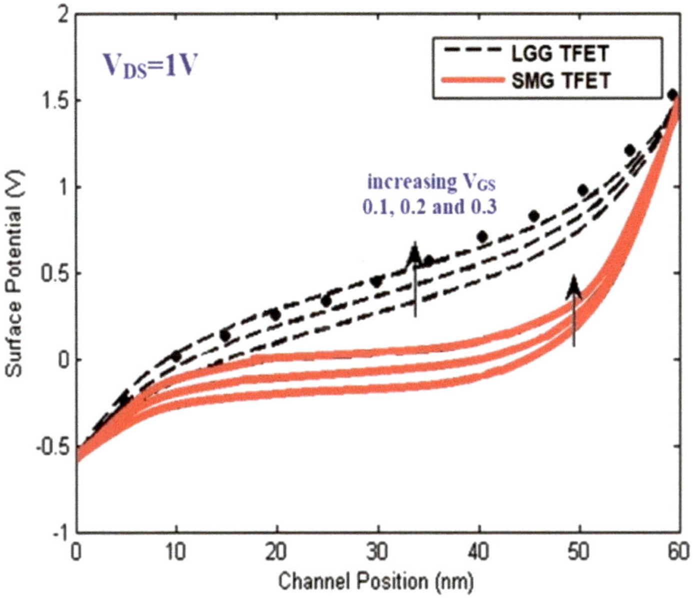

FIGURE 8.11 The surface potential profile comparison between SMG and LGG configuration for a fixed $V_{DS} = 1$ V with varying gate voltages.

By employing two different work-function in LGG configuration, the channel electrostatics in terms of electric field and also vertical electric field can be modified using this concept of "work-function engineering."[30] It is clear that a "x"-dependent term is coming into the equation of surface potential that ultimately provides a linear behavior of the surface potential in LGG configuration. Figure 8.12 shows the surface potential profile of LGG configuration for different drain biases and a fixed applied gate voltage. It is clear that the increase in V_{DS} does not create much effect at the source–channel junction and hence less DIBT can be achieved.[28]

Figures 8.13 and 8.14 show the electric field profile in lateral and transverse direction of the channel, respectively. In the ON state, when the sufficient gate voltage is applied then the band bending is more pronounced near the source–channel junction and in turn increases the electric field and hence carrier tunneling. It is clear from Figure 8.13 that the lateral electric field increases at the middle of the channel as compared to its source and drain side because of increase in its transverse component with applied gate voltage. In addition, it is clear from Figure 8.14 that the electric field near the drain region is reduced that ultimately helps to get smaller ambipolar current.

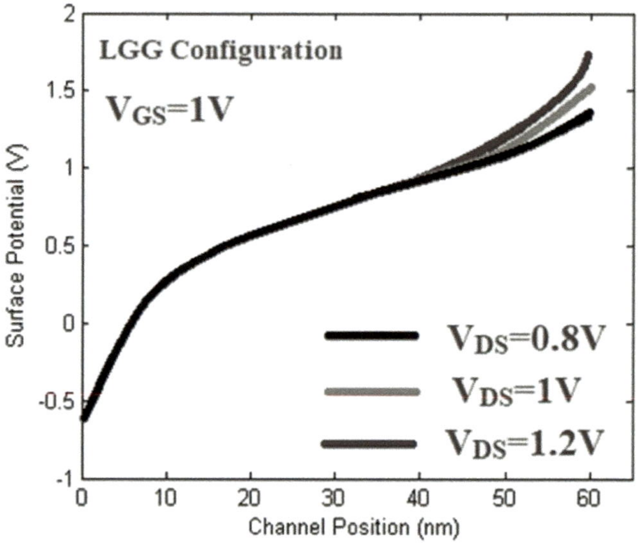

FIGURE 8.12 The surface potential profile for a fixed $V_{GS} = 1$ V with varying drain voltages.

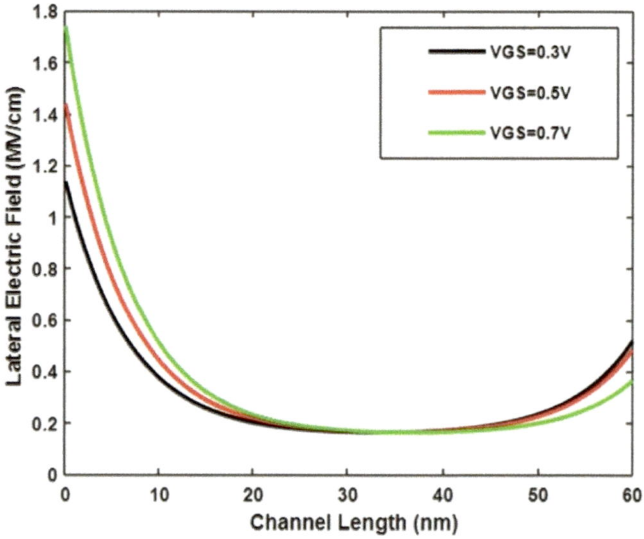

FIGURE 8.13 The lateral electric field profile along the channel length for a fixed V_{DS} = 1 V and varying gate voltages.

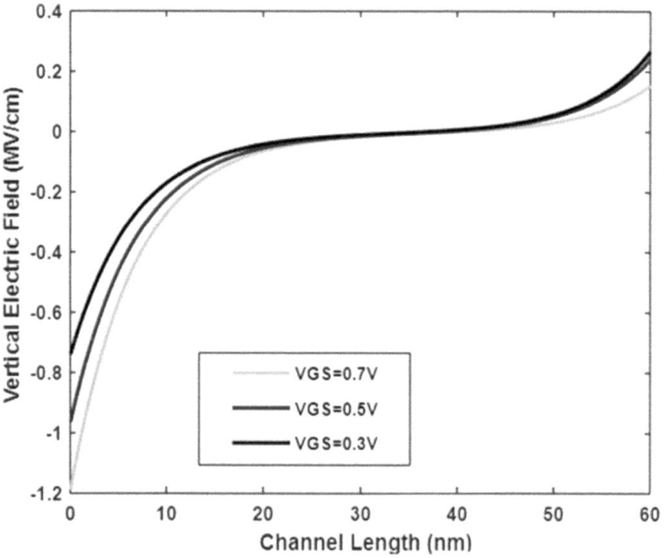

FIGURE 8.14 The vertical electric field profile with respect to channel length for a fixed $V_{DS} = 1$ V and varying gate voltages.

The resultant electric field can be plotted as a square root mean of lateral and vertical electric field that is the average effective field higher near the source–channel junction. In addition, the field at the drain–channel junction gets reduced due to proper choice of work-function in LGG configuration and in turn reduces the ambipolar current.

As the electric field in LGG configuration is achieved to be higher at the source–channel junction and in turn the tunneling current is higher as depicted from Figure 8.15. To see the impact of gate oxide permittivity, the transfer characteristics is plotted in Figure 8.16 for varying gate oxide permittivity. It is clear that higher the permittivity, higher will be the gate control and in turn higher the current. Similarly, the impact of gate oxide thicknesses is also studied in Figure 8.17 that shows that higher current can be achieved at smaller oxide thickness configuration.

8.4 CONCLUSION

In this chapter, work-function engineered tunnel FET structures have been analyzed based on analytical modeling and numerical simulations. We

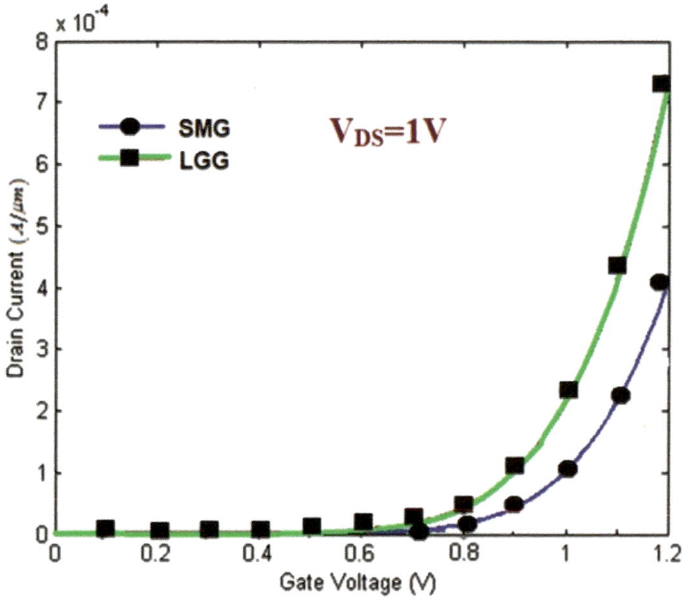

FIGURE 8.15 I_D–V_{GS} characteristics of structure S-B (LGG) in comparison with SMG configuration for a fixed V_{DS} = 1 V.

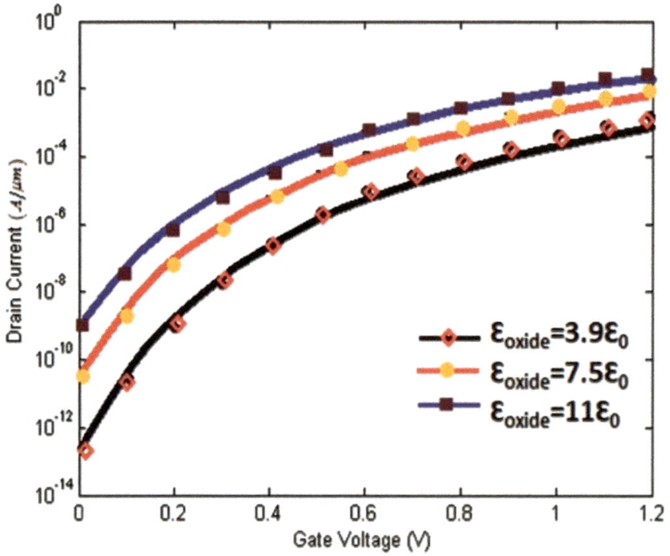

FIGURE 8.16 I_D–V_{GS} characteristics (log scale) of the structure S-B (LGG) for a fixed V_{DS} = 1 V with varying gate oxide permittivity.

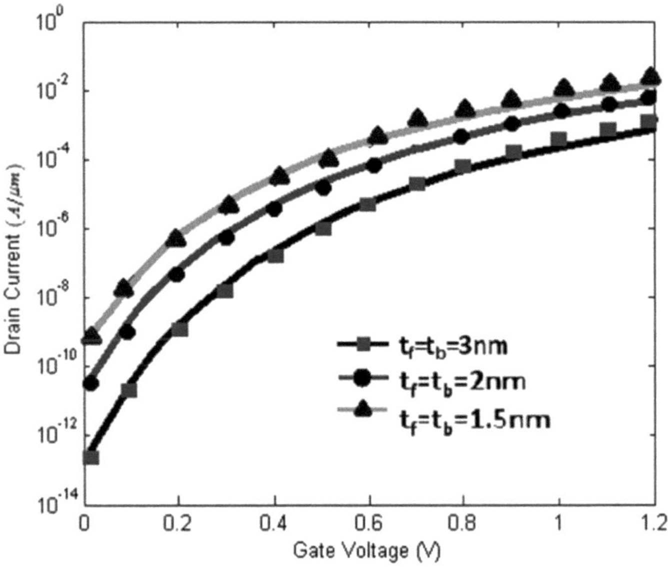

FIGURE 8.17 I_D–V_{GS} characteristics (log scale) of the structure S-B (LGG) for a fixed V_{DS} = 1 V with varying gate oxide thicknesses.

have discussed a general approach to model a TFET structure by using Poisson's equation and Kane's model with suitable boundary conditions. Using MATLAB, the analytical model equation has been plotted, and then a good fit is obtained using numerical simulation. We have considered two structures, viz., S-A (Double Gate Triple Metal TFET) and S-B (Double Gate Linearly Graded Gate TFET) to analyze how the work-function and geometrical modification of the structure helps to improve the device performance in terms of ON current and reduced ambipolar current. In structure S-A, three different metals are placed in cascade to form a gate electrode over which a potential barrier is formed in the channel by employing a proper choice of work-function in the gate electrode. This potential barrier obstructs the reverse tunneled carriers and in turn reduces the ambipolar current. Moreover, the choice of metal work-function placed near the source–channel junction enhances the electric field at the junction and results in higher ON current. However, the cascade connection of metals with unlike work-function causes an abrupt change in work-function between the two metals. To keep this concern in mind, another structure S-B has been studied thoroughly in which the work-function of two adjacent metals varies linearly over the channel. This helps in a smooth potential profile and enhanced electric field which in turn improves the device performances. In summary, an approach to solve the tunnel FET device physical phenomenon using a well-defined equations is discussed in this chapter. The general approach discussed here is useful to make a compact model of any device which in turn provides a bridge for device circuit co-design.

KEYWORDS

- **tunnel field effect transistor**
- **subthreshold slope**
- **Poisson's equation**
- **work-function engineering**
- **linearly graded gate**

REFERENCES

1. Skotnicki, T.; Beranger, C. F.; Gallon, C.; Boeuf, F.; Monfray, S.; Payet, F.; Pouydebasque, A.; Szczap, M.; Farcy, A.; Arnaud, F.; Clerc, S.; Sellier, M.; Cathignol, A.; Schoellkopf, J. P.; Perea, E.; Ferrant, R.; Mingam, H. Innovative Materials, Devices and CMOS Technologies for Low-power Mobile Multimedia. *IEEE Trans. Electron Devices* Jan. **2008**, *55* (1), 96–130.
2. Padilla, A.; Yeung, C. W.; Shin, C.; Hu, C.; Liu, T. K. Feedback FET: A Novel Transistor Exhibiting Steep Switching Behavior at Low Bias Voltages. *Proc. Intl. Electron Devices Meet.*, Dec. 2008, pp. 171–174.
3. Salahuddin, S.; Datta, S. Use of Negative Capacitance to Provide Voltage Amplification for Low Power Nanoscale Devices. *Nano Lett.* Dec. **2007**, *8*, 405–410.
4. Ionescu, A. M.; Riel, H. Tunnel Field-effect Transistors as Energy-efficient Electronic Switch. *Nature* Nov. **2011**, *479* (7373), 329–337.
5. Zhang, Q.; Zhao, W.; Seabaugh, A. Low-Subthreshold-Swing Tunnel Transistors. *IEEE Electron Device Lett.* Apr. **2006**, 27 (4), 297–300.
6. Tsvidis, Y.; McAndrew, C. *The MOS Transistor*, Third Edition; Oxford University Press, 2013.
7. Cheung, K. P. On the 60 mV/dec @300 K Limit for MOSFET Subthreshold Swing, *Proc. Intl Symposium on VLSI Tech. Sys. and App.*, June 2010, pp. 72–73.
8. Bagga, N. Investigation of Device Engineered Innovative Tunnel FETs for Improved Performances. Ph.D. Thesis, submitted at IIT Roorkee, 2019.

9. Appenzeller, J.; Lin, Y. M.; Knoch, J.; Avouris, P. Band-to-band Tunneling in Carbon Nanotube Field Effect Transistors. *Phys. Rev. Lett.* Nov. **2004**, *93*, 196–805.
10. Boucart, K.; Ionescu, A. M. Double-gate Tunnel FET with High-k Gate Dielectric. *IEEE Trans. Electron Devices* June **2007**, *54* (7), 1725–1733.
11. Khatami, Y.; Banerjee, K. Steep Subthreshold Slope n-and p-Type Tunnel-FET Devices for Low-Power and Energy-Efficient Digital Circuits. *IEEE Trans. Electron Devices* Nov. **2009**, *56*, 2752–2761.
12. Sarkhel, S.; Bagga, N.; Sarkar, S. K. A Compact Analytical Model of Binary Metal Alloy Silicon-on-nothing (BMASON) Tunnel FET with Interface Trapped Charges. *J. Comp. Electronics* Sep. **2017**, *16* (3), 704, 713.
13. Bagga, N.; Dasgupta, S. Surface Potential and Drain Current Analytical Model of Gate All Around Triple Metal TFET. *IEEE Trans. Electron Devices* Feb. **2017**, *64* (2), 606–613.
14. Bagga, N.; Kumar, A.; Dasgupta, S. Demonstration of a Novel Two Source Region Tunnel FET. *IEEE Trans. Electron Devices* Dec. **2017**, *64* (12), 5256–5262.
15. Bagga, N.; Chauhan, N.; Gupta, D.; Dasgupta, S. Novel Twofold Tunnel FET with Reduced Miller Capacitance: Proposal and Investigation. *IEEE Trans. Electron Devices* July **2019**, 1–7.
16. Bagga, N.; Kumar, A.; Bhattacharjee, A.; Dasgupta, S. Performance Evaluation of a Novel GAA Schottky Junction(GAASJ) TFET with Heavily Doped Pocket. *Superlatt. Microstruct.* May **2017**, *109*, 545–552.
17. Young, K. K. Short-Channel Effect in Fully Depleted SOI MOSFET's. *IEEE Trans. Electron Devices* **1989**, *36*, 399–402.
18. Kane, E.O. Zener Tunneling in Semiconductors. *J. Phys. Chem. Solides* Jan. **1960**, *12* (2), 181–188.
19. Kane, E.O. Theory of Tunneling. *J. Appl. Phys.* Jan. **1961**, *32* (1), 83–91.
20. Bagga, N.; Sarkar, S. K. An Analytical Model for Tunnel Barrier Modulation in Triple Metal Double Gate TFET. *IEEE Trans. Electron Devices* Jul. **2015**, *62* (7), 2136–2142.
21. Wang, H. et al. A Novel Barrier Controlled Tunnel FET. *IEEE Electron Device Lett.* Jul. **2014**, *35* (7), 798–800.
22. ATLAS User's Manual, Silvaco Inc., Santa Clara, CA, USA, 2016.
23. MATLAB Available Online: https://in.mathworks.com/products/matlab-online.html.
24. Sarkhel, S.; Manna, B.; Sarkar, S. K. Analytical Modeling and Simulation of a Linearly Graded Binary Metal Alloy Gate Nanoscale Cylindrical MOSFET for Reduced Short Channel Effects. *J. Comput. Electron.* Springer April **2014,** *13* (3), 599–605.
25. Manna, B.; Sarkhel, S.; Islam, N.; Sarkar, S.; Sarkar, S. K. Spatial Composition Grading of Binary Metal Alloy Gate Electrode for Short-Channel SOI/SON MOSFET Application. *IEEE Trans. Electr. Devices* **2012**, *59* (12), 3280–3287.
26. Verhulst, A. S.; Leonelli, D.; Rooyackers, R.; Groeseneke, G. Drain Voltage Dependent Analytical Model of Tunnel Field-effect Transistors. *J. Appl. Phys.* Jul. **2011,** *110*.
27. Van Zeghbroeck, B. Principles of Semiconductor Devices, chapter 6, P. Hall, Dec. 2009.
28. Martino, M. D. V.; Martino, J. A.; Agopian, P. G. D. Drain Induced Barrier Thinning on TFETs with Different Source/Drain Engineering, 2014 29th Symposium on Microelectronics Technology and Devices (SBMicro), Aracaju, 2014, pp. 1–4, doi: 10.1109/SBMicro.2014.6940092.

29. Sarkhel, S.; Bagga, N.; Sarkar, S. K. Compact 2D Modeling and Drain Current Performance Analysis of a Work Function Engineered Double Gate Tunnel Field Effect Transistor. *J. Computat. Electron.* **2016**, *15* (1), 104–114.
30. Bagga, N.; Sarkhel, S.; Sarkar, S. K. Exploring the Asymmetric Characteristics of a Double Gate MOSFET with Linearly Graded Binary Metal Alloy Gate Electrode for Enhanced Performance. *IETE J. Res.* **2016**, *62* (6), 786–794.

CHAPTER 9

Thin-Film Photovoltaic Devices with Asymmetric Heterocontact Geometries

GUFRAN AHMAD[1], SHRABANI GUHATHAKURATA[2], SHEETIKANTA MOHANTY[2], PRASHANT KUMAR SINGH[3], and SANDIPAN MALLIK[2*]

[1]*Department of Electrical Engineering, Dayalbagh Educational Institute, Agra 282005, India*

[2]*Department of Electronics Communication Engineering, National Institute of Science and Technology, Berhampur, Odisha 761008, India*

[3]*Department of Electronics and Communication Engineering, University College of Engineering and Technology, V.B.U., Hazaribag, Jharkhand 825301, India*

**Corresponding author. E-mail: sandi.iitkgp@gmail.com*

ABSTRACT

Silicon (Si) is the material of choice for the photovoltaic (PV) industry since the beginning (1954). Precisely, in most of the PV industries, Si solar cells (Si-SCs) are dominating. Si-SCs require nearly 10–15 kg of Si to produce per kilowatt-peak, while the traditional SCs use 200–300 μm thick Si layer for efficient photon absorption. So, highly purified raw materials are required for manufacturing Si-SC. Therefore, the cost of electricity produced by Si-PV solar cells cannot be much lower than that from fossil fuels. Comparing the active layer thickness of conventional Si-SCs with thin-film solar cells (TF-SCs), TF-SCs require a few μm, while the conventional Si-SCs require 200–300 μm. Hence, the electricity produced by TF-SCs might be cost-effective due to less material consumption. Additionally, TF-SCs are benefited by simplified processing steps and monolithic integration. Among TF-SC materials, (i) copper zinc tin selenide (CZTS) and copper indium gallium selenide (CIGS) have complex structures with high defect densities that limit their cost-competitiveness, efficiencies, and industrial uptake for decades; (ii) Perovskite suffers from various intrinsic instabilities (i.e., iodine diffusion) and degrade quickly in the presence of heat, UV light, and moisture; (iii) GaAs-based III-V single and multi-junction cells are too expensive to be deployed in large areas.

On the other hand, cadmium telluride (CdTe) TF-SC panel technology has outshined multicrystalline Si solar panels in terms of cost-efficiency. Although CdTe has a large and almost ideal bandgap (1.5 eV) that can utilize its energy at shorter wavelengths more efficiently compared to Si (1.1 eV). But the performance of CdTe solar cells is limited due to traditional use of cadmium sulfide (CdS) as a window layer that causes the parasitic absorption and recombination loss at CdS/CdTe junction along with the lack of an electron blocking layer at the traditional CdTe/metal back contact (e.g., Au) interface. While the parasitic absorption in CdS reduces the short-circuit current (J_{SC}) and the poor band alignment of CdS and Au with CdTe reduces the open-circuit voltage (V_{OC}), the fill factor (*FF*), and ultimately the efficiency. For such reasons, top priority is given to the solid research for having a cost-effective, easy-to-fabricate, robust, and scalable methods to improve the performance of CdTe PV technology. The large bandgap (i.e., low parasitic absorptive) transition metal oxide (TMO) dopant-free carrier transport layers have emerged as an alternative contact material that can be deposited at low temperatures to improve the overall performance and efficiency of CdTe solar cells. As it allows only one kind of carrier and blocking the another, thus, the probability of recombination decreases. As a result, there is a significant increase in V_{OC}, thus, increasing the overall efficiency of the solar cell as well. Therefore, this chapter will be focusing mainly on exploring the paths for obtaining high V_{OC}, which will perhaps yield high efficiency for TF-PV technology.

9.1 INTRODUCTION TO SOLAR CELL AND P/N-HOMOJUNCTION SOLAR CELLS

The conventional energy sources are not able to create a compatible and sustainable evolution for the human community due to global increase in pollution and energy consumption; hence, there is a high demand for energy. It is always believed that the usage of renewable energy sources is necessary for the growth of the human community. Solar energy is considered to be one of the most optimistic and trustworthy renewable energy sources. It is categorized into two types: (i) Thermal solar energy (an energy in which the solar energy is converted into thermal energy, which in turn used to power a turbine) and (ii) Photovoltaics (PV) (in PV, the energy of photons (radiated by sun) is directly converted into electricity). Hence, a solar cell is a device which directly converts light energy into electrical energy by PV effect. In 1839, a French physicist *Becquerel* discovered a current producing mechanism, when a material was exposed to sunlight. In 1876, *William G. Adams* and *Richard E. Day* reported the first PV effect in Selenium (a solid material).[1] In 1941, *Russel Ohl* patented the first Si solar cell[2] and later by 1954, the Si solar cell had an improved efficiency of 6%.[3]

With the knowledge of more than 70 years, solar cell technology is giving a tough combat to the established fossil fuel market, yet there are hurdles for the PV industry. If solar electricity becomes a large contributor to energy production, then several efforts must be made to lower the module cost. There are certain materials and technologies that directly convert light into electricity. In PV technology, one of the best suitable materials is silicon (Si), as it is nontoxic, abundantly, and economically available. Presently, the market of bulk wafer-based crystalline Silicon (c-Si) solar cells occupies the largest share (~90%).[4] The crystalline-based solar cells have high conversion efficiency and long durability under atmospheric exposure. Despite having high conversion efficiency, there are some drawbacks in this technology such as high energy payback time, low efficiency with increasing temperature and massive kerf loss resulting from ingot making, wafer swing, and polishing.[5–9] For producing electricity directly from the sun, second-generation solar cell technology (thin-film PV) provides a best cost-effective solution.[10] Thin-film (TF) solar modules have twofold shorter energy payback times (EPBT) as compared to wafer-based crystalline solar modules that use materials that are abundant and nontoxic.[6] EPBT is the period in which the solar cell can recover the same amount of energy that was consumed during its production.

In addition, TF technology has numerous advantages, such as (i) TF–PV technology consumes less material of lesser thickness as compared to bulk Si solar cells, which makes them lighter, less fragile, and significantly inexpensive to produce; (ii) TF–SCs are easily deposited on multiple substrates, such as glass, plastic, and various metal sheets because of their low fabrication temperature[11, 12]; (iii) using solution-based and gas phase deposition processes, large cell areas and modules can be made flexible, thereby making it possible to fabricate by roll-to-roll manufacturing process, which may drastically cut down the production costs.[13] Currently, Si TF, CdTe, and copper-indium-gallium-selenide (CIGS) are the three TF–PV technologies that are available in the market. The global PV production by technology in 2013 is shown in Figure 9.1a, and the diversification of the global solar PV market up to 2020 is shown in Figure 9.1b.

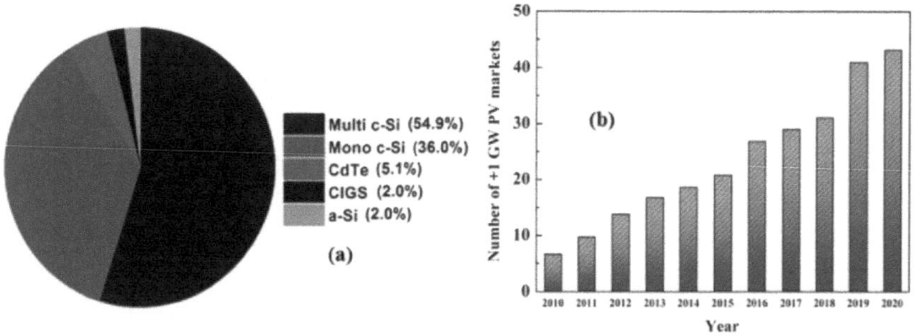

FIGURE 9.1 (a) Global PV production categorized by technology in 2013. Percent share is indicated in brackets. (b) The diversification of the global solar PV market (© IHS Markit).

In commercial PV production, CdTeTF solar cells are quite famous because of their low fabrication expense and high efficiency. CdTe has a 4.5 eV bandgap (direct). CdTe also possesses high intensity optical absorption coefficient in the solar spectrum (visible region) due to which the light is being easily absorbed by very few μm of CdTe, which ultimately reduces the material cost.[14] To fabricate a fine polycrystalline and chemically stable CdTe film, techniques involved are electrode position, sputtering, and vapor deposition techniques. CdTe solar cell has achieved efficiency ranging 19–20% for R&D and 10% or above in production and are now the front runners, in the production among TF solar modules.[15] However, some environmental issues are there with Cd emissions from CdTe PV modules because Cd is a heavy metal and potentially carcinogenic (causing cancer) that can accumulate in plants and animal tissues.[16, 17] Nonetheless, CdTe is a good option for fabricating a highly efficient, low cost and reliable thin-film module. Now,

CdTe on glass substrates has the conversion efficiency nearly 22%, and with substrates like polyamide or metallic foil, CdTe is making a good move toward flexible devices.

The standard Si solar cell is a homojunction of either p^+–n or n^+–p layer. Light absorption in the junction generates free electrons and holes, which proceed for diffusion. At the p–n junction, a space–charge region is formed and, thus an electric field exists. The electric field selectively sweeps holes one way and electrons the other, which causes the flow of photocurrent. In the traditional homojunction solar cells, an electric field separates the photogenerated carriers and drives carrier flow, while in the heterojunction solar cells, the band offsets at the selective contacts drives carrier gradient. Figure 9.2a and b shows the structure of homojunction solar cells, and their band diagram with photogenerated charge collection scheme, respectively.

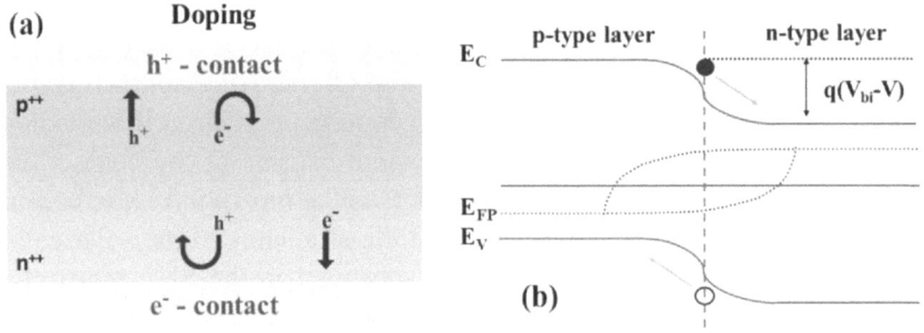

FIGURE 9.2 (a) Traditional p–n homojunction solar cell and (b) its equilibrium band diagram with photogenerated charge transport.

In commercial and laboratory Si solar cells, the base wafer is chosen to be thick for supporting top contacted p–n junctions. But many researchers have taken thin Si wafer to produce cost-effective and highly efficient solar cells. By controlling the dopant type and concentration, large asymmetries in the conductivity are presented to electrons and holes can be achieved. These heavily doped surface regions, commonly called homojunction, are typically formed by thermal diffusion, ion implantation or via rapid melt-recrystallization processes, most of which are inherited from the microelectronics industry. This approach has proven effective carrier separation for which the solar cell efficiency has reached up to ~20%.[18]

9.2 ALTERNATIVES TO CLASSICAL DOPED P–N HOMOJUNCTIONS: HETEROJUNCTION

The high doping concentration increases the thermal budget and introduces several fundamental losses and technological issues, which hinder the further reduction of the cost-to-performance ratio for the homojunction technology. Many recognized strategies are applied to remove these issues and progress forward to use heterocontacts to achieve carrier-selectivity. The modern archetype of the heterocontact concept is the Si heterojunction (SHJ) cell. Recently, the merits of heterocontact have been demonstrated conclusively, and it surpasses the homojunction technologies and claim for the world record efficiency for single junction c-Si cells.[19] Furthermore, the doped emitter and the Ohmic contacts are the two limitations of Si wafer-based solar cells in classical homojunction technology.[20, 21] Compared with the classical homojunction technology, the two parameters, contact passivation and open-circuit voltage (V_{OC}), are found quite better in the SHJ approach.[22] There are issues in the interfaces of Cu (In, Ga) Se_2 and CdTeTF-SC with the buffer layer and Ohmic back contact, respectively.[23, 24] Nowadays, the most notable hurdles in solar cell technology are separation and recombination of charge carriers at the contact. But among them, separation of charge carriers is the most vital issue because a perfect carrier separation can increase the efficiency of the solar cell to the theoretical benchmark. All the advancement in the SHJ approach is achieved by setting a high V_{OC} to reduce the contact resistance.[25] Since then, for collecting photogenerated charge carriers, the only condition that was taken into vision was selectivity.[26] Figure 9.3 shows the band diagram of heterojunction solar cell under illumination condition and splitting of Fermi level.

The efficiencies offered by heterojunction solar cells are quite close to conventional Si cells but with low thermal budget and uncomplicated fabricating process. Particularly, in this type of solar cells, the doped side of a p^+–n and n^+–p junction are changed by the carrier selective layer to collect a particular charge carrier from the absorber layer. The heterojunction solar cells have a unique feature that the electron and hole collecting layers can be formed by the passivation of the front and rear surface by deposition method at low temperature. The advanced heterojunction intrinsic thin layer (HIT) solar cell can be fabricated by depositing by doping a-Si:H layer on c-Si wafer by plasma-enhanced chemical vapor deposition (PECVD) that has achieved a power efficiency of 25.6% (Panasonic). Moreover, the Japanese company Kaneka has reported a record efficiency of 26.3% combining with the HIT and interdigitated back contact (IBC) structures. Therefore, heterojunction cell technology is highly promising and will compete with the hydrocarbon-based solar PV for electricity generation. In spite of the advantages in silicon hetero-junction (HIT) solar cells, there are fundamental challenges related to deposition of a-Si:H layer on wafer. But the HIT technology still relies on the PECVD process to deposit a-Si:H layer that requires a moderately high thermal budget; complicated deposition optimization yields undesired parasitic absorption loss in the layer. The other significant problems with a-Si:H material is their lower bandgap (~1.7), high defect density, and light-induced degradation effect, altogether limit the HIT solar cells performance.

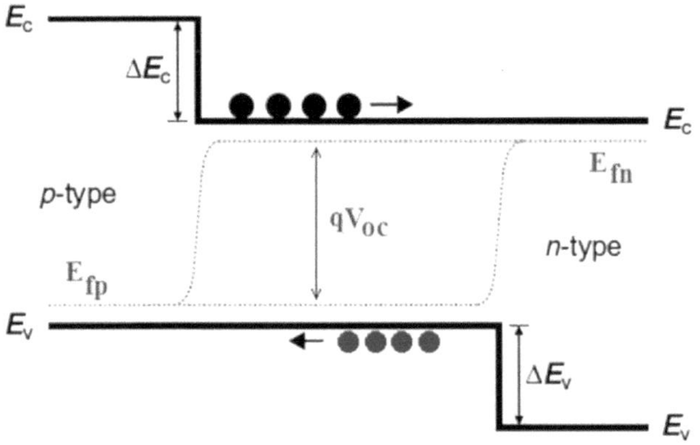

FIGURE 9.3 A schematic band diagram of a heterojunction solar cell under illumination at open circuit condition.

With continuous technology advancement and exploration of solar cells, they have a low cost per watt-peak. CdTe solar cells is the most competitive technology among the TF-PV technologies, and it has better cost-efficiency than c-Si solar panels. The most exciting benefit of CdTe PV technology is their ability to absorb sunlight close to an ideal wavelength. It can capture energy at shorter wavelengths than Si PV technology; therefore, it closely matches with the natural wavelengths of sunlight for optimal light to electricity conversion.

All processing technologies currently used for CdTe solar cell fabrication employed CdS/CdTe interface for carrier collection. The CdTe cell efficiency has reached up to ~ 22% due to ideal bandgap (~1.5 eV) and high absorption coefficient. However, in CdTe structure and choice of window layer still hinders the cell to reach the theoretical maximum efficiency of ~29%.

Thus, for improving the performance of CdTe solar cell, an alternative carrier selective layer is required to reduce the charge carrier extraction barriers and for efficient separation of electrons and holes at CdTe absorber interface. Thus, as an alternative to traditional impurity-doped carrier selective layer (Figure 9.4a), a metal oxide-based wide bandgap with optional work function material has drawn significant attention from the research community and commercial organizations to further achieve efficiency/cost ratio for large scale manufacturing of CdTe solar cells. The TMO at the CdTe interface forms an asymmetric band offsets for electrons and holes to establish carrier-selective contacts (CSC) (Figure 9.4b). The work function difference in the metal oxides and Fermi level of CdTe absorber has a broad energy band bending at the TMO/CdTe interface, thereby only one kind of photogenerated carrier can be allowed to transport which are either holes or electrons (Figure 9.4c).

The constructive carrier separation of electrons and holes is due to their respective conductivities which results in the substantial reduction of interfacial recombination loss and thus CdTe. PV devices may have significantly high V_{OC} as well as efficiency. High work function metal oxides are MoO_x, VO_x, WO_x, NiO_x, and CuO_x, which create a large conduction band offset to provide an excellent path for transportation of holes as they formed a heterocontact when deposited on CdTe, whereas the low work function metal oxides are TiO_x, MgO_x, and ZnO, which creates a small conduction band offset to provide an excellent path for transportation of holes when deposited on CdTe. Besides being used for carrier separation, these metal oxide layers also provide low contact resistance and absorption loss. Therefore, these alternating carrier selective materials with several key features such as low contact resistivity, low surface recombination velocities, and high built-in potential are vital for achieving high efficiency in CdTe solar cells. Moreover, TMOs can be deposited at low temperature (<200°C) by conventional deposition techniques resulting in the process at lower thermal budget. The schematic of the asymmetric heterocontact CdTe solar cell and its band diagram are shown in Figure 9.4d.

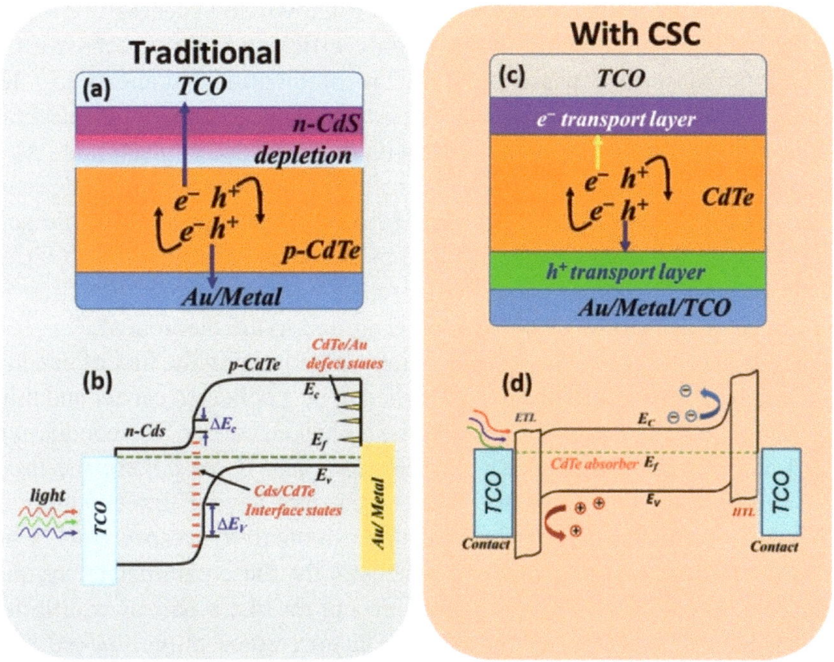

FIGURE 9.4 Different scheme of charge collection: (a) The conventional solar cell structure where carrier selectivity due to doping gradient, (b) charge collection due to asymmetric heterocontact formed by TMOs, (c) schematic of energy barrier of individual layer and formation of band offsets with CdTe absorber, and (d) proposed energy band diagram for TMO/CdTe heterojunction solar cell.

9.3 BASIC ASPECTS FOR CARRIER SELECTIVE CONTACTS

Carrier selective contacts (CSC) refer to the competency of contacts by enabling a distinctive kind of carrier to flow and forbidding the others. According to charge carrier transport theory, this phenomenon is based upon asymmetric resistance of one carrier above another, by the ratio of resistances between electrons and holes of the contact, which is precisely considered to be one of the essential parameters. CSCs are executed on the entire Si (lightly doped) surface by a facile fabrication method with a unidirectional pattern of carrier flow, instead of expensive fabrication techniques, providing exceptional surface passivation and fill factor (*FF*), hence, resulting in low carrier recombination, high V_{OC}, and low contact resistivity (ρ_c). Silicon heterojunction (SHJ) and tunnel oxide passivated contact (TOPCon) are the two most well-known and successful Si-based CSC technologies for designing highly efficient Si solar cells with an exceptional surface passivation (i.e. recombination parameter [J_o] less than 10 fA/cm^2) and low contact resistivity (i.e., less than 30 mΩ cm^2). The efficiencies of SHJ solar cells (with double-sided contact) and IBC Si solar cells are 25.1%[27] and 26.3%,[22] respectively, which is a world record. Nevertheless, the efficiency of 25.1% has been noted by a n-type solar cell having the front emitter diffused by boron and the rear with TOPCon, respectively.[28] Certainly, many researchers have calculated the efficiency of Si solar cells to be 29.4%, which is around to the theoretical.[29]

CSC comes up with a foremost aim of maximizing the flux of one kind of carrier across a virtual surface, named to be a collected carrier and minimizing the flux of another named to be a blocked carrier. The conductivity (σ) and potential gradient (η) for electrons and holes govern the fluxes in the contact. Therefore, the blocked carriers have a low conductivity, whereas collected carriers have a high conductivity, respectively, when they move across the contact. Experimentally, these parameters are quite tedious to calculate. Hence, results are put on like a pair of equilibrium parameters: (i) contact recombination factor (representing blocked carrier's flux) and (ii) contact resistivity (representing the interface resistance to collected carriers). These two parameters have a significant impact on carrier selectivity.

9.3.1 MAJORITY CARRIER CONDUCTANCE

Majority carrier conductance, generally named as low resistance contacts of solar cells, portrays the capability of contacts in transferring the majority carries constructively.[30] This phenomenon descends the Fermi level of majority carriers and equals the slope in the respective conduction band and the valence band of the contacts. In solar cells, surface passivation and carrier selectivity influence the open circuit characteristics whereas in majority carrier conductance, at the optimal power point, external current flow occurs. Hence, the efficiency of the solar cell greatly depends upon this specific parameter. In traditional solar cells, p–n junctions were used to collect and extract the carriers, instead of carrier selective contact. Previously, from the absorbers, Ohmic contact extracts the majority carriers and p–n junctions first collect the minority carriers and then transmute into majority carriers and later extracted by another Ohmic contact from either side of the junction. As a result, orthodox solar cells had theoretical and practical losses compared with thermodynamic ideals, and at the Fermi level, the carriers are extracted due to doping of the device rather than quasi Fermi level. Practically, in p–n junctions, it is taken that lowering the doping on either side results in lower V_{OC}. But the lifetime of minority carriers decreases on increasing the doping; hence, there is a trade-off between the performance and the design. In order to circumvent the trade-off, multiple solar cell structures were introduced with dissimilar doping at the heterocontacts, which further reduces the surface recombination in the contacts as well. Thermodynamically, the carriers generated are extracted at the energy equivalent to quasi Fermi level and no theoretical doping inherent dependency; hence, p–n junctions are not required.

Basically, the majority carrier conductance of Ohmic contact can be enhanced by doping or accumulation of carriers. On a silicon wafer, under low-temperature condition, a finalized device structure is fabricated using organic coating or thin-film deposition method for the epitaxial growth of Si. Since induced junctions such as p^+/n-Si and n^+/p-Si are formed, surface inversion conditions are utilized for minority carrier selectivity. In both the instances, the majority carriers are being repelled by band bending, due to which accumulation of minority carriers are being favored at metal–Si interface, for the extraction. A metal-insulator-semiconductor (MIS) pattern is found, whenever ultrathin (<3 nm) oxides are passivated at the interface. Therefore, MIS solar cells work on the principle, when photogenerated minority carriers are tunneled across the oxides while at the semiconductor surface, majority carriers are exhausted by the firm inversion layer.[31, 32] From the eq 9.1, using selective contact ($\Phi_{sel.cont.}$) and c-Si absorber (Φ_{c-Si}) work functions, induced band bending can be expressed as[33, 34]:

$$\Phi_{sel.cont.} - \Phi_{c-Si} = \Phi_s \approx V_{bi} \quad (\text{for } S_\Phi = 1) \tag{9.1}$$

where, φ_S is considered to be equivalent to junction's built-in potential (V_{bi}) that acts as the barrier potential for minority carriers (electrons) and provides a conductivity for majority carriers.

9.3.2 CHARGE TRANSPORTING LAYERS (CTL)

Charge transporting layers (CTL) perform a crucial role in the performances of TF-SCs like perovskite solar cells, CIGS, organic, CdTe, etc. CTLs withdraw an electron or hole and then infuse them into the electrodes, respectively. So, ideal CTLs should have the excellent energy layers matched with that of the perovskite layer, in order to expedite photogenerated charges and curtail charge losses.[35, 36] Further, high optical transmission is required to maximize the amount of light passing through and to reach the absorber layer. For the CTL, high charge mobility is necessary to transport the charges to their corresponding electrodes. Likewise, in order to curtail any recombination and to assure reasonable charge transportation, the compact layer with full coverage and pinhole-free layer is needed. CTLs must have useful interface contacts with absorber layers. CTLs must have an optimal thickness, which must be sufficiently thick in order to provide full coverage and sufficiently thin to minimize series resistance (R_s) that can further help in accomplishing high fill factor (*FF*) and consequently higher device performance. Furthermore, CTLs help in avoiding the metal-ion diffusion, moisture, and oxygen to the perovskites and make them substantial in air. In the recent past, different inorganic and organic CTLs, such as hole transporting layer and electron transport layer have been validated in the literature.[37, 38] In order to enhance the properties of CTLs and their device performance, different strategies have been used. Figure 9.5 shows the schematic energy level diagram of various metal oxide-based carrier selective contact layer for c-Si and thin-film solar cells.

9.3.2.1 ELECTRON TRANSPORT LAYER (ETL)

The creation of electron selective contacts with the absorber layers in order to intensify the electron extraction efficiency from light absorbing layers and hole blocking is the fundamental motive of the electron transport layer. This can help in enhancing the carrier separation and the device performance and also helps in reducing carrier recombination. The primary selection norm for the ETL includes an n-type semiconductor, having a large bandgap which is lucent to visible light having high carrier mobility and good electrical conductivity for effective electron transfer. Also, the semiconductor must have better band alignment with the materials so that it can be processed at a low temperature, ultimately, making it a cost-effective approach. Finally, in the ambient condition when it is exposed to the environment, the ETL needs to be stable. For the solar cells, many different inorganic and organic electron transport materials had been used.

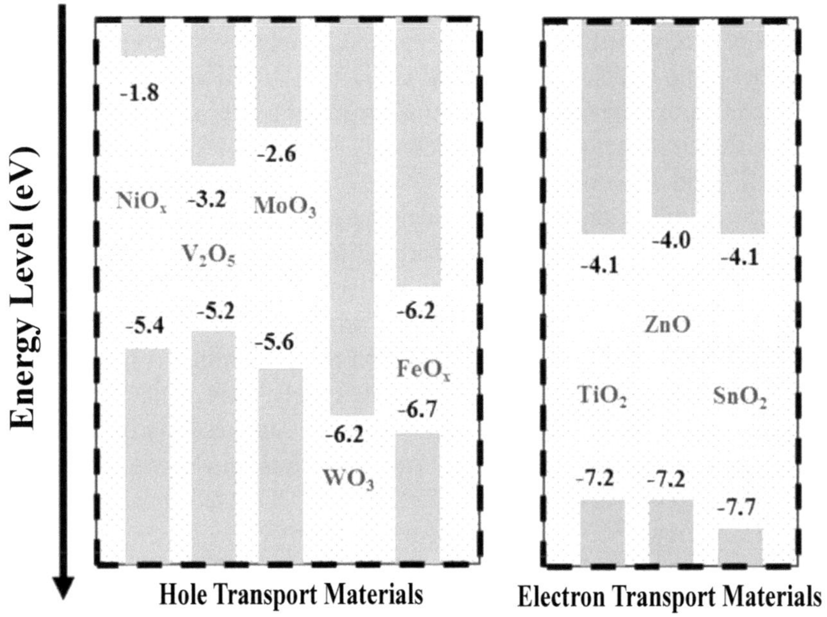

FIGURE 9.5 A schematic energy level diagram of various metal oxide-based charge transport layer.

9.3.2.1.1 Metal Oxides as Electron Transport Layer

The most challenging part while fabricating a solar cell is selecting a suitable electron selective contact. High-performance solar cells commonly use metal oxides such as titanium dioxide (TiO_2) and tin oxide (SnO_2) as electron transport layers.[39, 40] Other broad bandgap materials such as zinc oxide (ZnO),[41] aluminum oxide (Al_2O_3),[42] strontium titanium oxide ($SrTiO_3$),[43] silicon dioxide (SiO_2), and zirconium dioxide(ZrO_2)[44, 45] are also used as electron transport layers.

The most widely used transparent conducting oxide for electron selective contact is the n-type metal oxide TiO_2. High PV device performance was obtained using TiO_2 with 15.6% power conversion efficiency (PCE), and the use of TiO_2 nanoparticles as electron selective contact has improved the performance to 15.9%.[46] Processability of TiO_2 involves high temperature (>450°C), which makes it unsuitable for flexible devices and mass production.[47] In addition, the bulk electron mobility of TiO_2 (<1 $cm^2\ V^{-1}\ s^{-1}$)

is lower than most of its counterparts.[48] To get rid of these problems, other materials are doped in TiO_2, for example, graphene is incorporated with the TiO_2 nanoparticles[49] and is used as electron selective contact nanocomposite in perovskite solar cell. Alternative metal oxides such as ZnO, SnO_2, and ZrO_2 have also been used widely.

SnO_x with its non-ideal E_c (E_c = 4.5 eV) is still used as an ETL but with a slight modification of its electrical characteristics. Doping SnO_x aligns its bandgap with the adjacent absorbing layer as well as improves other electronic properties. Fullerene can be used as a surface coating on SnO_2 ETL to enhance its electron extraction process.[50] The main advantage of SnO_x film is low-temperature fabrication process, which is beneficial for application in flexible devices and tandem solar cells. ZnO owing to its superior electron mobility and high bandgap[51] has recently emerged as a promising candidate for ETL.[52] ZnO quantum dots which have much higher conduction property compared to ZnO film or TiO_2 have already been used as ETL.[53] Figure 9.6 shows the band offsets of various metal oxides and transparent conductive oxides used as a carrier selective contact layer for heterojunction solar cells. Performance of ZnO ETL can be further improved by doping. It is already reported that aluminum (Al), magnesium (Mg), and lithium (Li) can improve electron mobility in ZnO.[54–56] Tungsten oxide (WO_3) is another promising candidate as ETL because of their high electron mobility (10–20 $cm^2\ V^1\ s^{-1}$) and wide bandgap energy (2.7–3.9 eV). Further, its low-temperature processability and cost make it more advantageous in future solar cell fabrication. Increase in oxygen vacancy improves the electrical conductivity and photoelectric conversion efficiency of WO_3.[57, 58] Niobium-doped WO_x ETL on flexible solar cell showed 15.65% PCE.[59]

9.3.2.2 HOLE TRANSPORTING LAYERS (HTL)

The fundamental components for accomplishing high-performance Cd Te perovskite and organic solar cells are the hole transporting layers. In order to limit recombination of charge, HTL perform a vital role such as transportation and extraction of photogenerated holes from perovskite materials (light-absorbing) to back electrodes material.[60, 61] The generic requirements for the promising hole-transporting materials consist of enhanced energy level alignment with absorber perovskite materials, better conductivity, and immense hole mobility for effective charge transmission, and it must have better thermal stability, photostability, and low material costs. With the heterojunction contacts, one complication arises that is there is usually a trade-off among the necessity for the thicker layer in order to enhance contact selectivity and lower contact recombination, and at the same time thin layers are also required in order to reduce carrier extraction losses (specifically at the higher current density) and a parasitic absorption.

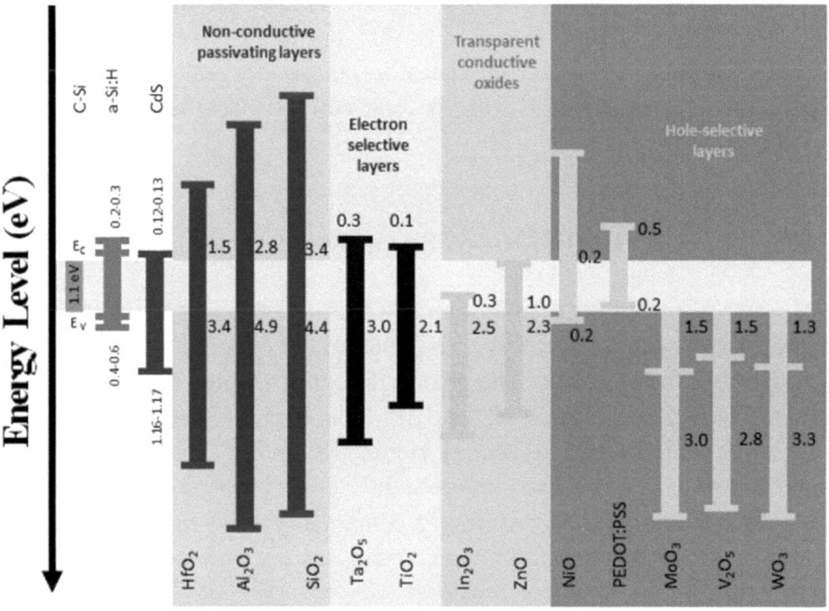

FIGURE 9.6 The energy level diagram to show the band offsets of different transition metal oxides and transparent conducting oxides used as a carrier selective layer heterojunction solar cells.

9.3.2.2.1 Metal Oxides as Hole Transport Layer

TMOs having work function higher than ~6 eV have been used as HTL that include molybdenum oxide (MoO_3),[62] vanadium pentoxide (V_2O_5),[63] and tungsten oxide (WO_3).[64] These materials have high hole mobility, higher conductivity, and better stability making them more promising for HTL. MoO_3 back contact significantly improves V_{OC} as reported by Paudel et al.[62] The efficiency of CdTe solar cell has reached 14.1% after using MoO_3/Au as back contact. V_2O_5 can be used as a buffer to reduce the Schottky barrier at the back contact and improve the device stability.[63] p-type nickel oxide (NiO) with its wide bandgap (3.6 eV) is used as a buffer layer in CdTe cells, and it is employed as back surface field (BSF) layer to reflect the electrons generated due to photons. As a result, the recombination at the back contact is decreased.[65] However, conductivity of NiO is poor and may lead to high series resistance which in turn decreases the fill factor of the cell. NiO efficiently accelerates charge extraction, which improves photovoltaic performance by shifting down the hole Fermi level. Cuprous oxide (2.1–2.6 eV) has been used as HTL because of its high hole mobility exceeding 100 cm^2 V^{-1} s^{-1}.[66, 67] Therefore, these inorganic metal oxides are potentially promising for HTL in solar cells to improve device stability and enhance efficiency.

9.4 SELECTIVE CONTACT STRATEGIES

A major setback for the evolution of solar cell contacts is the absence of measuring contact selectivity to a photovoltaic material that can be either perfect[26] or non-perfect. The two different strategies or models are (i) Resistive or Brendel and Peibst (BP) Model and (ii) Kinetic or REL (Roe, Egelhofer, and Lonergan) Model. Hence, let us first analyze the similarity and dissimilarities of these models. In BP model, it is assumed that the interface recombination current of a contact to a cell with V_{th} as the thermal voltage, V_i as the internal voltage, and $J_{s,o}$ as the interface recombination current at saturation is given by

$$J_S = J_{S,0}\left[\exp\left(\frac{V_i}{V_{th}}\right) - 1\right] \tag{9.2}$$

Meanwhile, an Ohmic resistance (R_C) in Ω cm^2 is represented by the contact to the majority carrier current, and thus the selectivity[68] can be written as

$$\sum\nolimits_r := \frac{V_{th}}{R_C J_{S,0}} \tag{9.3}$$

Now, in eq 9.3 majority carrier's R_c forms a relation with minority carrier's recombination resistance (zero bias) ($V_{th}/J_{s,o}$). At V_i, the absolute ratio of Ohmic and recombination resistance is specified by differential selectivity as

$$\tilde{\sum_r} := \frac{1}{R_C}\left(\frac{dJ_S}{dV_i}\right)^{-1} = \frac{Vth}{R_C J_{S,0}}\exp\left(\frac{-Vi}{V_{th}}\right) = \sum_r \exp\left(\frac{-Vi}{V_{th}}\right) \tag{9.4}$$

From eq 9.4, it is prominent that under bias condition, fraction of resistive and recombination losses is dependent on concentration of minority carriers. Equation 9.3 depicts when the selectivity is high, minority carrier recombination and series resistance are low, hence making it possible to extract all the majority carriers, with null voltage drop and null photo voltage due to minority carriers' recombination. On the other hand, if there is high series resistance and the recombination current leading to increase in voltage losses, it deteriorates the efficiency. Hence, there is a postulated divergence between the two type of carriers in eq 9.3. Therefore, the BP model is well known as a resistive model. In this model, the relation between current density (*J*) and external voltage (*V*) is written as

$$J = J_L - J_{S,0}\left[\exp\left(\frac{V - JR_C}{V_{th}}\right) - 1\right] \tag{9.5}$$

where J_L is represented as the current density of the photogenerated circuits. The assumptions taken in the REL model are quite different as compared to the BP model.[69] In this model, the respective carrier currents of major and minor carriers at the surfaces of PV absorbers with C_M and C_m as their respective concentrations are reported as

$$J^{(i)}_{M/m} = J^{(i)}_{M,0/m,0}\left[\frac{C^{(i)}_{M,0}}{C^{(i)}_{M,0/m,0}} - 1\right] \tag{9.6}$$

where $J^{(i)}_{\mathrm{M},0/\mathrm{m},0}$ is represented as majority/minority carrier current density (saturation) at corresponding contacts, let for $i = 1, 2$ and $C^{(i)}_{M,0/m,0}$ is represented as respective equilibrium carrier concentrations. At the two contacts, electrons and holes exchange their part as minority and majority carriers. In REL model, the selectivity is expressed as

$$\sum{}^{i}_{k} := \frac{J^{i}_{M,0}}{J^{i}_{m,0}} \tag{9.7}$$

where $J^{i}_{M,0}$ and $J^{i}_{m,0}$ are denoted as saturation current density of major and minor carriers, and *k* is denoted as the kinetic character of eq 9.6. It has been seen that the REL model stays numb for providing information on resistive losses (as in the BP model), but it portrays how the minority carriers are inversely dependent on the saturation current density with the BP model.

9.4.1 CONTACT AREA OPTIMIZATION

The picture of contact area optimization can be illustrated significantly with the concept of selectivity. For designing crystalline Si solar cells, few analytical approximations are taken to utilize this tool, outside the range of phenomenological relations.[68] After all approximations, the optimal output power (P) can be expressed as

$$P = J_{SC} V_{OC} FF_0 \left(1 - \frac{J_{SC} R_C}{V_{OC}}\right) \tag{9.8}$$

$$P = J_{SC} V_{th} \ln\left(\frac{J_{SC}}{J_{S,0}}\right) FF_0 - J_{SC}^2 FF_0 R_C \tag{9.9}$$

where FF_0 is called a fill factor without series resistance and defined as the function of V_{OC} to thermal voltage, which can be expressed as $FF_0 = FF_0(V_{oc}/V_{th})$. In ref.[70] using phenomenological equations, the effect of FF_0 and series resistance (Constant) is being described. Certainly, the optimization problem is being described in ref.[68], stating that the optimum area fraction of the solar cell surface is being confined to a small fraction *(f)*. For point contact, the characteristic length is small compared to minority carrier's diffusion length in absorber[71]; and along with perfectly passivated non-contact and non-conductive areas, hence, upon varying *f*, the selectivity remains unchanged. Nevertheless, on varying contact area fraction, there are high resistive losses against reduced recombination losses leading to a trade-off situation and which is desired for enhancing the efficiency.

9.4.2 SELECTIVITY AND PHOTOCURRENT COLLECTION EFFICIENCY

In Si-based solar cells, on measuring contact resistivity and interface recombination provide the experimental data for selectivity of contacts.[68] According to *Wong* and *Green,*[72] the selectivity closely corresponds to current transport efficiency, also known as photocurrent collection efficiency (f_{PC}), can be expressed as

$$f_{PC} := \frac{dJ}{dJ_L} \tag{9.10}$$

where, dJ is the derivative of the current collected with fixed external voltage (V) (at the solar cell terminal), and dJ_L is the derivative of the short circuit current, which is produced internally. Since f_{PC} is closely related to contact selectivity, certain scenario arises: (i) when contact is highly selective and applied voltage V is small, there is negligible mass recombination, thus $dJ = dJ_L$, hence, $f_{PC} = 1$, and (ii) when contact is less selective, charge carriers produced recombine at wrong contact, since short circuit current (JSC) which is generated internally may not increase the constant current density (J_{cont}). Again, Wong and Green theory[72] illustrated that the local photocurrent collection efficiency (f_{PC}^{loc}) can be defined as the ratio of differential change in local voltage across the junction (dV_j^{loc}) to the change in exterior voltage (dV), which can be expressed as

$$f_{PC}^{loc} := \frac{dV_j^{loc}}{dV} \tag{9.11}$$

Now, by optoelectronic reciprocity relation, f_{PC}^{loc} can be calculated using electro-modulated luminescence, since the junction voltage (local) depends on the luminescence emission.[73] Hence, the junction voltage (V_j) can be extracted from absolute electroluminescence intensity by electroluminescence images at different voltages (external) and calculated by the derivatives in eq 9.2. To perform electroluminescence, imaging systems are utilized which provide a local V_j, then the collection efficiency can also be obtained as a function of position in solar cells. Many researchers have demonstrated this opportunity on Si solar cells[74] and GaAs solar cell.[75] Ultimately, a global f_{PC} is obtained, after averaging the local f_{PC}^{loc} values over the solar cell area.[76]

9.5 ASSESSING THE SUITABILITY OF DIFFERENT CONTACT MATERIALS

It is very much essential to choose a particular contact material, i.e., hole transporting material (HTM) and electron transporting material (ETM) for maintaining charge selectivity and surface recombination to lower the energy losses at the heterocontact. In the recent past, variation of HTMs and ETMs has been examined for the achievement of higher efficiency TF-CIGS, CdTe, perovskite, and organic solar cells. There are three categories of most frequently used HTMs are small molecules, inorganic types, and organic polymers. In the small molecule category, Spiro-MeOTAD is the most commonly used HTMs in TF-PV devices for high efficiency. Presently, a conductive organic polymer named poly-Triarylamine (PTAA) is a shear replacement to Spiro-MeOTAD in perovskite devices.[40] However, these organic HTMs deliver a better carrier transport property which further escalates the performance and long-term stability. The main barrier for industrial applications is high material cost when compared with inorganic HTMs: CuSCN,[77] CuI,[78] NiO,[79] and Cu:NiO_x are responsible for low cost and stable performance. Till now in inorganic materials, the highest efficiency of 17.7% has been reported when Cu:NiO_x is used as HTM.[80] The metals with high work function such as gold (Au) and carbon (C) help in the extraction of holes from absorber layers and TFCdTe. Several research groups have validated the HTM-free perovskite solar cell with 5–12% efficiency,[81–83] and efficiency of ETM-free devices has been validated which is near ~14%.[84]

Presently, transition metal oxides (TMOs) are considered to be one of the best TFs and a perfect dopant-free CSCs for Si solar cells with high/low work functions and an acceptable band offset with Si. Some of the hole selective contacts with high work functions are, molybdenum oxide (MoO_x),[85] vanadium oxide (VO_x),[86, 87] and tungsten oxide (WO_x)[88, 89], and some contacts with small valence band offset with Si are nickel oxide (NiO_x)[14] and PEDOT:PSS [poly(3,4-ethylene-dioxythiophene):polystyr enesulfonate].[90]. Some of the electron selective contacts with low work functions are magnesium fluoride (MgF_2),[91] cesium carbonate ($CsCO_3$)[92], and lithium fluoride (LiF),[93] and contact with small conduction offset with Si is titanium dioxide (TiO_2).[94, 95] In comparison to doped-Si layer, the dopant-free CSCs can be easily deposited, at a cheaper cost by using spin-coating, atomic layer deposition (ALD), thermal evaporation, and even by printing and brush painting as it is universally validated by light-emitting diodes and organic solar cells.[96, 97]

Furthermore, dopant-free CSCs have a wide bandgap and is transparent, which make them more prominent as the window layer for the TF-SC and Si-based solar cells than the doped layer, which experiences the parasitic absorption loss in a blue wavelength range. When CSCs are installed in the device level, the prime drawback is their low surface passivation (with absorber surface) which readily reduces the efficiency of the cell. Generally, to enhance surface passivation on the solar cells, between the dopant-free contact layer and the absorber, a thin buffer layer must be inserted.

In a study on CdTeTF-SC, copper indium (CuI) TFs were used as a back-contact for the buffer layer, which results in reducing the energy barrier effectively at the back-contact, and it also improves CdTe solar cells V_{OC}.[98]

9.6 CONCLUSION

In the summary, we have observed the significance of heterocontacts in the emerging thin-film CdTe PV. We have seen that not only the two definitions related to contact selectivity are distinct but also the assumptions taken and areas of application are distinctive as well. The two approaches/models of carrier selectivity are discussed and described in a simpler way. We have seen that the carrier selective contacts (CSC) are one of the best heterocontact technologies to increase the efficiency of the solar cell. Coming to industrial feasibility and compatibility, TiO_2MoO_x, VO_x, WO_x, and NiO_x contacts (dopant-free) are the fore-runners as they have zero resistivity toward the base and can be easily inscribed on ultrathin substrates. Comparing with other technologies such as SHJ and TOPCon, which avails complex fabrication methods like PECVD and LPCVD, the transition metal oxide contacts can be easily fabricated by simple, low-cost techniques such as spin coating, sputtering, thermal evaporating, and screen printing. In addition to that the transition metal oxide contacts record an appreciable thermal stability and activation temperature (600°C) than SHJ (≤250°C) and TOPCon (~850°C). Therefore this chapter led to a better understanding of the fundamental of carrier selectivity in heterojunction solar cells, and with this in-depth understanding it would be possible to design the efficiency of heterojunction thin-film CdTe solar cells. The transition metal oxide based contact materials are promising candidates for a new generation of carrier-selective contacts for thin-film and silicon wafer-based solar cells.

ACKNOWLEDGMENTS

Sandipan Mallik and Shrabani Guhathakurata would like to sincerely thank the Department of Science and Technology-Science and Engineering

Research Board (DST-SERB) New Delhi, India for providing research funding [Project no: ECR/2018/002477] under DST-ECR scheme.

KEYWORDS

- **thin-film solar cells**
- **photovoltaic**
- **asymmetric heterocontact**
- **CdS**
- **CdTe**

REFERENCES

1. The Royal Society is Collaborating with JSTOR to Digitize, Preserve, and Extend Access to Proceedings of the Royal Society of London. ® Www.Jstor.Org. *Society*.
2. Ohl, R. S. Light-Sensitive Electric Device. United States Pat. Off. 2402662, 1941, 1–14.
3. Chapin, D. M.; Fuller, C. S.; Pearson, G. L. A New Silicon p-n Junction Photocell for Converting Solar Radiation into Electrical Power. *J. Appl. Phys.* **1954**, *25* (5), 676–677.
4. Battaglia, C.; Cuevas, A.; De Wolf, S. High-Efficiency Crystalline Silicon Solar Cells: Status and Perspectives. *Energy Environ. Sci.* **2016**, *9* (5), 1552–1576.
5. Wu, P.; Ma, X.; Ji, J.; Ma, Y. Review on Life Cycle Assessment of Energy Payback of Solar Photovoltaic Systems and a Case Study. *Energy Procedia* **2017**, *105*, 68–74.
6. Bhandari, K. P.; Collier, J. M.; Ellingson, R. J.; Apul, D. S. Energy Payback Time (EPBT) and Energy Return on Energy Invested (EROI) of Solar Photovoltaic Systems: A Systematic Review and Meta-Analysis. *Renew. Sustain. Energy Rev.* **2015**, *47*, 133–141.
7. Dupré, O.; Vaillon, R.; Green, M. A. Physics of the Temperature Coefficients of Solar Cells. *Sol. Energy Mater. Sol. Cells* **2015**, *140*, 92–100.
8. Chander, S.; Purohit, A.; Sharma, A.; Arvind; Nehra, S. P.; Dhaka, M. S. A Study on Photovoltaic Parameters of Mono-Crystalline Silicon Solar Cell with Cell Temperature. *Energy Rep.* **2015**, *1*, 104–109.
9. Solanki, C. S.; Singh, H. K. Anti-Reflection and Light Trapping in c-Si Solar Cells. 2017, No. Tanaka, 2010.
10. Green, M. A. Thin-Film Solar Cells: Review of Materials, Technologies and Commercial Status. *J. Mater. Sci. Mater. Electron.* **2007**, *18* (SUPPL. 1), 15–19.
11. Kazmerski, L. L. Solar Photovoltaics R&D at the Tipping Point: A 2005 Technology Overview. *J. Electron Spectros. Relat. Phenomena* **2006**, *150* (2–3), 105–135.

12. Koch, C.; Ito, M.; Schubert, M. Low-Temperature Deposition of Amorphous Silicon Solar Cells. *Sol. Energy Mater. Sol. Cells* **2001**, *68* (2), 227–236.
13. Cabarrocas, P. R. I. Deposition of Intrinsic, Phosphorus-Doped, and Boron-Doped Hydrogenated Amorphous Silicon Films at 50°C. *Appl. Phys. Lett.* **1994**, *65* (13), 1674–1676.
14. McCandless, B. E.; Sites, J. R. Cadmium Telluride Solar Cells, 2011.
15. Poplawsky, J. D.; Guo, W.; Paudel, N.; Ng, A.; More, K.; Leonard, D.; Yan, Y. Structural and Compositional Dependence of the CdTe x Se 1-x Alloy Layer Photoactivity in CdTe-Based Solar Cells. *Nat. Commun.* **2016**, *7*, 1–9.
16. Bakhiyi, B.; Labrèche, F.; Zayed, J. The Photovoltaic Industry on the Path to a Sustainable Future-Environmental and Occupational Health Issues. *Environ. Int.* **2014**, *73*, 224–234.
17. Aman, M. M.; Solangi, K. H.; Hossain, M. S.; Badarudin, A.; Jasmon, G. B.; Mokhlis, H.; Bakar, A. H. A.; Kazi, S. N. A Review of Safety, Health and Environmental (SHE) Issues of Solar Energy System. *Renew. Sustain. Energy Rev.* **2015**, *41*, 1190–1204.
18. Zhao, J.; Wang, A.; Green, M. A. Prog.; Photovoltaics: Res. Appl. 7, 471 (1999). Wiley Online Library| CAS| Web of Science® Times Cited.
19. Masuko, K.; Shigematsu, M.; Hashiguchi, T.; Fujishima, D.; Kai, M.; Yoshimura, N.; Yamaguchi, T.; Ichihashi, Y.; Mishima, T.; Matsubara, N. Achievement of More than 25% Conversion Efficiency with Crystalline Silicon Heterojunction Solar Cell. *IEEE J. Photovoltaics* **2014**, *4* (6), 1433–1435.
20. Lim, B.; Brendemühl, T.; Dullweber, T.; Brendel, R. Loss Analysis of N-Type Passivated Emitter Rear Totally Diffused Back-Junction Silicon Solar Cells with Efficiencies up to 21.2%. *IEEE J. Photovoltaics* **2016**, *6* (2), 447–453.
21. Brendel, R.; Dullweber, T.; Peibst, R.; Kranz, C.; Merkle, A.; Walter, D. Breakdown of the Efficiency Gap to 29% Based on Experimental Input Data and Modeling. *Prog. Photovoltaics Res. Appl.* **2016**, *24* (12), 1475–1486.
22. Yoshikawa, K.; Kawasaki, H.; Yoshida, W.; Irie, T.; Konishi, K.; Nakano, K.; Uto, T.; Adachi, D.; Kanematsu, M.; Uzu, H. Silicon Heterojunction Solar Cell with Interdigitated Back Contacts for a Photoconversion Efficiency over 26%. *Nat. Energy* **2017**, *2* (5), 17032.
23. Chirilă, A.; Reinhard, P.; Pianezzi, F.; Bloesch, P.; Uhl, A. R.; Fella, C.; Kranz, L.; Keller, D.; Gretener, C.; Hagendorfer, H. Potassium-Induced Surface Modification of Cu(In,Ga)Se_2 Thin Films for High-Efficiency Solar Cells. *Nat. Mater.* **2013**, *12* (12), 1107–1111.
24. Jaegermann, W.; Klein, A.; Mayer, T. Interface Engineering of Inorganic Thin-film Solar Cells–Materials Science Challenges for Advanced Physical Concepts. *Adv. Mater.* **2009**, *21* (42), 4196–4206.
25. Yoshikawa, K.; Yoshida, W.; Irie, T.; Kawasaki, H.; Konishi, K.; Ishibashi, H.; Asatani, T.; Adachi, D.; Kanematsu, M.; Uzu, H. Exceeding Conversion Efficiency of 26% by Heterojunction Interdigitated Back Contact Solar Cell with Thin Film Si Technology. *Sol. Energy Mater. Sol. Cells* **2017**, *173*, 37–42.
26. Würfel, P. Thermodynamic Limitations to Solar Energy Conversion. *Phys. E Low-dimensional Syst. Nanostructures* **2002**, *14* (1–2), 18–26.
27. Adachi, D.; Hernández, J. L.; Yamamoto, K. Impact of Carrier Recombination on Fill Factor for Large Area Heterojunction Crystalline Silicon Solar Cell with 25.1% Efficiency. *Appl. Phys. Lett.* **2015**, *107* (23), 233506.

28. Glunz, S. W.; Feldmann, F.; Richter, A.; Bivour, M.; Reichel, C.; Steinkemper, H.; Benick, J.; Hermle, M. The Irresistible Charm of a Simple Current Flow Pattern–25% with a Solar Cell Featuring a Full-Area Back Contact. In *Proceedings of the 31st European Photovoltaic Solar Energy Conference and Exhibition*; 2015; pp. 259–263.
29. Richter, A.; Hermle, M.; Glunz, S. W. Reassessment of the Limiting Efficiency for Crystalline Silicon Solar Cells. *IEEE J. Photovoltaics* **2013**, *3* (4), 1184–1191.
30. Berger, H. H. Contact Resistance and Contact Resistivity. *J. Electrochem. Soc.* **1972**, *119* (4), 507.
31. Green, M. A.; Godfrey, R. B. MIS Solar Cell—General Theory and New Experimental Results for Silicon. *Appl. Phys. Lett.* **1976**, *29* (9), 610–612.
32. Sze, S. M.; Ng, K. K. *Physics of Semiconductor Devices*; John Wiley & Sons, 2006.
33. Bivour, M. *Silicon Heterojunction Solar Cells: Analysis and Basic Understanding*; Fraunhofer Verlag, 2017.
34. Bisquert, J. *Nanostructured Energy Devices: Equilibrium Concepts and Kinetics*; CRC Press, 2014.
35. You, J.; Meng, L.; Song, T.-B.; Guo, T.-F.; Yang, Y. M.; Chang, W.-H.; Hong, Z.; Chen, H.; Zhou, H.; Chen, Q. Improved Air Stability of Perovskite Solar Cells via Solution-Processed Metal Oxide Transport Layers. *Nat. Nanotechnol.* **2016**, *11* (1), 75–81.
36. Lin, Q.; Armin, A.; Nagiri, R. C. R.; Burn, P. L.; Meredith, P. Electro-Optics of Perovskite Solar Cells. *Nat. Photonics* **2015**, *9* (2), 106–112.
37. Ke, W.; Fang, G.; Liu, Q.; Xiong, L.; Qin, P.; Tao, H.; Wang, J.; Lei, H.; Li, B.; Wan, J. Low-Temperature Solution-Processed Tin Oxide as an Alternative Electron Transporting Layer for Efficient Perovskite Solar Cells. *J. Am. Chem. Soc.* **2015**, *137* (21), 6730–6733.
38. Kim, H.-S.; Lee, C.-R.; Im, J.-H.; Lee, K.-B.; Moehl, T.; Marchioro, A.; Moon, S.-J.; Humphry-Baker, R.; Yum, J.-H.; Moser, J. E. Lead Iodide Perovskite Sensitized All-Solid-State Submicron Thin Film Mesoscopic Solar Cell with Efficiency Exceeding 9%. *Sci. Rep.* **2012**, *2* (1), 1–7.
39. Seo, J.; Uchida, R.; Kim, H.; Saygili, Y.; Luo, J.; Moore, C.; Kerrod, J.; Wagstaff, A.; Eklund, M.; McIntyre, R. Boosting the Efficiency of Perovskite Solar Cells with CsBr-Modified Mesoporous TiO_2 Beads as Electron-Selective Contact. *Adv. Funct. Mater.* **2018**, *28* (15), 1705763.
40. Yang, W. S.; Noh, J. H.; Jeon, N. J.; Kim, Y. C.; Ryu, S.; Seo, J.; Seok, S. Il. High-Performance Photovoltaic Perovskite Layers Fabricated through Intramolecular Exchange. *Science (80)* **2015**, *348* (6240), 1234–1237.
41. Son, D.-Y.; Im, J.-H.; Kim, H.-S.; Park, N.-G. 11% Efficient Perovskite Solar Cell Based on ZnONanorods: An Effective Charge Collection System. *J. Phys. Chem. C* **2014**, *118* (30), 16567–16573.
42. Lee, M. M.; Teuscher, J.; Miyasaka, T.; Murakami, T. N.; Snaith, H. J. Efficient Hybrid Solar Cells Based on Meso-Superstructured Organometal Halide Perovskites. *Science (80)* **2012**, *338* (6107), 643–647.
43. Bera, A.; Wu, K.; Sheikh, A.; Alarousu, E.; Mohammed, O. F.; Wu, T. Perovskite Oxide $SrTiO_3$ as an Efficient Electron Transporter for Hybrid Perovskite Solar Cells. *J. Phys. Chem. C* **2014**, *118* (49), 28494–28501.
44. Bi, D.; Moon, S.-J.; Häggman, L.; Boschloo, G.; Yang, L.; Johansson, E. M. J.; Nazeeruddin, M. K.; Grätzel, M.; Hagfeldt, A. Using a Two-Step Deposition Technique

to Prepare Perovskite ($CH_3NH_3PbI_3$) for Thin Film Solar Cells Based on ZrO_2 and TiO_2 Mesostructures. *Rsc Adv.* **2013**, *3* (41), 18762–18766.
45. Mei, A.; Li, 9.; Liu, L.; Ku, Z.; Liu, T.; Rong, Y.; Xu, M.; Hu, M.; Chen, J.; Yang, Y. A Hole-Conductor–Free, Fully Printable Mesoscopic Perovskite Solar Cell with High Stability. *Science (80)* **2014**, *345* (6194), 295–298.
46. Wojciechowski, K.; Saliba, M.; Leijtens, T.; Abate, A.; Snaith, H. J. Sub-150 C Processed Meso-SuperstructuredPerovskite Solar Cells with Enhanced Efficiency. *Energy Environ. Sci.* **2014**, *7* (3), 1142–1147.
47. Byranvand, M. M.; Kim, T.; Song, S.; Kang, G.; Ryu, S. U.; Park, T. P-Type CuI Islands on TiO2 Electron Transport Layer for a Highly Efficient Planar Perovskite Solar Cell with Negligible Hysteresis. *Adv. Energy Mater.* **2018**, *8* (5), 1702235.
48. Ponseca Jr, C. S.; Savenije, T. J.; Abdellah, M.; Zheng, K.; Yartsev, A.; Pascher, T.; Harlang, T.; Chabera, P.; Pullerits, T.; Stepanov, A. Organometal Halide Perovskite Solar Cell Materials Rationalized: Ultrafast Charge Generation, High and Microsecond-Long Balanced Mobilities, and Slow Recombination. *J. Am. Chem. Soc.* **2014**, *136* (14), 5189–5192.
49. Wang, J. T.-W.; Ball, J. M.; Barea, E. M.; Abate, A.; Alexander-Webber, J. A.; Huang, J.; Saliba, M.; Mora-Sero, I.; Bisquert, J.; Snaith, H. J. Low-Temperature Processed Electron Collection Layers of Graphene/TiO_2 Nanocomposites in Thin Film Perovskite Solar Cells. *Nano Lett.* **2014**, *14* (2), 724–730.
50. Ke, W.; Zhao, D.; Xiao, C.; Wang, C.; Cimaroli, A. J.; Grice, C. R.; Yang, M.; Li, Z.; Jiang, C.-S.; Al-Jassim, M. Cooperative Tin Oxide Fullerene Electron Selective Layers for High-Performance Planar Perovskite Solar Cells. *J. Mater. Chem. A* **2016**, *4* (37), 14276–14283.
51. Zhang, Q.; Dandeneau, C. S.; Zhou, X.; Cao, G. ZnO Nanostructures for Dye-sensitized Solar Cells. *Adv. Mater.* **2009**, *21* (41), 4087–4108.
52. Liu, D.; Kelly, T. L. Perovskite Solar Cells with a Planar Heterojunction Structure Prepared Using Room-Temperature Solution Processing Techniques. *Nat. Photonics* **2014**, *8* (2), 133–138.
53. Ameen, S.; Akhtar, M. S.; Seo, H.-K.; Nazeeruddin, M. K.; Shin, H.-S. An Insight into Atmospheric Plasma Jet Modified ZnO Quantum Dots Thin Film for Flexible Perovskite Solar Cell: Optoelectronic Transient and Charge Trapping Studies. *J. Phys. Chem. C* **2015**, *119* (19), 10379–10390.
54. Song, J.; Zheng, E.; Bian, J.; Wang, X.-F.; Tian, W.; Sanehira, Y.; Miyasaka, T. Low-Temperature SnO_2-Based Electron Selective Contact for Efficient and Stable Perovskite Solar Cells. *J. Mater. Chem. A* **2015**, *3* (20), 10837–10844.
55. Tseng, Z.-L.; Chiang, C.-H.; Chang, S.-H.; Wu, C.-G. Surface Engineering of ZnO Electron Transporting Layer via Al Doping for High Efficiency Planar Perovskite Solar Cells. *Nano Energy* **2016**, *28*, 311–318.
56. An, Q.; Fassl, P.; Hofstetter, Y. J.; Becker-Koch, D.; Bausch, A.; Hopkinson, P. E.; Vaynzof, Y. High Performance Planar Perovskite Solar Cells by ZnO Electron Transport Layer Engineering. *Nano Energy* **2017**, *39*, 400–408.
57. Li, W.; Da, P.; Zhang, Y.; Wang, Y.; Lin, X.; Gong, X.; Zheng, G. WO_3 Nanoflakes for Enhanced Photoelectrochemical Conversion. *ACS Nano* **2014**, *8* (11), 11770–11777.
58. Sahle, W.; Nygren, M. Electrical Conductivity and High Resolution Electron Microscopy Studies of WO_{3-x} Crystals with $0 \leq X \leq 0.28$. *J. Solid State Chem.* **1983**, *48* (2), 154–160.

59. Wang, K.; Shi, Y.; Gao, L.; Chi, R.; Shi, K.; Guo, B.; Zhao, L.; Ma, T. W(Nb)O_x-Based Efficient Flexible Perovskite Solar Cells: From Material Optimization to Working Principle. *Nano Energy* **2017**, *31*, 424–431.
60. Ameen, S.; Rub, M. A.; Kosa, S. A.; Alamry, K. A.; Akhtar, M. S.; Shin, H.; Seo, H.; Asiri, A. M.; Nazeeruddin, M. K. Perovskite Solar Cells: Influence of Hole Transporting Materials on Power Conversion Efficiency. *ChemSusChem* **2016**, *9* (1), 10–27.
61. Yu, Z.; Sun, L. Recent Progress on Hole-transporting Materials for Emerging Organometal Halide Perovskite Solar Cells. *Adv. Energy Mater.* **2015**, *5* (12), 1500213.
62. Paudel, N. R.; Compaan, A. D.; Yan, Y. Sputtered CdS/CdTe Solar Cells with MoO3−x/Au Back Contacts. *Sol. Energy Mater. Sol. Cells* **2013**, *113*, 26–30.
63. Shen, K.; Yang, R.; Wang, D.; Jeng, M.; Chaudhary, S.; Ho, K.; Wang, D. Stable CdTe Solar Cell with V2O5 as a Back Contact Buffer Layer. *Sol. Energy Mater. Sol. Cells* **2016**, *144*, 500–508.
64. Paudel, N. R.; Xiao, C.; Yan, Y. CdS/CdTe Thin-film Solar Cells with Cu-free Transition Metal Oxide/Au Back Contacts. *Prog. Photovoltaics Res. Appl.* **2015**, *23* (4), 437–442.
65. Galiana, B.; Rey-Stolle, I.; Baudrit, M.; García, I.; Algora, C. A Comparative Study of BSF Layers for GaAs-Based Single-Junction or Multijunction Concentrator Solar Cells. *Semicond. Sci. Technol.* **2006**, *21* (10), 1387.
66. Chu, C.-W.; Li, S.-H.; Chen, C.-W.; Shrotriya, V.; Yang, Y. High-Performance Organic Thin-Film Transistors with Metal Oxide/Metal Bilayer Electrode. *Appl. Phys. Lett.* **2005**, *87* (19), 193508.
67. Fortunato, E.; Barquinha, P.; Martins, R. Oxide Semiconductor Thin-film Transistors: A Review of Recent Advances. *Adv. Mater.* **2012**, *24* (22), 2945–2986.
68. Brendel, R.; Peibst, R. Contact Selectivity and Efficiency in Crystalline Silicon Photovoltaics. *IEEE J. Photovoltaics* **2016**, *6* (6), 1413–1420.
69. Roe, E. T.; Egelhofer, K. E.; Lonergan, M. C. Limits of Contact Selectivity/Recombination on the Open-Circuit Voltage of a Photovoltaic. *ACS Appl. Energy Mater.* **2018**, *1* (3), 1037–1046.
70. Green, M. A. Accuracy of Analytical Expressions for Solar Cell Fill Factors. *Sol. Cells* **1982**, *7* (3), 337–340.
71. Schöfthaler, M.; Rau, U.; Werner, J. H. Direct Observation of a Scaling Effect on Effective Minority Carrier Lifetimes. *J. Appl. Phys.* **1994**, *76* (7), 4168–4172.
72. Wong, J.; Green, M. A. From Junction to Terminal: Extended Reciprocity Relations in Solar Cell Operation. *Phys. Rev. B* **2012**, *85* (23), 235205.
73. Rau, U. Reciprocity Relation between Photovoltaic Quantum Efficiency and Electroluminescent Emission of Solar Cells. *Phys. Rev. B* **2007**, *76* (8), 85303.
74. Rau, U.; Huhn, V.; Stoicescu, L.; Schneemann, M.; Augarten, Y.; Gerber, A.; Pieters, B. E. Photocurrent Collection Efficiency Mapping of a Silicon Solar Cell by a Differential Luminescence Imaging Technique. *Appl. Phys. Lett.* **2014**, *105* (16), 163507.
75. Delamarre, A.; Lombez, L.; Watanabe, K.; Sugiyama, M.; Nakano, Y.; Guillemoles, J.-F. Experimental Demonstration of Optically Determined Solar Cell Current Transport Efficiency Map. *IEEE J. Photovoltaics* **2016**, *6* (2), 528–531.
76. Huhn, V.; Pieters, B. E.; Augarten, Y.; Gerber, A.; Hinken, D.; Rau, U. Imaging Photocurrent Collection Losses in Solar Cells. *Appl. Phys. Lett.* **2016**, *109* (22), 223502.

77. Qin, P.; Tanaka, S.; Ito, S.; Tetreault, N.; Manabe, K.; Nishino, H.; Nazeeruddin, M. K.; Grätzel, M. Inorganic Hole Conductor-Based Lead Halide Perovskite Solar Cells with 12.4% Conversion Efficiency. *Nat. Commun.* **2014**, *5* (1), 1–6.
78. Christians, J. A.; Fung, R. C. M.; Kamat, P. V. An Inorganic Hole Conductor for Organo-Lead Halide Perovskite Solar Cells. Improved Hole Conductivity with Copper Iodide. *J. Am. Chem. Soc.* **2014**, *136* (2), 758–764.
79. Xu, X.; Liu, Z.; Zuo, Z.; Zhang, M.; Zhao, Z.; Shen, Y.; Zhou, H.; Chen, Q.; Yang, Y.; Wang, M. Hole Selective NiO Contact for Efficient Perovskite Solar Cells with Carbon Electrode. *Nano Lett.* **2015**, *15* (4), 2402–2408.
80. Jung, J. W.; Chueh, C.; Jen, A. K. A Low-temperature, Solution-processable, Cu-doped Nickel Oxide Hole-transporting Layer via the Combustion Method for High-performance Thin-film Perovskite Solar Cells. *Adv. Mater.* **2015**, *27* (47), 7874–7880.
81. Etgar, L.; Gao, P.; Xue, Z.; Peng, Q.; Chandiran, A. K.; Liu, B.; Nazeeruddin, M. K.; Grätzel, M. Mesoscopic $CH_3NH_3PbI_3/TiO_2$ Heterojunction Solar Cells. *J. Am. Chem. Soc.* **2012**, *134* (42), 17396–17399.
82. Laban, W. A.; Etgar, L. Depleted Hole Conductor-Free Lead Halide Iodide Heterojunction Solar Cells. *Energy Environ. Sci.* **2013**, *6* (11), 3249–3253.
83. Shi, J.; Dong, J.; Lv, S.; Xu, Y.; Zhu, L.; Xiao, J.; Xu, X.; Wu, H.; Li, D.; Luo, Y. Hole-Conductor-Free Perovskite Organic Lead Iodide Heterojunction Thin-Film Solar Cells: High Efficiency and Junction Property. *Appl. Phys. Lett.* **2014**, *104* (6), 63901.
84. Ke, W.; Fang, G.; Wan, J.; Tao, H.; Liu, Q.; Xiong, L.; Qin, P.; Wang, J.; Lei, H.; Yang, G. Efficient Hole-Blocking Layer-Free Planar Halide Perovskite Thin-Film Solar Cells. *Nat. Commun.* **2015**, *6* (1), 1–7.
85. Geissbühler, J.; Werner, J.; Martin de Nicolas, S.; Barraud, L.; Hessler-Wyser, A.; Despeisse, M.; Nicolay, S.; Tomasi, A.; Niesen, B.; De Wolf, S. 22.5% Efficient Silicon Heterojunction Solar Cell with Molybdenum Oxide Hole Collector. *Appl. Phys. Lett.* **2015**, *107* (8), 81601.
86. Gerling, L. G.; Masmitja, G.; Voz, C.; Ortega, P.; Puigdollers, J.; Alcubilla, R. Back Junction N-Type Silicon Heterojunction Solar Cells with V_2O_5 Hole-Selective Contact. *Energy Procedia* **2016**, *92*, 633–637.
87. Gerling, L. G.; Mahato, S.; Morales-Vilches, A.; Masmitja, G.; Ortega, P.; Voz, C.; Alcubilla, R.; Puigdollers, J. Transition Metal Oxides as Hole-Selective Contacts in Silicon Heterojunctions Solar Cells. *Sol. Energy Mater. Sol. Cells* **2016**, *145*, 109–115.
88. Islam, R.; Shine, G.; Saraswat, K. C. Schottky Barrier Height Reduction for Holes by Fermi Level Depinning Using Metal/Nickel Oxide/Silicon Contacts. *Appl. Phys. Lett.* **2014**, *105* (18), 182103.
89. Bivour, M.; Temmler, J.; Steinkemper, H.; Hermle, M. Molybdenum and Tungsten Oxide: High Work Function Wide Band Gap Contact Materials for Hole Selective Contacts of Silicon Solar Cells. *Sol. Energy Mater. Sol. Cells* **2015**, *142*, 34–41.
90. Schmidt, J.; Zielke, D.; Gogolin, R.; Sauer, R.; Lövenich, W. Recent Advances in Polymer/Silicon Heterojunction Solar Cells. In *32nd European Photovoltaic Solar Energy Conf. and Exhibition*; 2016; pp. 85–88.
91. Wan, Y.; Samundsett, C.; Bullock, J.; Allen, T.; Hettick, M.; Yan, D.; Zheng, P.; Zhang, X.; Cui, J.; McKeon, J. Magnesium Fluoride Electron-Selective Contacts for Crystalline Silicon Solar Cells. *ACS Appl. Mater. Interfaces* **2016**, *8* (23), 14671–14677.

92. Zhang, Y.; Cui, W.; Zhu, Y.; Zu, F.; Liao, L.; Lee, S.-T.; Sun, B. High Efficiency Hybrid PEDOT: PSS/Nanostructured Silicon Schottky Junction Solar Cells by Doping-Free Rear Contact. *Energy Environ. Sci.* **2015**, *8* (1), 297–302.
93. Zhang, Y.; Liu, R.; Lee, S.-T.; Sun, B. The Role of a LiF Layer on the Performance of Poly (3, 4-Ethylenedioxythiophene): Poly (Styrenesulfonate)/Si Organic-Inorganic Hybrid Solar Cells. *Appl. Phys. Lett.* **2014**, *104* (8), 83514.
94. Avasthi, S.; McClain, W. E.; Man, G.; Kahn, A.; Schwartz, J.; Sturm, J. C. Hole-Blocking Titanium-Oxide/Silicon Heterojunction and Its Application to Photovoltaics. *Appl. Phys. Lett.* **2013**, *102* (20), 203901.
95. Yang, X.; Zheng, P.; Bi, Q.; Weber, K. Silicon Heterojunction Solar Cells with Electron Selective TiO_x Contact. *Sol. Energy Mater. Sol. Cells* **2016**, *150*, 32–38.
96. Greiner, M. T.; Lu, Z.-H. Thin-Film Metal Oxides in Organic Semiconductor Devices: Their Electronic Structures, Work Functions and Interfaces. *NPG Asia Mater* **2013**, *5* (7), e55–e55.
97. Duan, L.; Xie, K.; Qiu, Y. Progress on Efficient Cathodes for Organic Light-emitting Diodes. *J. Soc. Inf. Disp.* **2011**, *19* (6), 453–461.
98. Li, X.; Xiao, D.; Wu, L.; Wang, D.; Wang, G.; Wang, D. CdTe Thin Film Solar Cells with Copper Iodide as a Back Contact Buffer Layer. *Sol. Energy* **2019**, *185*, 324–332.

CHAPTER 10

Memristor: The Missing Fourth Circuit Component

SUDAKSHINA KUNDU[1*] and RAJESH DUTTA[2]

[1]*Department of Computer Science and Engineering, Maulana Abul Kalam Azad University of Technology, West Bengal, India*

[2]*Electronics and Communication Engineering, Maulana Abul Kalam Azad University of Technology, West Bengal, India*

[*]*Corresponding author. E-mail: skundu_fin@yahoo.co.in*

ABSTRACT

We are familiar with the two-terminal electrical components—the resistors, the capacitors and the inductors. According to Leon Chua there should be another relationship that includes non-linear systems and relates the change in electric charge moving through a circuit to the change of flux surrounding the circuit. This fourth non-linear circuit element is called Memristor—a memory-resistor as it promised to store information as a memory element. Memresistance is an intrinsic property of any electrical circuit observable only at scaled down dimensions. The first memristor device was realized in the HP labs in 2008. It can be designed from basic passive and active electronic circuit components to realize the electrical characteristics required. There are different models which have been used to account for the memristive behavior of this physical device. The memristors have great potential in the design of integrated circuits by addressing several important aspects of the VLSI chip designs and computational areas as Non-Volatile-Random-Access-Memory (NVRAM), associative memories, low-power memory as well as distributed state storage, dynamically adjustable load, programmable resistance, electronically adjustable filters and amplifiers. Memristors have great potential in image recognition and image processing applications and have been used in a switched-on radio frequency antenna.

10.1 INTRODUCTION

We are familiar with the two terminal electrical components–the resistors, the capacitors and the inductors. The four fundamental electromagnetic quantities, viz., the charge **q**, the electrical current **i**, the voltage **V**, and the magnetic flux **ϕ** are pair-wise related to one another by equations through the linear electrical circuit components of resistor, capacitor, and inductor through their properties of resistance (**R**), capacitance (**C**), and inductance (**L**) respectively. All of these relations are valid for the linear systems.

To elaborate on their interdependence, the relationships are as follows:

- The resistor (R) relates the current and the voltage as $\partial\mathbf{V} = R\partial\mathbf{i}$ that states that any change in current $\partial\mathbf{i}$ through an electrical circuit produces a corresponding change in voltage $\partial\mathbf{V}$ across the terminals. This is Ohm's Law. The constant of proportionality is called the resistance **R**. Alternately, any change in the voltage across any two terminals of an electrical circuit leads to a direct change in the current flowing through these terminals, that is, $\partial\mathbf{i} = (1/R)\partial\mathbf{V} = \mathbf{G}\partial\mathbf{V}$, where G = 1/R is the inverse of resistance. It is called conductance.
- The capacitor (C) relates the charge and the voltage as $\partial\mathbf{q} = C\partial V$; any change in voltage $\partial\mathbf{V}$ across a capacitor causes a corresponding change in the stored charge $\partial\mathbf{q}$. The constant of proportionality measures the capacitance **C** of the capacitor.
- The inductor (L) relates the magnetic flux and the current according to the equation $\partial\boldsymbol{\phi} = L\partial\mathbf{i}$ which implies that any change in magnetic flux $\partial\boldsymbol{\phi}$ across an electrical circuit leads to a change in the current $\partial\mathbf{i}$ through the circuit. The proportionality constant is the inductance L. There are two other relations that are governed by the basic Laws of Physics for Electricity and Magnetism that relate the time rate of change of one electromagnetic quantity to another.
- The rate of change of charge across a cross-section of an electrical circuit is equal to the current flowing through it [$\partial\mathbf{q} = i\partial t$].
- It is further noted that if the flux through any closed conducting loop changes it induces a voltage. This rate of change of flux $\partial\boldsymbol{\phi}$ with

time is equal to the induced voltage [mathematically $V = +\partial\boldsymbol{\phi}/\partial t$ is Lenz's Law; or $\partial\boldsymbol{\phi} = V\partial t$].

The mathematical relationship between the change in magnetic flux through a closed electrical circuit and the consequent work done by the induced electric field on the charges circulating is given by Faraday's Law. This is represented as one of the Maxwell's equations. If the change in the magnetic flux across the area of the loop is nonzero [$\partial\boldsymbol{\phi}/\partial t \neq 0$], then an electric field **E** circulates. If the loop is an electrical conductor then electrical charge **q** circulate around the loop and the actual work done on the charges by the induced electric field is $\int \mathbf{E}.\mathbf{dr} = V$. This work done per unit charge is called the electromotive force (EMF) and is expressed in Volts. Faraday's Law of electromagnetic induction gives the magnitude of the induced EMF while Lenz's law gives its direction of the induced current.

All these relationships are ***linear*** where the cause of the change and its effect are directly proportional and are related through a constant of proportionality that does not change with any change in the cause or the effect.

However, there is no pairing between the flux and the charge. According to Leon Chua, there should be another relationship, not restricted to linear systems, for relating the flux to the charge. The fourth relation is $d\phi = Mdq$ and it relates the change in electric charge dq moving through a circuit to the change of flux $d\phi$ surrounding the circuit (Figure 10.1).

This pairing was so far missing from the other pairing relations between the electromagnetic quantities till Leon Chua proposed the fourth circuit element in 1971 in his seminal paper[1] that related the charge moving through an electrical circuit to the magnetic flux that was surrounded by the circuit. This passive circuit element related the time integrals of the current and voltage across its two terminals and the resistance of the element depended on the charge flowing through the circuit. The resistance increased when current flowed in one direction and decreased if the flow of current was reversed. However, the value never reached zero but retained the value at the time of switching off the current. This gave this element the ability to retain some information, that is, a memory. Leon Chua named this fourth non-linear circuit element Memristor–a memory-resistor as it promised to store information as a memory element. According to Chua Memristor, this hypothetical device would link flux to charge in the same way as a nonlinear resistance would connect voltage to current. Chua was a great proponent of nonlinear electronics and believed in its great potential in various applications.

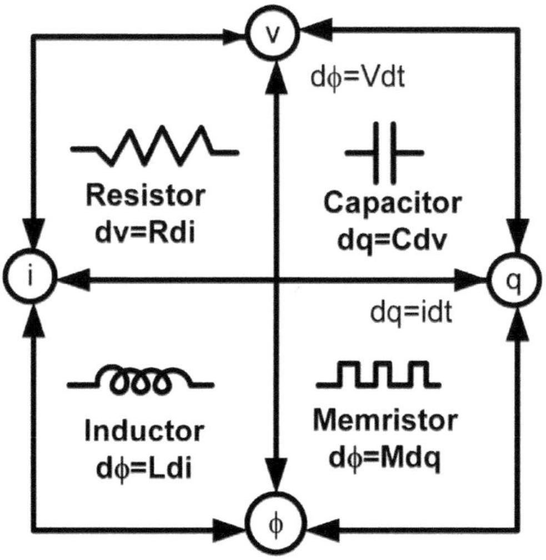

FIGURE 10.1 Relation between the electrical quantities of voltage V, charge q, flux ϕ, and current i with the four fundamental electronic components resistance R, capacitance C, inductance L, and memristance M.

The relationship between the change in flux surrounding a circuit $\partial\phi$ to the change of charge ∂q flowing through as proposed by Chua are related by the nonlinear circuit element Memristor M as $\partial\phi = M\partial q$. Initially, it has been a purely mathematical construct. This is the missing relationship between the pair-wise coupling of the four electrical quantities, the voltage V, the charge q, the magnetic flux ϕ, and the current i. Resistance couples voltage to current, capacitance relates charge to voltage, and inductance links magnetic flux to current. The memristor completes the picture by establishing a relationship between magnetic flux and charge.

The memresistance is an intrinsic property of any electrical circuit. But it is observable only at scaled down dimensions. So this property could not be detected in the past centuries when the electrical circuits were of macrodimensions. With the gradual scaling down of device size to the micro- and nanometer ranges over the past decades, the possibility of detecting memristor type behavior in nonlinear circuit increases.

The fundamental variables in electronic theories are flux and charge and not voltage and charge (Figure 10.2).

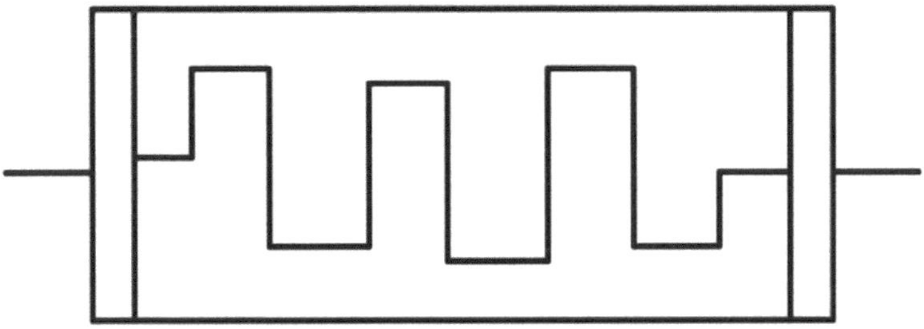

FIGURE 10.2 Symbol for memristor proposed by Chua.

10.2 HISTORICAL DEVELOPMENT

In 1967, Simmons and Verderber[2] observed hysteretic resistance switching effects by injecting gold ions in 20–300 nm thin film of silicon oxide. The observation was published as an article in the Proceeding of the Royal Society of London where the reason for the hysteresis was ascribed to electron trapping effects.

Leon Chua postulated a new two terminal circuit element in 1971 establishing the relationship between magnetic flux linkage and electrical charge.[3] This was declared as the fourth electrical circuit component memristor, in addition to the already known linear passive circuit components of resistance, capacitance, and inductance. This fourth circuit component was a nonlinear one and a purely mathematical abstraction at the time. By 1976, Chua and Sung Mo Kang, generalized the theory of memristor and memristive system in their publication.

The prediction of the fourth element by Leon Chua in 1971, following the symmetry arguments was simplistic and straightforward. However, the actual realization of the real memristor came much later as late as 2008.[4] There has been much research on resistance switching and circuits that partly realized the memristive behavior.

By 1986, a two-terminal reconfigurable resistance-switching array is patented by Johnson and Ovshinsky which is based on phase changing materials.[5] Although this array is distinct from basic memristive behavior, but some of its elements like the crossbar architecture and two terminal resistance switch are later made use of in designing the memristor.

An electrically reprogrammable variable resistance device made of tungsten oxide is proposed by Thakoor, Moopenn, Daud, and Thakoor in

1990. Although it is referred to as a memristor, it is not clear whether this is an implementation of the memristor proposed by Chua in 1971. There is no mention of the paper by Chua in this publication.[6]

In 1993, two-terminal organic thin film resistance switches forming artificial neural functions are patented by Nichogi, Taomoto, Asakawa, and Yoshida of the Matsushita Research Institute.[7] These appear to possess properties similar to those suggested by Chua for his memristor. However, there is no specific mention of the element memristor.

Buot and Rajagopal demonstrated a memristor like storage of binary information due to the bow-tie current–voltage characteristics of AlAs/GaAs/AlAs quantum-well diodes with specially doped spacer layers.[8] This publication in 1994 too appears to have no bearing to the nonlinear memory element proposed by Chua.

The year 1998 has seen a number of research publications that have proposed devices that have similar behavior as the memristor of Chua. These include the programmable metallization cell of Kozicki and West,[9] the ionic programmable resistance device by Swaroop, West, Martinez, Kozicki, and Akers,[10] the defect tolerant computer architecture of Heath, Kuekes, Snider, and Williams of Hewlett-Packard Laboratories. Although none claim any connection to the memristor but all have similar functionalities.

Beck, Bednorz, Gerber, Rossel, and Widmer working in the Zurich Research Laboratory of IBM report reproducible resistance switching effects by using thin oxide films[11] in 2000. In the same year, Kuekes, Williams, and Heath of Hewlett-Packard Laboratories have patented a two-terminal nonlinear resistance switch made of nanoscale crossbars using rotaxane molecular structures.[12] The bearing of these devices to the memristor proposed by Chua has not yet been established.

The importance of oxide bilayers in achieving high-to-low resistance ratios is shown by Liu, Wu, Chen, and Ignatiev of the University of Houston that can be used as non-volatile memory elements in a conference held in San Diego in 2001. They demonstrate the zero-crossing Lissajous curves suggested by Chua and Kang without relating this behavior to the memristor. The underlying mechanism has not been explained.[13]

A reversible two-terminal metal oxide-based resistance switching device is designed by Rinerson, Chevallier, Longcor, Kinney, Ward, and Hsia in 2005.[14] Lan, Bill, and VanBuskirk receive a US patent in the same year for their resistance switching memory cells by combining layers of organic material and metal oxides/sulfides. They all have similar I–V characteristics as that of the memristor but the new element is not mentioned in the documents.[15]

The original paper of Chua is cited for the first time by Ovshinsky while applying for the patent for a neural synaptic system. This system is based on phase change material that acts as a two-terminal resistance switch. He receives the US patent in 2006.[16]

A number of US patents are granted in 2007 for devices and systems that have memristic behavior. These include the programming of ionic complex resistance switches for acting as fuse or antifuse by Bulovic, Mandell, and Perlman,[17] memristor like two-terminal resistance switches for reconfigurable circuit architectures by Snider[18] and two-terminal resistance switching materials of Mouttet for various applications in signal processing, control systems, communication circuits, and pattern recognition.[19] Snider emphasizes the importance of memristive devices for pattern recognition and reconfigurable circuits.[20]

The year 2008 saw a flood of research activities related to memristors.[22–30] These centered around the nanoscale two-terminal cross-bar arrays for neural networks, the link between Chua's memristor and the two-terminal resistive switching behavior of nanoscale systems, use of two-terminal resistance switching materials in analogue computational architectures, polymeric and titanium dioxide memristors, the behavior of metal oxide–metal nanodevices, spin-memristive systems, etc. Many of these were awarded US patents and several presented in journals and conferences. The two-terminal hysteresis resistance patented was useful in image processing and pattern recognition. However, there was no direct correlation of the new devices and their applications to the memristor proposed by Leon Chua, although they possessed similar characteristics.

The first memristor device was realized in the HP labs along the same principles as suggested by Leon Chua in 2008. It was the first device of its kind that was grown on a simple material or a physical structure that had all the properties of a memristor predicted by Leon Chua. It was made of a simple thin two-layer TiO_2 sandwiched between platinum electrodes. This device owes its performance to the physical characteristics of TiO_2.[4]

Since then many research workers are working in different laboratories to design and develop memristive devices and their applications in various circuits of interest.

10.3 MEMRISTOR PROPERTIES

Leon Chua originally connected the flux linkage $\phi(t)$ to the amount of electrical charge q(t) that has flown through the circuit by the nonlinear functional relationship

$$f\,[\phi(t), q(t)] = 0, \tag{10.1}$$

where both the flux $\phi(t)$ and the charge q(t) are time-varying functions; the flux $\phi(t)$ being the time integral of voltage and charge q(t) is that of the electric current.

$$\phi(t) = {}_0\!\int^t V(\tau)\, d\tau \tag{10.2}$$

and

$$q(t) = {}_0\!\int^t i(\tau)\, d\tau \tag{10.3}$$

The variable flux linkage is a generalized term, not necessarily the magnetic flux alone. It is a generalization from the circuit characteristics of the inductor.

The memristor is therefore defined mathematically as the rate of change of flux linkage with charge, where the flux linkage itself varies with the charge.

$$M = d\phi(q)/dq \tag{10.4}$$

This can alternatively be written as

$$M[q(t)] = [d\phi(q)/dt]/[dq/dt] = R_M(q) \tag{10.5}$$

where $R_M(q)$ is the small signal memresistance defined at an operating point. This memresistance is a charge-controlled or current-controlled memristor.

If the order is reversed, it becomes a flux-controlled or voltage-controlled memristor as shown below:

$$M[\phi(t)] = [dq/dt]/[d\phi(q)/dt] = G_M(\phi) \tag{10.6}$$

The above relationships make it possible to design the memristor.

When the resistance of the classical resistor is made to vary according to the time-domain integral of the current flowing through it, it may be made to behave as a current- or charge- controlled memristor.

Similarly, when the conductance of a classical conductor is made to vary according to the time-domain integral of the terminal voltage, then it behaves as a flux- or voltage-controlled memristor.

The current–voltage characteristic of a memristor is like a Lissajous figure, that is, it has a pinched hysteresis loop. This is shown in Figure 10.3.

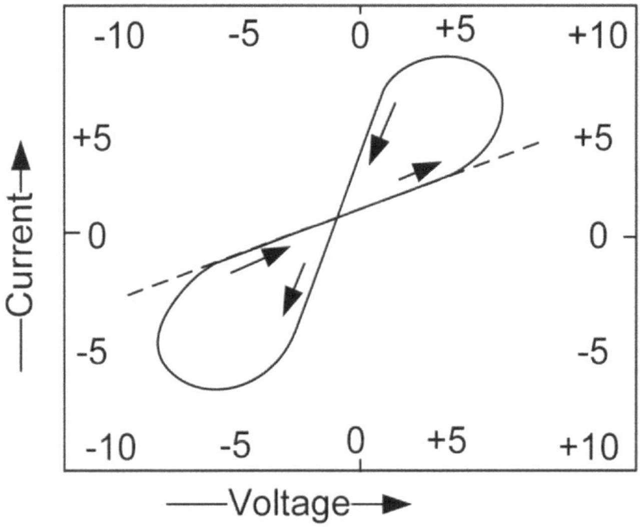

FIGURE 10.3 Lissajous figure–Memristic characteristic; x-axis represents voltage (in Volts) and y-axis represents current (in Amperes) in appropriate scales.

The pair-wise relationship between the four electromagnetic variables current (i), charge (q), flux (ϕ), and voltage (V) and their corresponding constants of proportionalities, which represent the different passive circuit components are shown in Table 10.1.

TABLE 10.1 Different Passive Circuit Components.

Device	Unit	Relation	Nature
Resistor	Ohm	$R = dV/di$	Linear
Capacitor	Farad	$C = dq/dV$	Linear
Inductor	Henry	$L = d\phi/di$	Linear
Memristor	Ohm	$M = d\phi/dq$	Nonlinear

The current-controlled resistor or the voltage-controlled conductor can be realized by designing a circuit consisting of an electronically controlled resistor and an integrator.

Therefore, the equation connecting the voltage and current through the memristor can be written as

$$V(t) = M[q(t)].i(t) \tag{10.7}$$

When the current i(t) is zero, the voltage is also zero, and M(t) is constant. This is the memory effect. The power consumed by the device resembles the square-law characteristic of a constant resistor,

$$P(t) = V(t).\ i(t) = i(t)^2.\ M[q(t)] \tag{10.8}$$

When an a.c. current flows through the circuit, there is very little change in the memristor value and the latter appears as a simple resistor. For very rapid variations in the value M[q(t)], the current will stop quickly and so will the power consumption.

The memristor can be designed from basic electronic circuit components, both passive and active, to realize the electrical characteristics required. The first circuit made up of passive and active electronic components to behave as a memristor was suggested by Leon Chua.[1] This circuit is shown in Figure 10.4.

10.4 THE FIRST MEMRISTOR DESIGNED

The memristor of the HP Laboratories[4] has been grown on a two-layer thin film of TiO_2, which is sandwiched between two platinum electrodes. The TiO_2 layer is 5 nm thick with an oxygen deficient n-type semiconducting layer interfaced with an un-doped dielectric layer of much higher resistance. The oxygen vacancies behave as a +2 positively charged carriers (Figure 10.5).

To understand the behavior of the device, we consider the two important properties of TiO_2 that have been utilized in the design of the device:

1. The conductivity of TiO_2 is sensitive to oxygen depletion. The material is usually an insulating material. But when oxygen vacancies are created, it starts behaving as a semiconductor. It is then represented as $TiO_{2-10.}$
2. The oxygen vacancies have a tendency to migrate within the material by drifting when an electric field is applied. This is called anion migration.

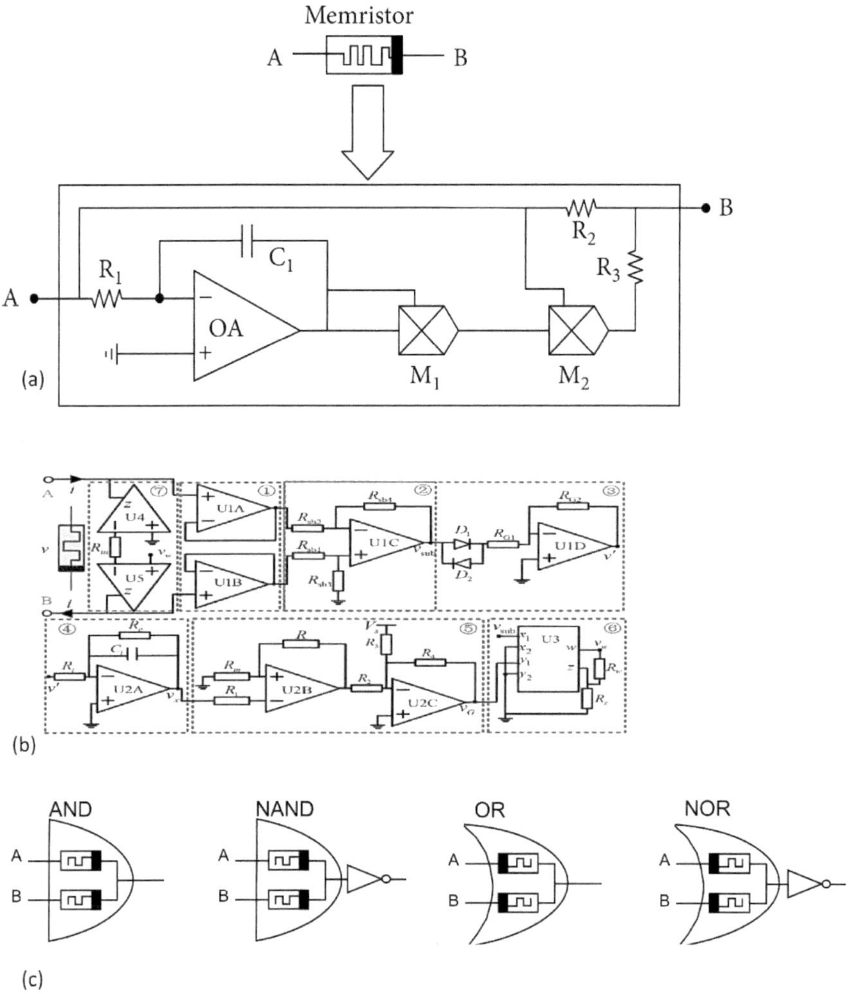

FIGURE 10.4 The memristor circuit.

One of the two layers is doped with oxygen vacancies so that it behaves as an n-type semiconductor while the other layer is kept un-doped to serve as an insulator. The total length of the device between the two terminals is D and the doped layer of length W behaves as the semiconductor while the rest of the un-doped region of length (D-W) behaves as an insulator of high resistance.

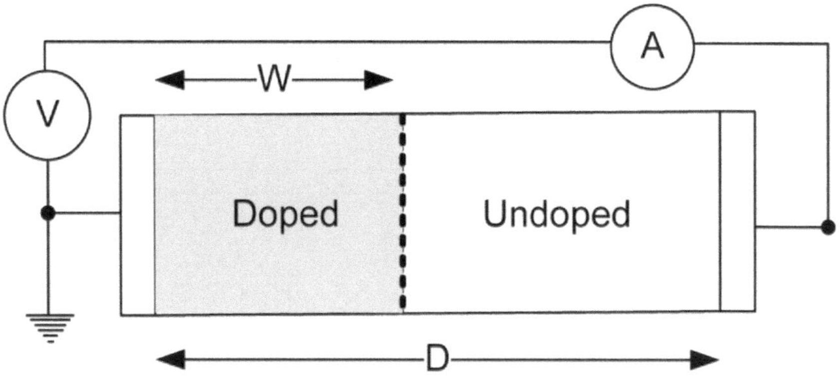

FIGURE 10.5 Memristor grown in the HP Labs for the first time.
Source: Courtesy Reference 4.

When an electric current passes between the terminals of the device, it causes the oxygen vacancies to migrate and the length of the doped layer increases so that the total resistance of the device decreases. When the entire width D is filled up with the oxygen vacancies, that is, W = D, the resistance of the device is low and the device is turned on. This resistance is denoted by R_{ON}. It is the resistance of the device if the entire length is doped and this resistance is of the order of 1 kΩ.[31]

If an opposite current flows between the terminals, the oxygen vacancies are swept away and the length of the doped region decreases. When all the oxygen vacancies are swept away, the entire un-doped layer of width D has a very large resistance R_{OFF} and the device turns off. R_{OFF} is the resistance of the device if the entire TiO_2 is un-doped.

When W(t) ≠ D, and there are two distinct regions in the TiO_2 layer, then the effective resistance of the device is[32]

$$R_M(t) = V(t) / i(t) = \{R_{ON}[W(t)/D] + R_{OFF}[1-W(t)/D]\} \quad (10.9)$$

This is a general equation and is valid even if W(t) = 0 or W(t) = D.

For TiO_2, $R_{OFF}/R_{ON} \approx 10^2 >> 1$. When $R_{ON} << R_{OFF}$, then $\Delta R = (R_{OFF} - R_{ON}) \approx R_{OFF}$.

The current i(t) due to a voltage V(t) are related by $V(t) = R_M(W).\ i(t)$ according to Kirchoff's law. In this case, W is a function of the current that causes a change in width W(t) and hence W is also a function of the time. In the constant linear-drift model, the boundary between the doped and the un-doped regions shift making the time rate of change W(t) a nonzero quantity and defined by the relation

$$dW(t)/dt = v_d = [\eta\mu_D R_{ON}/D]\ i(t) \quad (10.10)$$

where η is the memristor polarity (η may be +1 or -1, depending on whether there is expansion or contraction of the doped region), μ_D is the drift mobility of the vacancies and the current 9(t) corresponds to the uniform electric field $E = [R_{ON}/D]\ i(t)$.

So the actual width of the doped region is

$$W(t) = W_0 + [\eta\mu_D R_{ON}/D]\ q(t) \quad (10.11)$$

where W_0 is the initial value of the width of the doped region in the absence of any migration of the vacancies. The width of the doped region changes linearly with the charge passing through it.

The total amount of charge that is required to pass through the device in order to spread the doped region over the entire length D of the device is Q_0

$$Q_0 = D^2/(\mu_D R_{ON}) \quad (10.12)$$

It is the amount of charge required to extend the doped region by an amount $(D\text{-}W_0)$ and take up the entire device length. Using eqs 10.9, 10.11 and 10.12, the memristor is given by the expression

$$M(q) = R_M(t) - (\eta.q.\Delta R)/Q_0 \quad (10.13)$$

where $R_M(t)$ is given by eq 10.9. R_0 is found from eq 10.9 by putting $W = W_0$.

For an ideal circuit consisting of a memristor and a voltage source, the current–voltage equation is

$$[R_M(t) - (\eta.q(t).\Delta R)/Q_0]dq/dt = d[R_M(t)q(t) - \tfrac{1}{2}(\eta.q(t)^2.\Delta R)/Q_0]\ /dt = V(t) \quad (10.14)$$

The solution to this equation yields the values of charge and current as

$$q(t) = (Q_0.R_0/\Delta R)[\ 1 - \sqrt{\{1 - 2\eta\Delta R\ \phi(t)/(Q_0.R_0^2)\ \}}] \quad (10.15)$$

$$i(t) = V(t)/[R_0\sqrt{\{1 - 2.\eta.\Delta R\ \phi(t)/(Q_0.R_0^2)\}}] = V(t)/M(q(t)) \quad (10.16)$$

Here $\phi(t) = \int_0^t V(\tau)d\tau$ is the magnetic flux associated with the voltage.

The charge is shown to be a function of the magnetic flux in eq 10.15 in conformity with the fourth symmetry relation $d\phi = M(q)\times dq$. Equation 10.15 is based on the linear-drift model. Therefore, it will be valid only when the charge flowing through the memristor q(t) is less than two extrema - $q_{max}(t) < Q_0(1 - W_0/D)$ for $\eta = +1$ and for $q_{max}(t) < QW_0/D)$ for $\eta = -1$.

Equation 10.16 ensures that the memristor does not introduce any phase change between the current and voltage. The current i = 0 only when the voltage V = 0. However, the memristor is a purely dissipative element. It also follows from the eq 10.16 that current is a multivalued function of the voltage V. This is due to the fact that when the voltage is alternating, that is, $V(t) = V_0.\sin(\omega t)$, then the flux linkage is $\phi(t) = V_0.[1 - \cos(\omega t)]/\omega$, but when $V(\pi/\omega - t) = V(t)$, we have $\phi(\pi/\omega - t) \neq \phi(t)$. So the current i(t) will be a multivalued function of the voltage V(t). This indicates the presence of hysteresis.

It is to be further noted that the flux linkage ϕ is inversely related to the frequency or $\phi \propto 1/\omega$. Memristive behavior becomes prominent at low frequencies, $\omega \leq \omega_0$ where $\omega_0 = 2\pi/t_0$ and the constant $t_0 = D^2/\mu_D V_0$ is the time needed by the vacancies to move over a distance D under a constant voltage V_0. For a typical memristive device discussed here, $t_0 \approx 0.1$ m sec and $\omega_0 \approx 50$ k Hz.

10.5 TOWARD REALIZING A MEMRISTOR

Let us consider the memristor designed from a thin two-layer TiO_2 placed between two platinum electrodes, the first ever realization of Leon Chua's memristor as a simple physical device.[4]

The resistance of the device is R_{ON} when the entire width D is filled up with the oxygen vacancies. Here, W(t) = D and $R_M(t) = R_{ON}$.

When W(t) = 0, the device is off with a very large resistance $R_M(t) = R_{OFF}$.

The edge of the doped region shifts depending on the direction of flow when a current passes through the device. When $W(t) \neq 0$, or $W(t) \neq D$, the effective resistance of the TiO_2 layer for any intermediate state is given by eq 10.9 which is valid even if W(t) = 0 or W(t) = D.

When $R_{OFF} >> R_{ON}$, the memresistance is given by

$$R_M(t) = R_{OFF}\,[1 - [(\mu_D R_{ON})/D^2].q(t)]. \qquad (10.17)$$

This can be implemented in P-Spice.

There are different models which have been used to account for the memristive behavior of this physical device. The Linear Boundary Model assumes that TiO_2 has a high resistive un-doped layer and a low resistive layer doped with oxygen vacancies. The positively charged oxygen vacancies can act as donors of electrons and make the layer more conductive.

The boundary between the two layers drifts with an applied voltage depending on the direction of the bias. The overall resistance of the device that defines its memresistive state changes accordingly. However, this model does not take into consideration of the fact that the drift speed of the oxygen vacancies becomes zero at the two ends of the device.

This has been addressed by others[33, 34] by using a 'window function' so that the speed of the vacancies becomes zero at the terminals.

Further, the change of the conductive state of the memristor is also dependent on the change in the nature of the metal–semiconductor junction. The change of the overall resistance of the TiO_2 layer and the change in the nature of the metal–semiconductor contacts are considered by two different theories. One of the two platinum electrodes at the terminals acts as an Ohmic contact since there is always a huge population of oxygen vacancies in its neighborhood. Therefore, the current is controlled by the other metal–semiconductor junction that is a Schottky contact. It acts as a rectifying contact but can also be turned into an Ohmic one by the application of a voltage of proper polarity that crowds that junction with more oxygen vacancies.

The current–voltage relationship has two terms, one is responsible for a tunnel dominant conduction with the current component given by $i_T(t) = w^n\beta \sinh [\alpha V(t)]$, where β and n are fitting parameters and w is a normalized state variable having values between 0 and 1 and the other accounts for the rectifying conduction of the device. The second term is $i_D(t) = \chi(\exp[\gamma V(t) - 1)$, where γ and χ are fitting parameters.

The current therefore becomes

$$I(t) = i_T(t) + i_D(t) = w^n\beta \sinh [\alpha V(t)] + \chi[\exp\{\gamma V(t)\} - 1]. \quad (10.18)$$

Although the current–voltage relation is simple and is capable of reproducing the memristor characteristic curves, yet, it does not involve any device physics to explain the details of working.

Lehtonen et al. have improved upon the model to define the state-derivative equation as

$$dw/dt = a\, V(t)^{\varrho}\, f(w) \quad (10.19)$$

where a is a fitting constant, ϱ is an odd integer and f(w) is the window function.

Simmons' Tunnel Barrier Model[35] provides for the tunnel current in TiO_2 memristor but the equations are complicated making the simulations and computations difficult. Two other models, namely, the TEAM model[36] and VTEAM model[37] are excellent in terms of simplicity and accuracy but none discuss the underlying aspects of the device physics. A model based on the theory of Schottky diodes will go a long way to explain the behavior of the TiO_2 memristor.

10.6 APPLICATION AREAS OF MEMRISTORS

The memristors have great potential in the design of integrated circuits by addressing several important aspects of the VLSI chip designs and computational areas.[38] There has been intense research work aimed at utilizing this unique device in improving the performance of the integrated circuit chips and the applications they are put to.

The most obvious application lies in the design of the **nonvolatile random access memory** (NVRAM).[39] Memristor has the advantage of acting as a memory without continuously drawing power. It also takes up very little area/space. Therefore, the memristor can act to improve upon the design of the digital memory applications.

The memristor also has a great possibility as **associative memories.**[39] The associative memories can map an input pattern to an output pattern depending on the similarities between the input pattern and that stored in the memory.

There is a prospect of using the memristor for **low power memory** as well as **distributed state storage.** The maximum resistance value of the memristor is used for its digital memory applications.

Memristors can be used in analogue circuits as well. It can be used as a **dynamically adjustable load** in electrical and electronic circuits. It can also act as **programmable resistance.**[40]

It has been proposed that it can be used to design **electronically adjustable filters** and **amplifiers**.[41]

The memristor finds extensive possible utility in **image recognition and image processing** applications by making use of the edge detection when the large changes in the digital image are identified. The light intensities are converted into voltages that are applied to the points of a grid of memristors. The resulting changes in the memristor resistance or memresistance are recorded in the frames and the edge-detected images can be recovered by a measurement of the resulting resistance values acting as memories.[42]

Memristor has been used in a **switched-on radio frequency antenna.**[43] The possibilities are endless.

10.7 FUTURE SCOPE

Modern nanoscale technologies can facilitate the growth of memristors which can find extensive applications in circuit design and integrated circuits for computers, especially in the design of memory elements. The application areas include not only non-volatile memories but encompass analog electronics, image processing, and reconfigurable antenna in the RF range since memristors have the ability to store both analog and digital data in simple and power efficient way.

KEYWORDS

- **memristor**
- **memristance**
- **hysteresis loop**
- **magnetic flux**
- **linear boundary model**

REFERENCES

1. Chua, L. Memristor –The Missing Circuit Element. *IEEE Trans. Circuit Theory* **1971**, *18* (5), 507–519.
2. Simmons, J. G.; Verderber, R. R. New Conduction and Reversible Memory Phenomena in Thin Insulating Films, *Proceedings of the Royal Society of London*, October 3, 1967, pp. 1–37.
3. Chua, L. O.; Kang, S. M. Memristive Devices and Systems. *Proc. IEEE* February **1976**, *64* (2).
4. Strukov, D.; Snider, G.; Stewart, D.; Williams, R. Missing Memristor Found. *Nature* **2008**, *453* (7191), 80–83.
5. Johnson, R.; Ovshinsky, S. US Patent 4,597,162.

6. Thakoor, S.; Moopenn, A.; Daud, T.; Thakoor, A. P. Solid-state Thin-film Memistor for Electronic Neural Networks. *J. Appl. Phys.* **1990**, *67*, 3132–3135.
7. Nichogi, K.; Taomoto, A.; Asakawa, S.; Yoshida, K. Matsushita Research Institute, U.S. Patent 5,223,750, 1993.
8. Buot, F. A.; Rajagopal, A. K. Binary Information Storage at Zero Bias in Quantum-well Diodes. *J. Appl. Phys.* **1994**, *76*, 5552.
9. Kozicki, M.; West, W. Assigned to Axon Technologies Corp. and the Arizona Board of Regents, U.S. Patent 5,761,115, June 2, 1998.
10. Swaroop, B.; West, W.; Martinez, G.; Kozicki, M.; Akers, L. Programmable Current Mode Hebbian Learning Neural Network Using Programmable Metallization Cell, *Proceedings of the IEEE International Symposium on Circuits and Systems*, June 3, 1998, Vol. 3, pp. 33–36.
11. Heath, J.; Kuekes, P.; Snider, G.; Williams, S. A Defect-Tolerant Computer Architecture: Opportunities for Nanotechnology. *Science* June 12, **1998**, *280* (5370), 1716–1721.
12. Beck, A.; Bednorz, J. G.; Gerber, C.; Rossel, C.; Widmer, D. Reproducible Switching Effect in Thin Oxide Films for Memory Applications. *Appl. Phys. Lett.* July 3, **2000,** *76*, 2749.
13. Kuekes, P.; Williams, S.; Heath, J. Assigned to Hewlett-Packard, U.S. Patent 6,128,214, October 3, 2000.
14. Liu, S.; Wu, N.; Chen, X.; Ignatiev, A. A New Concept for Non-Volatile Memory: The Electric Pulse Induced Resistive Change Effect in Colossal Magnetoresistive Thin Films. Non-volatile Memory Conference, San Diego, CA, November 6–7, 2001.
15. Rinerson, D.; Chevallier, C.; Longcor, S.; Kinney, W.; Ward, E.; Kuo-Ren Hsia, S. Assigned to Unity Semiconductor, U.S. Patent 6,870,755, March 22, 2005.
16. Lan, Z.; Bill, C.; VanBuskirk, M. A. Assigned to Advanced Micro Devices, U.S. Patent 6,960,783, November 1, 2005.
17. Ovshinsky, S., U.S. Patent 6,999,953, 2006.
18. Bulovic, V.; Mandell, A.; Perlman, A. Assigned to Spansion, U.S. Patent 7,183,141, February 27, 2007.
19. Snider, G. Assigned to Hewlett-Packard, U.S. Patent 7,203,789, April 10, 2007.
20. Snider, G. Self-organized Computation with Unreliable, Memristive Nanodevices. *J. Nanotechnol.* August 10, **2007.**
21. Mouttet, B. U.S. Patent 7,302,513, November 27, 2007.
22. Greg Snider of HP Labs receives U.S. Patent 7,359,888 (assigned to Hewlett-Packard) April 15, 2008.
23. Strukov, D.; Snider, G.; Stewart, D.; Williams, S. The Missing Memristor Found. *Nature* May 1, **2008**, *453*, 7191.
24. Mouttet, B. Logicless Computational Architectures with Nanoscale Crossbar Arrays, *2008 NSTI Nanotechnology Conference and Trade Show*, Boston, June 1–5, 2008.
25. Yang, J. J.; Pickett, M. D.; Li, X.; Ohlberg, D. A. A.; Stewart, D. R.; Williams, S. R. Memristive Switching Mechanism for Metal/Oxide/Metal Nano-devices. *Nat. Nanotechnol.* July 15, **2008.**
26. Mouttet, B. Proposal for Memristors in Signal Processing. Proceedings of Nano-Net 2008, *Nanotechnology Conference*, Boston, September 14–16, 2008.
27. Pershin, Y. V.; Di Ventra, M. Spin Memristive Systems: Spin Memory Effects in Semiconductor Spintronics, UC San Diego. *Phys. Rev. B* September 23, **2008**, *78*, 159905.

28. Stewart, D.; Beck, P.; Ohlberg, D. Assigned to Hewlett-Packard, U.S. Patent 7,443,711.
29. Mouttet, B. U.S. Patent 7,447,828, November 4, 2008.
30. Mouttet, B. U.S. Patent 7,459,933, December 2, 2008.
31. Marani, R.; Gelao, G.; Gina Perri, A. A Review on Memristor Applications.
32. Joglekar, Y.; Wolf, S. The Elusive Memristor: Properties of Basic Electrical Circuits. *Eur. J. Phys.* **2009**, *30*, 661–675.
33. Lehtonen, E.; Laiho, M. CNN Using Memristors for Neighborhood Connections, *Proc. Int. Workshop Cellular Netw. Appl.*, Feb. 2010, pp. 1–4.
34. Pickett, M. D.; Strukov, D. B.; Borghetti, J. L.; Yang, J. J.; Snider, G. S.; Stewart, D. R.; Williams, R. S. Switching Dynamics in Titaniumdioxide Memristive Devices. *J. Appl. Phys.* Oct. **2009**, *106* (7), 1–6.
35. Biolek, Z.; Biolek, D.; Biolkova, V. Spice Model of Memristor with Nonlinear Dopant Drift. *Radioengineering* June **2009**, *18* (2), Part 2, 210–214.
36. Kvatinsky, S.; Friedman, E. G.; Kolodny, A.; Weiser, U. C. TEAM: ThrEshold Adaptive Memristor Model. *IEEE Trans. Circuits Syst. I: Regular Papers* January **2013**, *60* (1), 211–221.
37. Kvatinsky, S.; Ramadan, M.; Friedman, E. G.; Kolodny, A. VTEAM–A General Model for Voltage Controlled Memristors. *IEEE Trans. Circuits Syst. II: Express Briefs* August **2015**, *62* (8), 786–790.
38. Prodromakis, T.; Toumazou, C. A Review on Memristive Devices and Applications, *Proceeding of 17th IEEE International Conference on Electronics, Circuits, and Systems (ICECS 2010)*, 2010, pp. 934–937.
39. Yuanfan, Y. Matching in Memristor Based Auto-associative Memory with Application to Pattern Recognition, *Proceedings of 12th International Conference on Signal Processing (ICSP)*, 2014, pp. 1463–1468.
40. Sangho, S.; Kyungmin, K.; Sung-Mo Kang, S. Memristor Applications for Programmable Analog ICs. *IEEE Trans. Nanotechnol.* **2011**, *10* (2), 266–274.
41. Ascoli, A. Memristor-based Filtering Applications, *Proceedings of 14th Latin American Test Workshop (LATW)*, 2013, pp. 1–6.
42. Gregory, M. D.; Werner, D. H. Application of the Memristor in Reconfigurable Electromagnetic Devices. *IEEE Antennas Propag. Magazine* **2015**, *57* (1), 239–248.

CHAPTER 11

Technological Development of Graphene and Graphene Nanocomposite-Based Supercapacitor Electrode

MILAN DHARA[1] and SUNIPA ROY[2*]

[1]*Department of Applied Electronics & Instrumentation Engineering, Future Institute of Engineering and Management, Kolkata, India*

[2]*Department of Electronics & Communication Engineering, Guru Nanak Institute of Technology, Kolkata, India*

[*]*Corresponding author. E-mail: sunipa_4@yahoo.co.in*

ABSTRACT

Graphene and graphene-based nanomaterials are captivating immense attention nowadays due to their novel properties, like highly stretchable specific surface area, phenomenal electrical conductivity, high transparency, remarkable chemical stability, and brilliant mechanical behavior. Utilizing these remarkable properties, graphene acts as a prominent nanomaterial for supercapacitor electrodes and various energy storage devices. Moreover, its admirable mechanical properties with immense tensile strength, Young's modulus and stretching capability (elasticity) make it greatly applicable electrode materials to fabricate ultrathin, flexible, safe energy storage devices in wearable and handy electronic products, biointegrated devices, recreational sports gadgets, medical utensils, telecommunication, automobile, space satellites, and hybrid electric vehicles. Here, we have emphasized on the recent developments of graphene and various modified graphene-based nanocomposites as a promising electrode material for supercapacitor applications depending on the various nanostructures formed by various synthesis techniques, to improve the specific capacitance, voltage window, cyclic stability, charging/discharging time, electrical conductivity, and specific energy and power density. But in the last few decades, researchers devoted themselves to formulate a graphene and/or its derivatives-based supercapacitors with high gravimetric energy density while retaining the intrinsic properties of high specific power density and long shelf life. Extensive research is going on for development of various graphene-based materials, improvization of their compositions and dimensions at varying scale using effective, low-cost synthesis techniques with enhanced electrochemical properties. In this review, classification of supercapacitors are studied based on their charge storage phenomenon (electrostatic or non-Faradaic, Faradaic, or combination of two) as well as functional active electrode materials (symmetric and asymmetric) and the latest development and perspective of graphene or graphene-based materials as supercapacitor electrodes have been critically analyzed.

11.1 INTRODUCTION

In the 21st century, one of the key challenges facing our society is the rapid curtailment of fossil fuels with increasing demand for energy. Also, unlimited fossil-fuel consumption appear as a serious concern to the world ecology and economy. Hence, energy needs to be used efficiently along with looking for low-cost renewable and environmentally friendly energy generation systems (such as piezoelectric nanogenerators, solar cell, wind energy, etc.) that can complement the fossil fuels to shrink environmental issues, such as global warming due to greenhouse effect, acid rain, and air pollution.[1, 2] But, there is a limitation of continuous availability of natural resources for nonconventional energy sources. Hence, energy storage devices need to be incorporated with this type of green cell for efficient, clean, and versatile use of energy. Therefore, research interest is appreciably increased toward the advanced technologies for high-performance energy storage (e.g., batteries, fuel cells, capacitors, and supercapacitors) and conversion (e.g., photovoltaic cells) schemes.[3] Among storage devices, Supercapacitors or ultracapacitors are increasing popular than the conventional available energy storage systems, because of their high power densitics, short charging time, long life cycle (>100,000 cycles), long cell life, high reliability and flexibility, and low maintenance type.[7–12] As compared with the commercial available energy storage devices (rechargeable battery), supercapacitor is a device which is capable of delivering high power density (hundred to thousand times) in the same volume by disregarding slower charge transfer reactions. But, they are unable to store equal amount of charge as batteries can (3–30 times lower). Supercapacitor has longer cycle life due to its charge storage mechanism which involves storing of charge physically at electrode–electrolyte interface through non-Faradic electrostatic way. Unlike battery, this reversible electrostatic charge storage process does not alter the electrode volume, thus eliminates swelling caused by reversible chemical charge transfer reaction. On the other hand, operating voltage of supercapacitor cell should be maintained at a low value in order to eliminate chemical decomposition of solvent. Unlike battery, supercapacitors can operate at a lower temperature (~–40°C) with reliable performance.[2–7] Hence, supercapacitors are widely used in portable electronic devices, consumer electronics, low-emission hybrid cars, buses, and trucks, memory backup systems, military devices, and process power plants. By bridging with batteries, supercapacitors can deliver high power required during acceleration and can recover the energy during braking in hybrid electric vehicles (HEVs).[13] Overall, the distinctive characteristics of supercapacitor (Table 11.1) often are used as complement for batteries and fuel cells.[14]

But, low energy density (~10 Wh/kg) of the commercially available supercapacitors is one of the serious concerns for further widening of its applications. Hence, improving energy density with allowable power density of supercapacitor is still a great challenge for the researchers. The energy density of a supercapacitor cell is calculated by the equation $E = 1/2\ CV^2$, where C is the cell capacitance and V is the cell potential. Thus, energy density of the cell can be escalated by either enhancing cell capacitance with novel electrode materials or rising cell voltage, which is usually limited by the breakdown strength of dielectric solvent.

Porous carbon-based materials, such as activated carbon (AC) and carbon nanotubes (CNTs) are preferably used as active electrode elements for supercapacitors due to their high gravimetric surface area, brilliant electrical conductivity, controllable porosity, nontoxicity, easy processing, good stability, and reasonable cost. AC, when used as electrodes, has high gravimetric surface area of 3000 m^2/g, but has limited per unit mass capacitance of 200 F/g due to inaccessibility of charge storage ions into its micropores and interior atoms at higher scan rate. On the other hand,

TABLE 11.1 Comparisons of Different Energy Storage Devices.

Parameters	Conventional capacitor	Supercapacitor	Battery
Storage Mechanism	Electrostatic	Electrostatic/Faradic with Capacitive	Chemical
Specific Energy (Wh/kg)	<0.1	1–10	~20–150
Specific Power (W/kg)	>10,000	500–10,000	<1000
Cell Voltage (V)	6–800	1–2.75	1.2–4.2
Discharge Time	10^{-6} to 10^{-3} seconds	Few Seconds to few Minutes	0.3–3 h
Charging Time	10^{-6} to 10^{-3} seconds	Few Seconds to few Minutes	1–5 h
Coulombic Efficiency (%)	About 100	85–98	70–85
Cycle Life (cycles)	Almost Infinite	>500,000	About 1000
Cell life	>>10 years	>10 years	~3 years (depending upon applications)
Operating temperature	− 20 to +100 °C	− 40 to +85 °C	− 20 to +65 °C
Lifetime	>100 k cycles	>100 k cycles	150 to 1500 cycles

CNTs which can be categorized as single-walled CNTs (SWCNTs) and multiwalled CNTs (MWCNTs) have smaller specific surface about <500 m^2/g and smaller specific capacitance about <100 F/g than AC due to the availability of only the outer layer of CNTs that are stacking into bundle, which is of concern for energy density.[15–17] Furthermore, CNTs are expensive, and moreover, it is very difficult to produce and handle vertically aligned CNTs. Other carbon-based materials with well-defined physical and chemical properties, such as fullerene, graphite, templated porous carbon, functionalized porous carbon, AC fibers, carbon aerogels, carbon composites, etc., are being continuously explored for supercapacitors.[18–22]

Graphene, the most unique material developed in the field of nanomaterial science for supercapacitor electrode, is a single-layered (with a thickness of only ~0.34 nm) two-dimensional sp^2 bonded (covalently bonded to three other atoms) carbon allotrope consisting of a hexagonal (or honeycomb) lattice structure with interatomic distance of 0.142 nm. It has strikingly strong mechanical strength and Young's modulus, large specific surface area, remarkable thermal and electrical conductivity, immense optical transmittance, highly stretchability along with high chemical stability, and superhydrophobicity (Table 11.2) which makes it a replacement of silicon for both research fields and further applications.[23–25]

This 2D planar single sheet of sp^2 hybridized carbon material is the fundamental structure of all other important carbon allotropes.[26] Each of these carbon allotropes has characteristics more or less equivalent to graphene, but the latter exceed in excellence. Graphene monolayer sheets can be enfolded to form a sphere, called zero-dimensional (0D) fullerenes, rolled up with respect to its axis to form one-dimensional (1D) cylindrical structure, called carbon nanotubes; two-dimensional (2D) structured with one or few layers, called graphene; and stacked by a weak van der Waals force to form three-dimensional (3D) graphite structures as shown in Figure 11.1.[29] Depending upon the number of layers in a framework, graphene nanomaterials may be regarded as mono-, bi-, and tri-layer graphene. If the number of graphitic layers increases more than these, it is called few-layer graphene (~5–10 graphitic layers) or multi-layer or thick graphene or nanocrystalline thin graphite (~20–30 layers).[28]

Again, due to the remarkable properties, its application is profoundly surging in the fields ranging from research to various upcoming electronics and photonics devices of real world.

TABLE 11.2 Properties of Graphene.

Properties of graphene	Value
Carrier mobility at room temperature	~200,000 cm^2/Vs
Theoretical specific surface area	2630 m^2/g
Optical transparency	97.70 %
Resistivity	10^{-6} ohm-cm
Young's Modulus	~ 1 TPa
Thermal conductivity	3000–5000 W/m-K
Saturation velocity	4–5 × 10^7 cm/sec

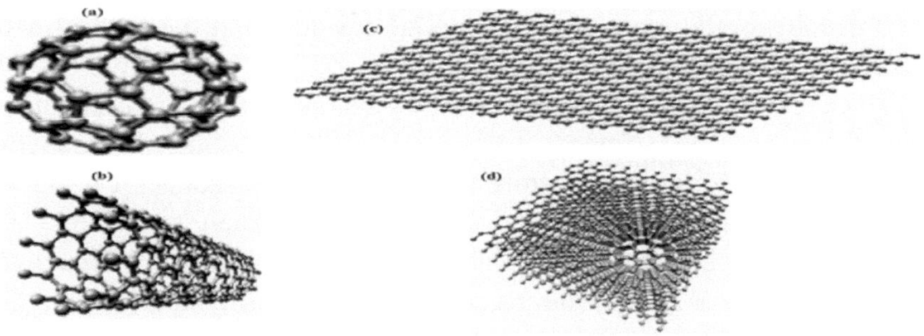

FIGURE 11.1 Mother of all graphene forms. (a) 0D fullerene, (b) 1-D CNT, (c) 2D graphene, (d) 3D graphite.

Source: Reprinted with permission from Ref. [29]. © 2018 STAM Open Access. https://creativecommons.org/licenses/by/4.0/

Recent researches regarding energy production and storage with the goal of emergent high performance and environment-friendly energy storage devices incurred immense interest in the graphene-based electrodes due to its promising characteristics in fuel cells, lithium-ion batteries, and supercapacitors.[30–36] The surface area of single sheet graphene is 2630 m^2/g, which is substantially greater than that of carbon which makes graphene and graphene-based materials as active electrodes in the energy storage applications. Moreover, the excellent conductivity of 10^6 S/m and specific capacitance of graphene along with relatively low production cost acted as added reasons for its use as electrodes.[37] In this review paper, we have critically analyzed the recent researches and developments of graphene and graphene nanocomposite-based materials with various synthesizing techniques as supercapacitor electrodes.

11.2 CLASSIFICATION OF SUPERCAPACITOR

Supercapacitor, an electrochemical energy storage device is constructed with two electrodes (anode and cathode) and an ion permeable separator that usually operates on the accumulation of ions on the exterior of the electrode. Depending on the charge storage principle, there are three classes of supercapacitors, namely, electric double-layer capacitor (EDLC), pseudocapacitor and hybrid supercapacitor as shown in Figure 11.2.

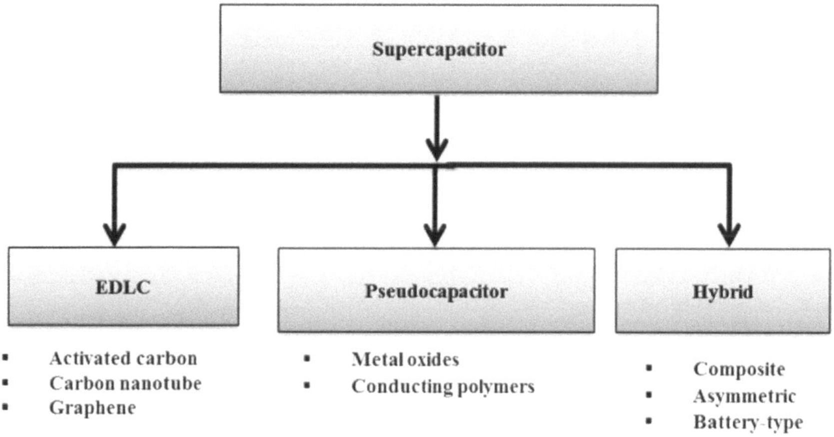

FIGURE 11.2 Classification of supercapacitor on charge storage mechanism.

Again, depending upon the material used for the electrodes, supercapacitors may be symmetric or asymmetric type (Figure 11.3). Symmetric type supercapacitors use the same type of electrodes for cathode and anode. But, in the case of asymmetric type, anode and cathode are made up of dissimilar type of electrode materials.[38]

11.2.1 ELECTRIC DOUBLE-LAYER CAPACITOR (EDLC)

Electric double-layer capacitor (EDLC) is an energy storage device that stores energy by creating double layers of charge at the interface between electrode and the electrolyte, where charges are accumulated through electrostatic attraction as shown in Figure 11.4a. Like conventional capacitor, charge is stored electrostatically due to repeatable adsorption and desorption of ions of the electrolyte onto the electrochemically stable active electrode material. Unlike conventional capacitor, charges within the double-layer are separated through a very small partition (0.3–0.8 nm) that depends on the electrolyte concentration and size of the ions. This energy storage mechanism involves only physical re-arrangement of charges instead of charge transfer chemical reactions and crystallographic phase transition at the interface as in the case of pseudocapacitor or battery and hence the capacitance arises due to true capacitance effect.[39, 40]

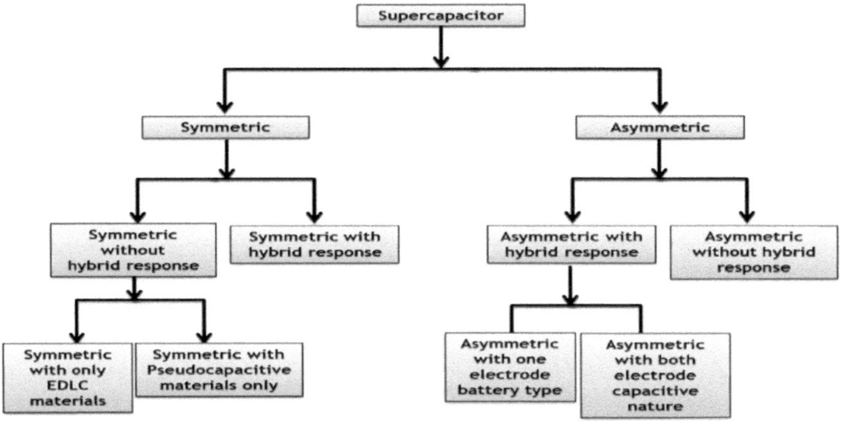

FIGURE 11.3 Classification based on electrode used in supercapacitor cell.

Typically, an EDLC comprises of two electrodes, electrolyte, current collectors, and an ion permeable separator, which prevents the electrical short circuit between the electrodes. During charging of EDLC (Figure 11.4a) with electrical potential between the electrodes, electrons are forcefully dragged from positive electrode to negative electrode through external circuit. As a result of deficiency and excess of charge creation at the electrode surface, the dissolved and solvated cations and anions of electrolyte move toward the surface of negatively charged electrode and positively charged electrode, respectively to compensate the external charge unbalance. This initiates two electric double-layers (EDLs) near the surface of electrode within the electrolyte which is analogous to conventional capacitor. Since, the separation between the active electrode material and counter ions adsorbed to its surface is very small (usually in an atomic range), electric field is very strong (Figure 11.4c) within the

EDL. As a result, each EDL stores enormous amount of energy. Again, the electrical energy is stored in an EDLC through the accumulation of charges near the surface of electrode. Large surface area with high pore volume is a crucial condition for high-performance supercapacitor electrode.[8, 12]

During discharging process (Figure 11.4b) by connecting a load externally to the supercapacitor cell, electrons are flow from negative electrode to positive through external load and hence giving discharge current in opposite direction. This discharge current will continue to flow until the cell is fully discharged, that is, surface adsorbed electrolyte ions are mixed again (Figure 11.4d). The mobility of ions into the pores is greatly influenced by the pore size. The pores which are inaccessible to the ions do not contribute any double-layer capacitance. From various report, it was seen that the pores below 0.5 nm are not accessible to hydrated ions (<1 nm for organic electrolytes).[42–45]

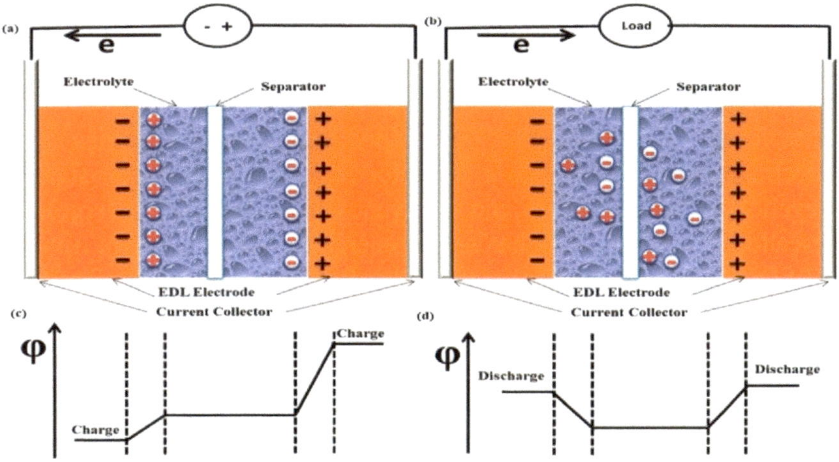

FIGURE 11.4 Schematic diagram of EDLC charge storage mechanism: (a–b) charging-discharging process of EDLC and (c–d) electric field across the electrode during charging and discharging.

The above concept of EDL was first depicted and modeled by von Helmholtz in the 19th century. This simple Helmholtz EDL model was again renovated by Gouy and Chapman when they included diffusion layer originated due to continuous distribution of electrolyte ions driven by thermal motion. Further, Stern merged the above two models to represent the ions distribution at the electrode and electrolyte interface in the EDL mechanism. The model is composed of two layers of ion allocation based on the binding of ions at the inner and outer section termed as compact layer (also called stern layer) and diffuse layer, respectively. The compact layer is formed specifically by adsorbed ions (generally, anions irrespective of the charge nature of the electrode, called inner Helmholtz plane (IHP) lie in close proximity to the core of the electrode) and nonspecifically adsorbed counter ions called outer Helmholtz plane (OHP) which are in contact with the electrolyte. As the voltage window of the electrodes is limited either by the solvent (1.23 V) or organic (3.5 V) electrolyte decomposition, the energy storage capacity of EDLC can be enhanced by adopting appropriate electrode materials having high surface area, such as porous graphene and its derivatives.[46, 47] Thus, EDLC is associated with better power performance, long life cycle, and elimination of damage of electrodes due to the absence of chemical reactions, but with restricted energy density.[48]

11.2.2 PSEUDOCAPACITOR

The name "pseudocapacitor" originated due to the properties of the electrode that behaves like a capacitor in its electrochemical signature (cyclic voltammogram and galvanostatic charge–discharge). In pseudocapacitors, the mechanism for energy storage is different from that of EDLC. The pseudocapacitors involve fast and reversible surface redox Faradaic reaction between electrode and electrolyte ions (Figure 11.5), in which reduction and oxidation reactions occur during charge and discharge phases and thus generating voltage-dependent pseudocapacitance ($C = dQ/dV$), that is,. linear variation of the charge stored with changing potential within the voltage window. Due to the different oxidation state of the electrode material, there occurs a movement of electrons across the double layer that develops Faradaic current through the supercapacitor.[39, 41] Cyclic voltammetry response signature is similar to that of the EDLC which is quasi-rectangular in shape with no or very small redox peaks. In this type of supercapacitor, charge storage can take place both at the surface and the bulk of the electrodes. Compared with the EDLC, these types of supercapacitors possess higher gravimetric capacitance and energy densities due to additional charge transfer within the defined potential. It is reported that pseudocapacitor can have 10–100 times more capacitance than EDLC.[49] Because of slower Faradaic reaction, the power density is much lower than EDLC. Again, they typically have inferior cycle life due to active material degradation as a result of Faradaic reactions.[50, 51] The most commonly researched classes of pseudocapacitive materials having quick and reversible Faradaic reaction, low cost and relatively longer cyclic life include transition metal oxides (ruthenium oxide, manganese oxide, and vanadium oxide), metal sulfides (nickel sulfide, copper sulfide, zinc sulfide, and cobalt sulfide) and conducting polymers, such as polyaniline, polypyrrole or derivatives of polythiophene.

11.2.3 HYBRID SUPERCAPACITOR

Hybrid supercapacitors are supercapacitors in which the devices pair with two electrodes with different charge storage mechanisms, that is, one capacitive (electrostatic) and other Faradaic (redox type with or without capacitive), that is, the better power performance of EDLC with high gravimetric capacitance of pseudocapacitors (Figure 11.5) and the resultant device is in-between a supercapacitor and a battery.[52] Their structure is asymmetric in nature with the combination of one electrostatic electrode and another electrochemical electrode. As a result of hybridization, operating potential window of the cell is escalated due to different potential window of the two electrodes.[53] Based on the electrode materials employed, hybrid cells are classified into three different categories: asymmetric hybrids, composite hybrids, and battery-type hybrids. Asymmetric hybrid supercapacitor combines EDLC-type electrode against pseudocapacitive type electrode. For example, Cao et al. has developed an asymmetric hybrid supercapacitor based on MnO_2 as anode and graphene as cathode with 1M Na_2SO_4 aqueous electrolyte. This asymmetric cell has operated on 2.0 V, giving an energy density of 25.2 Wh/kg at a power density of 100 W/kg.[54] Other examples are CNT//CNT-MnO_2 SC,[54] AC//MnO_2 SC,[55] etc. Due to superimposition of Faradaic and non-Faradaic charge storage process in asymmetric hybrids, better cycling stability than pseudocapacitors and high power density and energy density than EDLC are achieved. Composite hybrid supercapacitors are constructed with composite materials as electrodes which are the integration of carbon materials with either metal oxides or conducting polymers. Example of this kind of category includes MnO_2-CNT//MnO_2-CNT SC,[57] RuO_2-rGO//

RuO_2-rGO SC,[58] and V_2O_5-CNT//AC.[59] Lastly, battery-type hybrids are constructed using two different electrodes. The battery-type electrodes (redox, intercalation/conversion) that store charge by noncapacitive Faradaic manner, are combined with supercapacitive electrodes that store charge either by electrostatic or capacitive Faradic way so as to obtain high power density with allowable energy density. For example, MnO_2-CNT// $LiMn_2O_4$ SC,[60] AC//PbO_2 SC,[61] MoO_2-C nanofilm//$NiCo_2O_4$ SC[62] etc. Therefore, this asymmetric structure leads to increase in cell voltage, thus increasing power and energy densities, but the use of Faradic electrode degrades the stability in life cycle. Again, the device plotted nonlinear charge/discharge characteristics (*V* vs. *Q*).

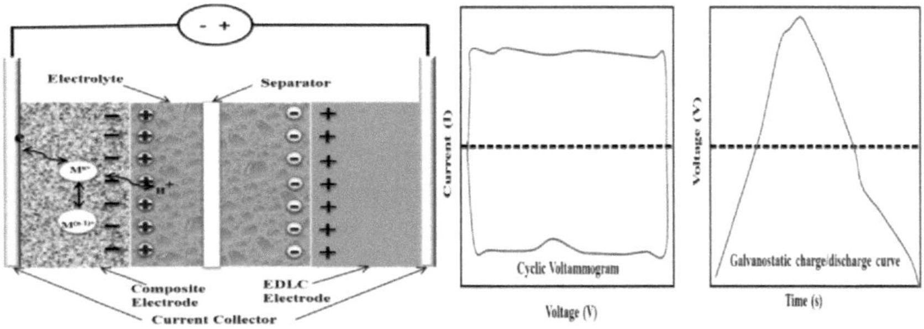

FIGURE 11.5 Pseudocapacitor charge storage mechanism in hybrid electrode and its CV and GCD (galvanostatic charge-discharge) curve.

11.2.4 SYMMETRIC AND ASYMMETRIC SUPERCAPACITOR

Symmetric supercapacitors use same electroactive material in both electrodes (i.e., anode and cathode) within a single cell. In this configuration, either double-layer type or pseudocapacitive type electroactive material are coated to form anode and cathode of the cell. As a result, energy is stored through the same charge storage mechanism in both of the electrodes having equal mass for charge balance. In EDL type symmetric supercapacitors, carbon-based materials (activated carbon, porous carbon, carbon nanotubes graphene, and graphene oxide) are used for developing both anode and cathode of a cell to store charges only due to the surface adsorption of ions from the electrolyte due to the electrostatic attraction. For examples, graphene//graphene SC, AC//AC SC, rGO//rGO SC. But, the symmetric device which is assembled with a pseudocapacitive material coated anode against the same as cathode, yields pseudocapacitor type symmetric cell. Some metal oxides (ruthenium oxide, manganese oxide, vanadium pentoxide, etc.) and conducting polymers (PANI, PEDOT, PTh, and PPy) are generally employed as pseudocapacitive materials to store charge in a capacitive Faradaic process. Example of these types of supercapacitors are RuO_2-rGO// RuO_2-rGO SC,[63] MnO_2-graphene//MnO_2-graphene SC,[64] MnO_2-CNT// MnO_2-CNT SC,[65] etc. In order to escalate the cell operating voltage and hence the energy density, supercapacitor can be prepared by anode and cathode having dissimilar electroactive material. This class is termed as asymmetric supercapacitor. Both electrodes of this kind may have either same charge storage mechanism with different electroactive material or different charge storage mechanisms. The hybrid configuration uses two electrodes having different energy storage mechanism in a single cell. When EDLC-type electrode is used against pseudocapacitive kind in a single cell, it is referred to as asymmetric hybrid with capacitive nature. For example, CNT//CNT-MnO_2 SC,[55] AC//MnO_2 SC,[59] V_2O_5-CNT//AC SC,[66] etc. A supercapacitor that is assembled with two dissimilar EDL-type electrodes or two different types of conducting polymer as anode and cathode but used same energy storage mechanism, is called asymmetric without hybrid configuration. This category includes GC//CNT SC, rGO//AC SC etc. Another asymmetric hybrid has been investigated; where a high energy battery-type electrode is assembled with a high power capacitive-type electrode (generally EDLC electrodes) in a single cell. This configuration improves the electrochemical performance of the cell. It is termed as battery-type asymmetric hybrid, for example, MnO_2-CNT//$LiMn_2O_4$ SC,[67] AC//PbO_2 SC,[68] MoO_2-C nanofilm//$NiCo_2O_4$ SC[69] etc. Finally, a symmetric device made up of carbon-based materials with aqueous electrolyte containing electrochemically active molecule can initiate a redox reaction on the surface of one of the electrodes. This is called symmetric hybrid.

11.3 GRAPHENE-BASED NANOMATERIALS FOR SUPERCAPACITOR ELECTRODE

The performance of supercapacitor depends on electrochemical properties and kinetic features of electrode materials, electrolytes, and voltage range used. For the enrichment of power density, it is required to escalate the kinetics of ions and electron transport toward the electrodes. Concurrently, specific energy of the supercapacitors can be enlarged by optimization of the electrode materials.[51, 54] Generally, in real system, the storage mechanism for EDLC and pseudocapacitors occurs concurrently and are inseparable. But for convenience, the classification is considered depending on dominant storage mechanisms. In EDLC, the gravimetric capacitance is dependent on charge accumulation on the exterior of the electrode. Hence, it is proportional to the specific surface area and conductivity of the electrode, which should be increased. In most cases, EDLCs are constructed using different carbon-based materials, but graphene shows excellence due to its intrinsic physical, mechanical, and chemical properties. Only drawback associated with graphene-based nanomaterials is self-aggregation of graphene layers and lower energy density due to their charge storage mechanism.[50] On the other hand, pseudocapacitor store energy via fast and repeatable surface or near surface redox reaction. The most widely analyzed materials used for pseudocapacitors include (i) transition metal oxides (such as RuO_2, MnO_2, Co_3O_4, Fe_3O_4, ZnO, NiO, etc.), metal hydroxides (such as $Ni(OH)_2$, $Co(OH)_2$, Ni-$Co(OH)_2$), or sulfides (like CoS_2, CuS, NiS, $NiCo_2S_4$, etc.) (ii) conducting polymers (CPs) like polyaniline (PANi), polypyrrole (PPy), and polythiophene (PTh), and (iii) materials retaining oxygen- and nitrogen-enclosing surface functional groups.[75–85] Though, conductive polymers (CPs) and transition metal oxides (TMOs) have remarkable conductivity and high theoretical gravimetric capacitance respectively, but poor long-term stability of conducting polymers and low conductivity of metal oxides make barrier for the future energy storage applications. Hence, now a days, researchers are focusing on the improvizations of graphene-based electrode materials (Figure 11.6) to yield better electrochemical performances by utilizing varying structural benefits and varying synthesis approaches.[25, 27]

Depending upon the number of layers in a framework, graphene nanomaterials may be regarded as monolayer graphene, few-layer graphene (FLG) which have ~5–10 graphitic layers, and multilayer graphene (MLG) consisting of ~20–30 graphitic layers. Again, depending on the structural variations, graphene-based materials can be grouped as 0-dimensional termed as free-standing graphene dots, one-dimensional termed as graphene yarns or fibers, two-dimensional termed as graphene films/sheets, and three-dimensional termed as graphene foam/hydrogel/aerogel.[70] Finally, graphene may be modified to graphene oxide (GO) and reduced graphene oxide (rGO) based on carbon-to-oxygen ratio in its structure.

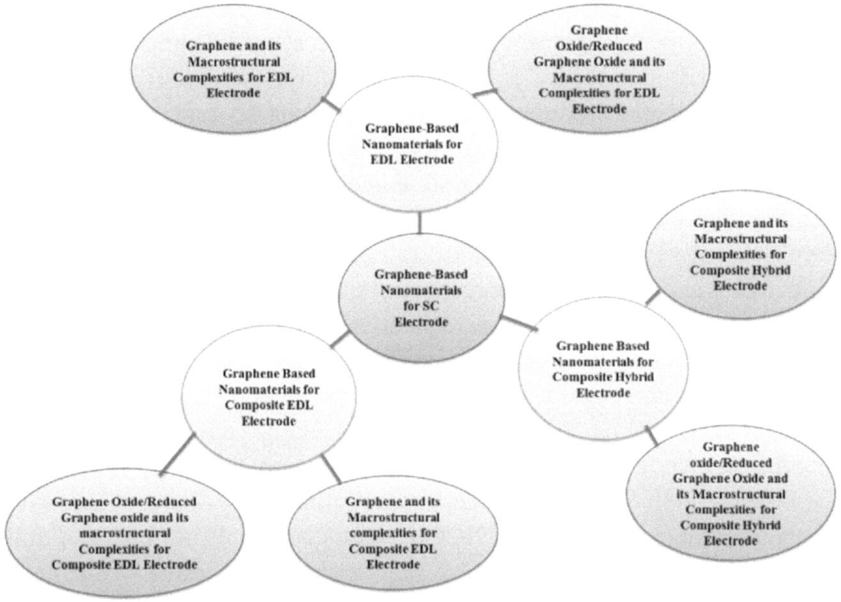

FIGURE 11.6 Classification of graphene-based nanomaterials for SC electrode.

11.3.1 GRAPHENE-BASED NANOMATERIALS FOR EDL ELECTRODE

The predecessors of graphene used for supercapacitor electrodes include carbon-based materials processed by different strategies. Activation of carbon introduces uncontrollable pore structures accompanied by oxygenated functionalities. Though, it possesses large surface area, but inaccessible pores, unstable oxygenated groups, and high resistance limit their performance. The template-guided synthesis method can utilize either hard (zeolite) or soft (polymers) template to prepare porous carbon or carbon aerogels. The porous template carbon provides good performance in supercapacitors due to controllable porosity, but hard template method exhibits high cost and elaborateness. Carbon aerogels, conversely, exhibit lesser specific capacitance, but are cheaper. CNT or porous carbon nanomaterials are effective materials, but possess lower specific capacitance than graphene in some cases.[72–74]

11.3.1.1 GRAPHENE AND ITS MACROSTRUCTURAL COMPLEXITIES FOR EDL ELECTRODE

The graphene has been remarkably accepted as supercapacitor electrodes because of its ultraporous structure, high specific surface area, and unusual electron mobility. In addition to theoretical specific surface area of 2630 m^2/g, the monolayer graphene exhibits specific capacitance of nearly 550 F/g upon utilization of full area. But, practically, the capacitive value is much less due to irreversible agglomerations of graphene layers via van der Waals force during generation and applications. Hence, the key challenge is to enhance the overall properties of graphene to make it effectively applicable for supercapacitor electrodes.[71]

Stoller et al. devised a supercapacitor based on chemically modified graphene electrode that exhibited specific capacitance of 135 F/g and 99 F/g with aqueous and organic electrolytes, respectively.[86] Graphene quantum dots (GQDs), owing to their nanosize, excellent conductivity, quantum confinement, edge effect, and water solubility, are impressed for potential persistence in the field of supercapacitor applications. Zhang et al. derived a gravimetric capacitance of 296.7 F/g with good cycle performance and a satisfactory energy density of 41.2 Wh/kg with GQD-based EDLC cell in aqueous electrolyte.[87] A symmetric GQD-based microsupercapacitor cell demonstrated by Liu et al. showed an outstanding areal capacitance of 486.1 $\mu F/cm^2$ in aqueous solution. In ionic electrolyte, it showed a wide potential window of 2.7 V.[88]

Fiber shape graphene is promising as an electrode for wearable and portable energy storage devices due its amazing mechanical flexibility and excellent conductivity.[89–91] Li et al. fabricated graphene fiber fabrics (GFFs) with increased specific surface area through hydrothermal activation approach as shown in Figure 11.7a. TEM images show that crumpling of graphene sheet within graphene fiber (Figure 11.7b, c). The electrode showed areal capacitance of 1060 mF/cm^2 at 1 mA/cm^2, long cell life (50,000 GCD cycles), and ultrahigh energy density of 23.5 $\mu Wh/cm^2$ at 1 A/cm^2 and a power density 26.3 mW/cm^2 at 200 mA/cm^2 and good rate capability (Figure 11.7c).[92] The effective surface area of graphene not only depends on the distribution of pores at solid state but also highly on the number of graphene layers with less agglomeration. Wang et al. prepared graphene-based materials through reduction of GO using gas-based hydrazine reduction at room temperature. As-prepared electrode showed less agglomeration than the chemically modified graphene (CMG), yielding excellent specific capacitance of ~205 F/g with 90% of capacitance retention, and corresponding specific energy of 28.5 Wh/kg at 10 kW/kg.[93]

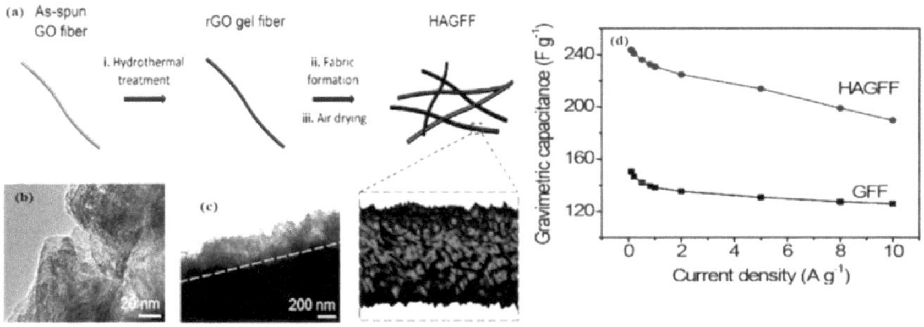

FIGURE 11.7 (a) Schematic illustration of the hydrothermal activation process for HAGFF (hydrothermally activated graphene fiber fabrics) with a hierarchical structure, TEM images of HAGFF at (b) low and (c) high magnification, and (d) gravimetric capacitance at different current densities for HAGFF and GFF. Reprinted with permission from Ref. [92].©2017 American Chemical Society.

The 2D graphene nanomaterials are emerging as unique morphology carbon materials for supercapacitor energy storage applications due to their good chemical stability, excellent carrier mobility, and large surface area. This class has different morphologies, such as nanofilms, nanosheets, nanopaper, nanocoatings, and nanolayers.[94] Graphene surface functionalities have an immense effect on the capacitive value of the electrode. Fang et al. developed functionalized graphene through acid-assisted method, showing a gravimetric capacitance of 400 F/g.[95] In another way, Du et al. prepared functionalized graphene sheets from graphite oxide by low-temperature thermal exfoliation approach followed by carbonization in N_2 with improved BET surface area of 404 m^2/g. As-prepared material indicated a gravimetric capacitance of 264 F/g in 2 mol/L KOH aqueous electrolyte and 73 F/g in organic $LiClO_4$ electrolyte at 5 mV/s.[96] Another low-temperature (as low as 200 °C) exfoliation (Chemical) was proposed by Lv et al. with the help of vacuum pressure to prepare graphene from graphite oxide which offered gravimetric 264 F/g in aqueous electrolyte and 122 F/g in organic electrolyte at 100 mA/g.[97] In another way, Yu et al. achieved gravimetric capacitance of 135 F/g at 0.75 A/g in aqueous 2M KCL solution in transparent graphene film prepared through vacuum filtration.[98]

The performance of graphene materials as supercapacitors can be enhanced by increasing the number of layers and the associated surface area leading to the development of graphene-based macrostructures with 3D networks, that is, graphene aerogels, graphene foams, and graphene sponges. These 3D graphene materials have unique structure morphology with micro-, meso-, and macro-interconnected pores, tunable surface area, abundant active sites, multidimensional electron transportation, and fast electrolytic ion diffusion which enhance the electrochemical activities of the material, thus they improve energy and power density, and overall supercapacitance performance.[71] Lee et al. achieved a large gravimetric capacitance of 206 F/g while retaining 96% of capacitance after 10000 cycles at a higher current density in mesoporous graphene nanoball (MGB).[99] Xu et al. prepared a 3D graphene hydrogel-based solid-state supercapacitor, yielding a high gravimetric capacitance of 186 F/g at 1 A/g and good cyclic performance with 8.4% reduction of capacitance after 10,000 GCD cycles.[100] Xu et al. further improved the electrochemical performance of 3D graphene through incorporating hydroquinones into it to prepare functionalized graphene hydrogel (FGH), yielding a specific capacitance of 441 F/g at 1 A/g as compared with nonfunctionalized graphene hydrogel (211 F/g). Further, FGH demonstrated high rate capability (80% capacitance retention at 20 A/g) and good cyclic response (only 14% reduction of capacitance value even after 10, 000 GCD cycles).[101] Laser scribing is another simple, catalyst-free, eco-friendly technique for the preparation of graphene. Zhang et al. prepared 3D porous graphene from carbon nanosphere through single-step CO_2 laser scribing process that offered areal capacitance of 0.941 mF/cm^2 in 0–2 V window, an ultrahigh volumetric power density of 28 W/cm^3, and long-term cycling stability (98.6% capacitance retention value even after 12,000 GCD cycles) in symmetric cell.[102] An ultrahigh gravimetric capacitance of 366 F/g at 2 A/g was obtained in graphene aerogel–nickel foam (GA@NF) hybrid material prepared by Ye et al. through freeze-dying and followed by thermal annealing process.[103]

11.3.1.2 GRAPHENE OXIDE/REDUCED GRAPHENE OXIDE AND ITS MACROSTRUCTURAL COMPLEXITIES FOR EDL ELECTRODE

Graphene oxide (GO) is an another important element of the graphene family with layered structure and surface-related properties and is covalently bonded to oxygen functional groups, such as hydroxyl (C–OH), epoxy (C–O), and carboxyl (C=O) on the basal planes and edges. It is an interesting form of graphene nanomaterials due to its high conductivity after reduction and selectivity after functionalization.[104] They are identical for the supercapacitor electrodes as their graphene sides could be exposed to electrolyte with possibly high surface area, yielding high specific capacitance.

Rajagopalan et al. introduced chemically modified graphene oxide to show the effect of potassium hydroxide (KOH) treatment of GO on supercapacitor performance by giving gravimetric capacitance of 253 F/g at 0.2 A/g in aqueous solution and outstanding cyclic stability.[105] Aboutalebi et al. synthesized flexible, porous GO and rGO fibers and yarns through wet-spinning strategy for supercapacitor that responded a remarkable gravimetric capacitance of 409 F/g at 1 A/g in aqueous solution.[106] Fan et al. fabricated 1D GO nanoscrolls (GONS) via rolling GO with assistance in liquid nitrogen through shock cooling technique as shown in Figure 11.7a–c. 2D GO wrinkled into thin and curled 1D tubular structures (Figure 11.7d, e). As prepared electrode exhibited propitious gravimetric capacitance of 156 F/g at 1 A/g in aqueous KOH solution with long cycle life.[107] Down et al. proposed the GO-based supercapacitor via a facile screen-printing technique that offered a highly competitive capacitance of 423 F/g with excellent power handling capability of up to 13.9 kW/kg and an energy density of 11.6 Wh/kg.[108]

Although the theoretical gravimetric capacitance value of 2D graphene is about 550 F/g, but the capacity of most of the practical solid-state supercapacitors lie in the range between 80 and 118 F/g due to reduction of active surface area and diffusion rate on agglomeration. Recently, graphene-based macrostructures with 3D networks, such as graphene aerogels, foams, and sponges, has been attracted toward the high-performance solid-state supercapacitor due to tunable electroactive surface, free-active sites, multidirectional electron transportation, and fast ion diffusion.[109] Another binder-free defect-etching technique was suggested by Xu et al. to prepare solution dispersible holey graphene oxides (HGO) with excellent in-plane nanopores for improved ion diffusion. This HGO electrode showed an ultrahigh gravimetric capacitance of 283 F/g at 1 A/g in 1M H_2SO_4 aqueous electrolyte and good cyclic routine, yielding 94% capacitance retention over 20,000 GCD cycles at 20 A/g.[110] Karthika et al. proposed the functionalized exfoliated graphene oxide (EGO) for supercapacitor electrodes by simple chemical methods, yielding a ultrahigh gravimetric capacitance of 146 F/g and outstanding energy density of 20 Wh/kg as compared with pure, lower level oxidized and reduced electrode.[111]

Due to the oxygen functionalities, conductivity of GO is heavily reduced causing shutting down the charge transportation. But, reduction by chemical or thermal process leads to a structure close to graphene with high electrical conductivity.[112, 113] Vivekchand et al. prepared thermally reduced graphene oxide for supercapacitor electrode that showed a high capacitance of 117 F/g and operating potential window of 3.5 V in an ionic liquid.[114] Zhang et al. developed conductive, free-standing, and flexible porous carbon films by chemical activation of rGO paper using KOH, yielding high gravimetric surface area of 2400 m^2/g. Symmetric cell prepared using these electrodes exhibited a gravimetric capacitance of 120 F/g at 10 A/g and specific power of 500 kW/kg at 26 Wh/kg in organic electrolyte.[115] The charge transfer characteristics of GO was greatly enhanced by removing oxygen functional groups to form rGO through electrochemical method as proposed by Yang et al. The electrode offered a high gravimetric capacitance of 223.6 F/g at 5 mV/s in aqueous electrolyte.[116] (Figure 11.8).

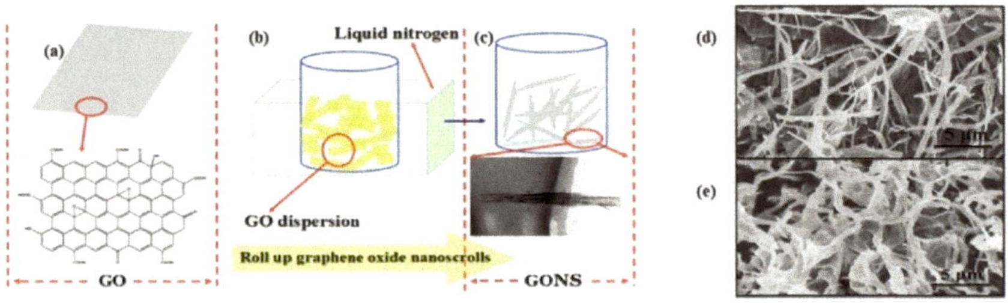

FIGURE 11.8 Schematic illustration of the process for the rolling up formation of GONS. (a) GO dispersed in water and (b) rolling up graphene oxide nanoscrolls with liquid nitrogen, (c) the rolled up GONS, (d) SEM images of GONS, and (e) SEM images of GNS. Reprinted with permission from Ref. [107]. ©2015 Springer Open Access.

To prevent self-aggregation of graphene sheets, they are assembled into 3D architecture which effectively improves the surface area and thus giving excellent capacitive performance. To emphasize this, Ye et al. proposed the 3D graphene aerogel–nickel foam hybrid EDLC electrode, yielding gravimetric capacitance of 366 F/g 2 A/g.[117] Electrochemical reduction is a green synthesis method performed in a room temperature using nonhazardous chemicals. Yu et al. showed a high gravimetric capacitance of 246 F/g in aqueous electrolyte in electrochemically reduced graphene oxide.[118] Chen et al. utilized mild reducing agent, such as hydrobromic acid to prepare partially RGO which retained some oxygen functional group, leading to the additional pseudocapacitance for the RGO electrodes and thus showed an unusual capacitance value of 348 F/g at 0.2 A/g in aqueous electrolyte.[119]

11.3.2 GRAPHENE-BASED NANOMATERIALS FOR COMPOSITE EDL ELECTRODE

Graphene nanomaterials with different micro and macro structures are emerging as energy storage electrodes for supercapacitor applications due to their large surface area, high conductivity, good chemical stability, and remarkable electrons mobility. But, the self-agglomeration or irreversible re-stacking of graphene sheets causes a significant deterioration in surface area and diffusion rate via π-π stacking interaction and van der Waals force between graphene sheets. Hence, even if graphene nanomaterials have theoretical capacitance of 550 F/g, but their practical value is greatly reduced due to the unsuccessful utilization of full surface area.[120, 121] Several techniques, such as hydrothermal reduction, incorporating reducing agents, and adding spacer between graphene nanosheets have been successfully utilized to encounter these problems. In this way, however the surface area of the highly conductive graphene-based materials can be increased, but the low contact resistance between graphene and metal current collector is another substantial challenge for the development of high-performance all-solid-state supercapacitor. In this regard, several strategies, such as electrophoretic deposition, chemical vapor deposition, vacuum filtration, etc. have been imposed for the preparation of binder-free electrodes.[122, 123]

11.3.2.1 GRAPHENE AND ITS MACROSTRUCTURAL COMPLEXITIES FOR COMPOSITE EDL ELECTRODE

The introduction of spacers or stabilizers into the graphene layers is an effective way to increase the specific capacitance of the graphene. Further, this technique not only increases the accessibility of electrolyte ions, but also improves the utilization of both surfaces and nanochannels of graphene sheets.[125] Wang et al. introduced inexpensive carbon black (CB) as spacer into graphene paper (GP) through vacuum filtration to develop binder-free flexible pillared GP electrode in which CB eliminates the restacking of graphene sheets. This pillared GP-based supercapacitor exhibited excellent gravimetric capacitance of 138 F/g and 83.2 F/g in aqueous and organic electrolyte, respectively and good cyclic performance with 3.85% (in aqueous electrolyte) and 4.35% (in organic electrolyte) capacitance degradation values.[126] Again, Yan et al. proposed the graphene nanosheets/CB prepared by sonication and in situ reduction, exhibiting a gravimetric capacitance of 175 F/g at 10 mV/s in 6M KOH aqueous electrolyte.[127] Adding AC into graphene improves the surface area, packing density (~0.3 g/cm^3), diffusion pathway, and conductivity. Zheng et al. proposed the graphene/AC nanosheets composite through hydrothermal carbonization using glucose and chemical activation by KOH (Figure 11.9), which offered a maximum gravimetric capacitance of 210 F/g in aqueous electrolyte and a capacitance of 103 F/g in organic electrolyte was achieved which are far higher than the commercial AC (76 F/g at 1 mV/s).[128]

Among the heteroatoms, nitrogen is proved to be efficient, as incorporation of nitrogen into graphene can gain wettability of the contact surface and accessibility of the ions.[130] Li et al. prepared 3D interconnected porous nitrogen-enrich graphene (NG) hydrogel through hydrothermal treatment using ethanediamine as reducing and doping agent. This 3D porous structure enhances the ion accessibility during EDL mechanism. As-prepared binding-free electrode exhibited per unit mass capacitance of 316.8 F/g at 0.5 A/g.[131] Si et al. proposed the platinum-graphene composite electrode in which former nanoparticles act as spacer for restacking of graphene resulting an improvement of specific capacitance from 14 F/g to 269 F/g.[132] Chen et al. developed 1D silver nanowire-doped graphene through filtration process. As-prepared Ag NW/G paper was flexible and conductive with amazing potential response for energy conversion and energy storage devices.[133]

Zhi et al. prepared Ag nanowires/3D-graphene foam/ordered mesoporous carbon (Ag-GF-OMC)-based electrode in which OMC was used as flexible current collector, whereas Ag increases the conductivity of the composite. The Ag-GF-OMC electrode achieves a high capacity of 213 F/g in 6M KOH aqueous solution, shows excellent stability with more than 90% capacitance retention after 10,000 cycles, and specific energy of 4.5 Wh/kg was achieved with this electrode in supercapacitor applications.[134] Cherusseri et al. derived vertically aligned graphene–carbon fiber hybrid electrode with improved electrolyte–ion diffusion by electrophoretic process. As-prepared fiber electrode exhibited a remarkable gravimetric capacitance of 333.3 F/g with 100% capacitance value even after 100000 GCD cycles.[135] Kou et al. proposed a wet-spinning assembly strategy to spin graphene-carbon nanotube core-sheath fibers for yarn supercapacitor electrode which showed amazing capacity of 269 mF/cm^2 and 177 mF/cm^2 and specific energy of 5.91 μWh/cm^2 and 3.84 μWh/cm^2 in liquid and solid electrolyte, respectively.[136] Cheng et al. used CVD process to grow CNT along with graphene fiber for flexible supercapacitor electrode due to ultrahigh area-specific capacitance of 1.2–1.3 mF/cm^2 with good cyclic performance.[137]

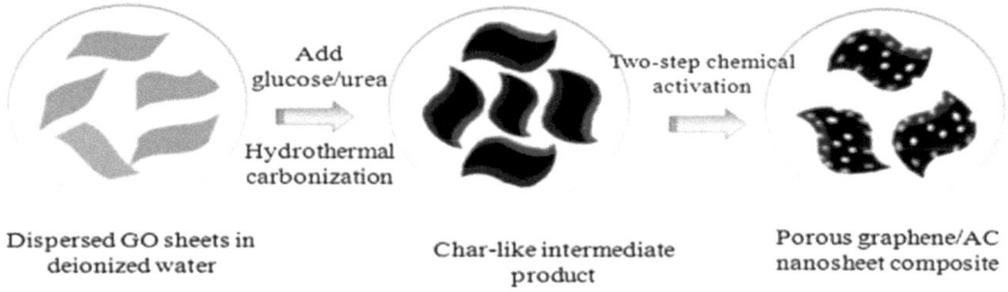

FIGURE 11.9 Schematic illustrations showing the experimental steps of preparing porous graphene/AC nanosheet composite. Reprinted with permission from Ref. [128]. © 2014 Elsevier B.V.

11.3.2.2 GRAPHENE OXIDE/REDUCED GRAPHENE OXIDEAND ITS MACROSTRUCTURAL COMPLEXITIES FOR COMPOSITE EDL ELECTRODE

Nonconductivity of GO due to the presence of the oxygen-groups on the carbon layers is eliminated by thermal reduction or chemical reduction. Moreover, its pore volume and self-agglomeration are improved with the incorporation of hetero-atoms or other carbon materials, such as AC, CB, or CNT. Due to the similarities in structural and physical properties, the carbon nanotubes (CNTs) may be used as a spacer to avoid the agglomeration of graphene film in the hybridized assembly.[138] Due to the strong π-π interactions between graphene sheets and CNTs, mechanical and electronic properties of the combined structure are greatly influenced. To emphasize this concept, Qiu et al. used hybridization technique to form free-standing

GO/CNT films through supermolecular interaction followed by vacuum filtration that showed an ultrahigh per unit mass capacitance of 140 F/g at 0.1 A/g in 1M H_2SO_4 when used as supercapacitor electrode.[139] Again, supercapacitive behavior of GO electrodes is strongly influenced by the reduction process and the amount of CNT incorporated with it. A partial reduction of GO papers using hydrazine or annealing at a low temperature of 220°C in air yields excellent capacitive performance than the heavy reduction or high temperature treatment. Huang et al. proposed the partial reduction of GO with addition of small amount of CNT (12.5 wt. %) to form thick GO/CNT sandwich papers that exhibited an outstanding gravimetric capacitance of 151 F/g at 0.5 A/g with 86% capacitance retention value even after 6000 cycles at 5 A/g.[140]

Zhu et al. improved the electrochemical performance of graphene by incorporating high capacity of AC with excellent rate capability of 3D graphene. They prepared AC/rGO aerogel through hydrothermal treatment followed by freeze-drying process as shown in Figure 11.10a. Due to the AC particles encapsulated on a 3D network (Figure 11.10b, c), a maximum gravimetric capacitance of 294 F/g (144 F/g for nonaqueous) in aqueous electrolyte at 0.05 A/g and good rate capability with 70.4% capacitance retention value were achieved with this electrode (Figure 11.10d).[129]

Supercapacitor electrodes based on graphene doped with heteroatoms show high electronic conductivity, surface wettability, and improved capacitance in addition to pseudoresponse. Nitrogen is one of the dopant elements introduced into graphene structure to form strong atomic bond. Xu et al. prepared a highly N-doped graphene electrode (NrGO/GQDs) for supercapacitor energy storage applications which exhibited excellent per unit mass capacitance of 344 F/g at 0.25 A/g, ultrahigh energy density of 43 Wh/kg at a power density of 417 W/kg, and good cyclic performance (82% capacitance remains even after 3000 GCD cycles).[141] Kumar et al. synthesized simple, low-cost rGO/MWCNT flexible nanocomposite film electrodes via chemical route that showed 200 F/g specific capacitance at 0.25 A/g in 2M KOH aqueous electrolyte.[142] Again, Yu et al. described a process for the synthesis of carbon microfiber made of single-walled carbon nanotube (SWCNT) and nitrogen-doped rGO that showed remarkable electrical conductivity and high packing density while maintaining a large ion-accessible surface area. The volumetric capacitance and energy density were 300 F/cm^3 and ~6.3 mWh/cm^3, respectively in PVA/H_3PO_4 gel electrolyte.[143] Surfactants could also be intercalated onto GO sheets, followed by reduction to produce graphene electrode. Zhang et al. intercalated graphene oxide with different surfactants, such as tetrabutylammonium hydroxide (TBAOH), cetyltrimethylammonium bromide (CTAB), and sodium dodecylbenzene sulfonate (SDBS) to prepare surfactant-stabilized graphene materials (GTR, GCR, and GSR) followed by reduction using hydrazine, yielding a maximum gravimetric capacitance of 194 F/g in 2M H_2SO_4 aqueous electrolyte and 144 F/g in ionic liquid at current density of 1 A/g with TBAOH.[144]

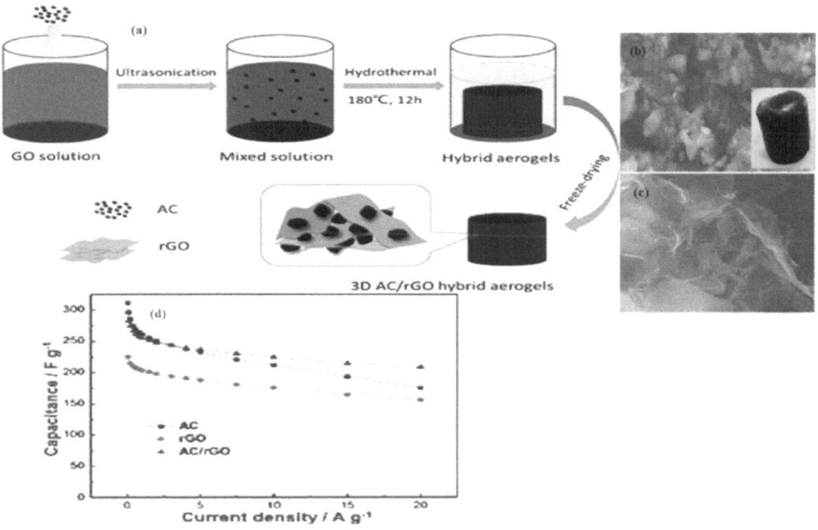

FIGURE 11.10 (a) Schematic diagram for the preparation of the 3D AC/rGO hybrid aerogels, (b) SEM images of AC/rGO aerogels, (c) magnified SEM image of AC/rGO aerogels, and (d) plot of specific capacitance at various current densities. Reprinted with permission from REF. [129]. ©2017 Wiley-VCH Verlag GmbH & Co.

11.3.3 GRAPHENE-BASED NANOMATERIALS FOR COMPOSITE HYBRID ELECTRODE

The large surface and good conductivity of graphene results in enormous charge accumulations at electrode–electrolyte interface. However, the actual capacitive behavior of bare graphene nanomaterials is not at satisfactory level due to agglomeration causing the reduction of surface area and hence capacitance. Recently, the idea that graphene incorporation with pseudocapacitive materials has emerged as an alternative way for energy storage candidates. Nanocomposites of graphene and metal oxides/hydroxides/nitrides/sulfides or conducting polymers have attracted wide attention for supercapacitor applications due to synergistic effect of redox reaction in pseudocapacitive material with high surface area/conductivity of graphene material. In this case, graphene improves the overall conductivity by creating charge transfer channel, while pseudocapacitance appears from the metal oxides/hydroxides/nitrides/sulfides or conductive polymers.[91]

11.3.3.1 GRAPHENE AND ITS MACROSTRUCTURAL COMPLEXITIES FOR COMPOSITE HYBRID ELECTRODE

11.3.3.1.1 Transition Metal Oxides (TMOs) Composite with Graphene and Its Macrostructural Complexities

Amorphous ruthenium oxide (RuO_2) is an outstanding candidate for pseudocapacitive electrode due to its excellent conductivity, highly reversible Faradic reactions, and ultrahigh capacitance value (720 F/g). But, its practical application was limited because of high cost, low porosity, low rate capability, and rapid depletion of power density at high charge/discharge rate. These drawbacks are eliminated by incorporating RuO_2 into carbon-based materials.[145]

Lin et al. has proposed ruthenium oxide/graphene sheet (RuO_2/GS) composite by hydrothermal and low-temperature annealing treatment in which paper-like graphene sheet enabled pore making. This composite electrode showed specific capacitance of 551 F/g at 1 A/g and its capacitance was retained up to 94.3% after 2000 cycles at 5 A/g which was more than the GS.[146] With the rapid development of wearable electronic devices (WEDs), the demand for graphene-based fiber electrodes is increasing due to their flexibility, wearability, and energy harvesting capability. However, bare graphene-based fiber electrode suffers from low capacity and limited potential applicability. Inserting spacers like CNTs into graphene fiber to expand its layer distance is an effective strategy to synthesize porous graphene fibers. But, high production cost of CNTs and nonscalable fabrication process of graphene fibers limit their application.[147, 148] Hence, developing pseudocapacitive material into fibrous graphene is an effective way to improve the capacitance of the electrodes. Li et al. incorporated graphene hydrated RuO_2 @graphene in situ on flexible carbon fibers through vapor-phase hydrothermal method that showed high volumetric capacitance of 210.14 F/cm^3. Their symmetric cell has excellent cyclic performance with 98.40% capacitance retention value after 50,000 continuous GCD cycles.[149] Again, Lin et al. hydrothermally incorporated RuO_2 particles on the surface of paper-like graphene sheets (GS) to fabricate RuO_2/GS composite film, exhibiting a specific capacitance of 551 F/g at 1 A/g in 1M H_2SO_4 electrolyte.[150] Further, Thangappan hydrothermally prepared ruthenium oxide nanoparticles anchored on graphene nanosheets (RuO_2/G), yielding a specific capacitance of 441.1 F/g at 0.1 A/g, operating voltage window of 1.6 V, and very small amount of capacitance (6% of its maximum value) was degraded after 1000 GCD cycles. The energy density was calculated as 61.2 Wh/kg at a power density of 183.8 W/kg in 1M Na_2SO_4 electrolyte.[151] In another way, Wang et al. developed a 3D hydrous ruthenium oxide (RuO_2)-anchored graphene and CNT hybrid foam (RGM) (Figure 11.11a) porous composite-based symmetric cell that showed a maximum gravimetric capacitance of 502.78 F/g with 106% capacitance retention after 8100 GCD cycles due to uniform incorporation of RuO_2 on hybrid structure (Figure 11.11b, c).[152]

Manganese dioxide (MnO_2) is getting importance as electrode material due to its low cost, environmental friendliness, a wide potential window, and large specific capacitance.[153] So, the incorporation of MnO_2 layer with graphene fiber is an effective way to improve the capacitive performance of the hybrid materials. But high mass loading of MnO_2 will increase the electrode resistance and utilization rate resulting in the reduction of specific capacitance. Wet-spinning is a most widely utilized method for graphene fiber production to encounter these limitations. Ma et al. wrapped graphene nanosheets on MnO_2 nanowires through wet-spinning process to inhibit self-aggregation and facilitating high mass loading and electrochemical utilization rate of MnO_2. Its symmetric cell with PVA/H_3PO_4 electrolyte showed a high volumetric capacitance of 66.1 F/cm^3, energy density (5.8 mWh/cm^3), and power density (0.51 W/cm^3).[154] Chen et al. used MnO_2-modified graphene sheets on graphene fiber (MnO_2/G/GF) that offered area-specific capacitance of 9.1–9.6 mF/cm^2 (length-specific capacitance of 143 μF/cm and gravimetric capacitance of 34–36 F/g) which was more than other fiber-shaped supercapacitor.[155] Yan et al. synthesized graphene-MnO_2 composite film by microwave irradiation that exhibited high specific capacitance of 310 F/g at a scan rate of 2 mV/s.[156] Planar supercapacitors have recently attracted the researchers for stretchable/flexible electronic devices due to the promising flexibility and shorter ion transportation path due to 2D ions transportation. The SC device expands along the 2D horizontal planes making the device compact. Peng et al. reported hybrid nanostructures of qasi-2D ultrathin MnO_2/graphene nanosheets by chemical integration of δ-MnO_2 nanosheets on graphene by strong electrostatic interaction in the solution treatment process which gives a high specific capacitance of 267 F/g at 0.2 A/g, excellent rate capability, and cycling stability with 92% of the capacitance was retained after 7000 cycles.[157] Again, Sadak et al. proposed the electrochemical deposition of MnO_2 nanoflowers on graphene paper (GrP) yielding excellent capacitive performance with specific capacitance of 385.2 F/g at 1 mV/s in 0.1M Na_2SO_4 electrolyte.[158]

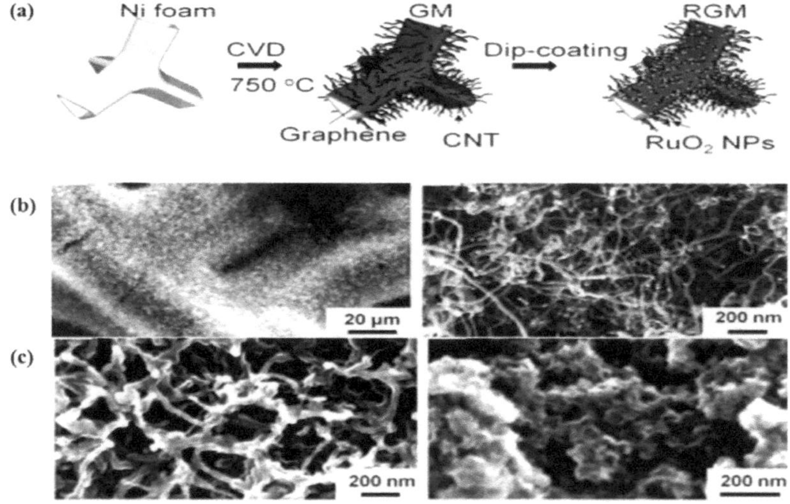

FIGURE 11.11 (a) Schematic illustration of the preparation process of RGM nanostructure foam and (b–c) SEM images of as-grown GM foam and lightly loaded RGM and heavily loaded RGM.

Source: Reprinted with permission from Ref. [152]. © 2014 Springer.

But, for high power applications like hybrid electric vehicles or fuel cell electric vehicles, the requirement of high mass loading of MnO_2 in hybrid electrodes reduces the specific capacitance due to close-packed structures resulting in limited electroactive surface area. This limitation was eliminated by He et al. who prepared a flexible supercapacitor electrode using 3D graphene network loaded with MnO_2 (containing ~92.9% of mass of the entire electrode) by electrodeposition, resulting in a high area capacitance of 1.42 F/cm^2 at a scan rate of 2 mV/s. Furthermore, capacitance has been maximized to 130 F/g by the optimization of electrodeposition process.[159] Graphene hydrogels (GHs), one of the members of 3D graphene family, can be developed through the reduction of GO. Several reducing agents, such as HI, NaI, $N_2H_4.H_2O$, ascorbic acid, etc. have been utilized for GHs preparation. But, among all of them, glucose is proved to be a low-cost and environment-friendly reducing agent for the fabrication of GHs.[160] Meng et al. prepared 3D MnO_2/graphene hydrogel composite through one-spot hydrothermal process utilizing glucose as a sort of green and effective reductant, and hence, showing a maximum capacitance of 200.6 F/g at 1 A/g.[161]

Vanadium pentoxide (V_2O_5) is also a most promising used as electrode materials for supercapacitor energy storage applications because of its low-cost, high capacity, which exhibits variable redox oxidation states. Only the limitations associated with V_2O_5 are low conductivity and poor cycling stability that are extensively improved by V_2O_5-based composite hybrid electrode.[162] Ndiaye et al. synthesized 3D vanadium pentoxide/graphene foam composite (3D V_2O_5/GF) by hydrothermal and freeze-drying process. The synthesized composite when combines with carbon-based positive electrode, showed a high operating cell voltage of 1.6 V in 1M KOH electrolyte and delivered a maximum energy density of 39 Wh/kg and a power density of 947 W/kg at current density of 1 A/g.[163] He et al. used an in situ hydrothermal technique to deposit cobalt oxide (Co_3O_4) on graphene to form Co_3O_4@GE to deliver an ultrahigh gravimetric capacitance of 415 F/g at 3 A/g in aqueous electrolyte.[164] In spite of excellent supporting backbone of GNSs, strong π-π interactions and van der Waals force between the planar basal planes cause their layers to restack. These drawbacks can be overcome by employing a vertically aligned graphene nanosheets (VAGN). Liao et al. hydrothermally deposited Co_3O_4 nanoparticles on VAGN and carbon fabric composite to fabricate Co_3O_4/VAGN/CF hybrid structure which exhibited a high specific capacitance of 3480 F/g. A symmetric supercapacitor was developed using this sample which resulted maximum energy density of 80 Wh/kg at power density of 20 kW/kg.[165] In another way, Dong et al. incorporated Co_3O_4 nanoparticles into three-dimensional graphene foam formed by in situ synthesis in graphene foams by hydrothermal method with a gravimetric capacitance of 1100 F/g at discharge current density of 10 A/g.[166] Again, Bao et al. descried another way to prepare 3D graphene frameworks/Co_3O_4 composites by the thermal explosion that showed a high specific capacitance of 1765 F/g.[167]

11.3.3.1.2 Conducting polymers Composite with Graphene and Its Macrostructural Complexities

Recently, conducting polymers (CPs) like polyaniline (PANI), polypyrrole (PPy), and polythiophene (PTh) have emerged as electrode materials for supercapacitor energy storage applications owing to their flexibility, low cost, fast pseudo-Faradaic charge transfer, and high conductivity in doped states with little environmental hazards. However, the main drawback of CPs is their poor cycling capability due to structural instability during charge–discharge process. Again, large volumetric swelling and shrinking due to doping and dedoping lead to poor lifetime.[168] Nowadays, CPs composite with graphene materials are drawing the massive attention in which additional pseudocapacitance arises from CPs and graphene materials and enhances their mechanical stability and life cycle.

Among conductive polymer, PANI is considered to be most promising electrode materials due to its low cost, easy synthesis, relatively high conductivity, and fast and repeatable surface redox reaction through continuous movement (insertion/extraction of electron in π-orbital) of electrons across the electrode–electrolyte interface. Its capacitive performance is further extended by considering nanometer-sized polymer particles with high active region and short diffusion path. These are essential for the access of electrolyte to the electrode in both EDL and redox mechanism.[169] Hence, when they are composite with high conductive graphene electrode, enormous capacitance is generated, leading to the high energy density for commercial energy storage applications.

Mondal et al. prepared GQDs-doped polyaniline composite electrode materials through chemical oxidation of aniline which exhibited wonderful gravimetric capacitance of 1044 F/g at 1 A/g with a high energy density of 117.45 Wh/kg at power density of 448.8 W/kg in 0.5 M H_2SO_4 aqueous electrolyte.[170] Fiber-based SC coated with 1D PANI not only increases the specific capacitance but also induces pores onto the structure, giving quick ions transportation inside the electrode resulting in excellent rate capability.[171–174] Therefore, Zhang et al. developed a core-sheath porous structure by incorporating polyaniline nanorods on graphene fibers (GF@

PANI) through chemical polymerization, yielding ultrahigh capacitance of 357.1 mF/cm^2, with energy density of 7.93 μWh/cm^2 at power density of 0.23 mW/cm^2.[175] Again, Zhang et al. prepared chemically modified graphene and PANI nanofiber composite through in situ polymerization, showing capacitance as high as 480 F/g at 0.1 A/g.[176] In a similar manner, a graphene–PANI composite was developed by Meng et al. by a polymerization method. The structure exhibited specific capacitance of 1046 F/g and 96% of capacitance value was maintained even when the current density increased from 10 to 100 A/g.[177] A maximum gravimetric capacitance of 1046 F/g at 1 mV/s was obtained in graphene nanosheets/PANI composite prepared by Yan et al. through in situ polymerization. In this composite, GNS not only offers highly conductive support but also provides large surface area for the well-dispersed deposition of nanoscale PANI particles.[178] Another facile method was reported by Kulkarni et al. to prepare a self-supported polyaniline nanofiber (PANI) on 3D graphene framework (GF) which delivered a maximum specific capacitance of 1024 F/g in 1M H_2SO_4 aqueous electrolyte at 10 mV/s (1002 F/g at 1 mA/g) with a capacitance retention value of 86.5% after 5000 GCD cycles.[179] Further, composite films of chemically converted graphene (CCG) and polyaniline nanofibers were prepared by Wu et al. through vacuum filtration in which PANI–NFs are sandwiched between CCG layers. The composite electrode containing 44 wt% showed a high electrochemical capacitance of 210 F/g at a discharge rate of 0.3 A/g and retained about 94% of this value when the current density was increased to 3 A/g.[180] Sawangphruk et al. proposed a composite by incorporating silver nanoparticles (AgNP) as co-oxidizing agents into the composite through spray-coating to form AgNP–PANI–graphene nanocomposite, yielding a maximum specific capacitance of 828 F/g at 1.5 A/g with 97.5% capacity retention value after 3000 cycles.[181]

Polypyrrole (PPy) is another most promising electrode material for supercapacitor due to its intrinsic electrical conductivity, low cost, simple synthesis, and high capacitance. But it shows poor stability and rate capability which limits its wide application. To improve electrochemical properties, graphene can be incorporated with PPy to form a composite electrode in which graphene not only increases the electrical conductivity but also effectively improves the stability of the conducting polymer during energy storage process.[182] Above strategy was reported by Zhang et al. to prepare PPy/graphene composite sheets using in situ polymerization method, showing a high specific capacitance of 482 F/g at 0.5 A/g.

Excellent cycling response was reported as there was a little attenuation (less than 5%) of specific capacitance after 1000 cycles.[183] Another facile and simple wet-spinning approach was suggested by Ding et al. to prepare graphene/polypyrrole (G/PPy) composite fiber with controlled diameter (15–80 μm) having area-specific capacitance of 115 mF/cm^2 at current density of 0.2 mA/cm^2.[184] On the other hand, Zhang et al. synthesized a graphene-based film by intercalation of PPy spheres between graphene layers, thus forming a hollow structure, exhibiting specific capacitance of 500 F/g at 5 A/g.[185]

11.3.3.2 GRAPHENE OXIDE/REDUCED GRAPHENE OXIDE AND ITS MACROSTRUCTURAL COMPLEXITIES FOR COMPOSITE HYBRID ELECTRODE

11.3.3.2.1 Transition Metal Oxides (TMOs) Composite with Graphene Oxide/Reduced Graphene Oxide and Its Macrostructural Complexities

Graphene-based electrodes usually have high power density and long cell life. But their energy storage capacity is limited due to their physical electrostatic charge storage mechanism. On the other hand, pseudocapacitor stores charge through fast and repeatable surface redox reactions which leads to the 10–100 times specific capacitance than the EDLC. Hence, extensive researches have been carried out to bridge various pseudocapacitive materials into graphene-based materials.

The ruthenium oxide (RuO_2) is an outstanding pseudocapacitive material due to its ultrahigh theoretical specific capacitance (~1400 F/g), good electrical conductivity (~10^5 S/cm), wide potential window, fast charging–discharging rate, and high chemical stability.[186] Zhai et al. hydrothermally prepared nano-RuO_2 decorated holey graphene composite fiber-based microsupercapacitor with 42.5 wt% Ru that delivered excellent volumetric capacitance of 1054 F/cm^3 at 2 mV/s in 1M H_2SO_4 aqueous electrolyte. Again, in symmetric cell with PVA/H_3PO_4 gel electrolyte, it attained energy density of 27.3 mWh/cm^3 at power density of 2954.1 mW/cm^3 with excellent cycling performance.[187] The uniform anchoring of RuO_2 nanoparticles onto the surface of RGO sheets yields excellent conductive path for the electron transport to and from the current collector.

To emphasize this, RGO/ruthenium oxide composite film was developed by Chen et al. through sol–gel synthesis followed with hydrothermal and annealing process that showed a specific capacitance of 471 F/g at 0.5 A/g in 1M H_2SO_4 aqueous electrolyte with 45 wt% of RuO_2 and retained 92% of the capacitance value after 3000 cycle.[188] In a same way, Amir et al. prepared RGO–RuO_2 composite via in situ sol–gel route. Due to the uniform dispersion of ultrafine, hydrated amorphous RuO_2 nanoparticles (1.0–2.0 nm) on RGO sheets, RGO–RuO_2 composite showed a maximum capacitance of 500 F/g at a current density of 1 A/g.[189] Amorphous hydrous RuO_2 is an outstanding candidate for supercapacitors due to its proton–electron mixed conductive nature and high capacity. Charge storage arises from Faradaic reaction involving the simultaneous insertion of an electron and proton. However, self-agglomeration and low conductivity of amorphous RuO_2 nanoparticles need to be modified through hybridization. Yang et al. prepared a 3D porous framework of RuO_2/reduced graphene oxide hydrogel (RuO_2/RGOH) by one-step hydrothermal self-assembled route without using any organic solvent or surfactant, yielding gravimetric capacitance of 345 F/g at 1 A/g.[190]

Due to high cost of RuO_2, manganese oxide materials have emerged as attractive pseudocapacitive elements due to their structural flexibility and wide applications in catalysis, ion exchange, molecular adsorption, biosensors, and energy storage. But, poor electrical conductivity and compact structure limits ions diffusion and rate capability. So, MnO_2, when combines with graphene with 3D structure, improves electrochemical performances. Li et al. hydrothermally synthesized the reduced graphene/ MnO_2 (HRG/MnO_2) composite which exhibited an excellent gravimetric capacitance of 211.5 F/g at 2mV/s with only 25% of capacitance decay after 1000 charge/discharge cycles.[191] Again, Hallow micro/nanostructured materials having large surface area offer high electrochemical response in supercapacitor applications due to less self-aggregation with porous structure which improves the accessibility of electrolyte ions toward the active materials.

Chen et al. reported rGO–MnO_2 hollow sphere hybrid electrode materials using a solution-based ultrasonic co-assembly process which showed a maximum specific capacitance of 578 F/g at 0.5 A/g.[192] In another way, Chen et al. fabricated a nanocomposite of graphene oxide with needle like MnO_2 nanocrystals by a simple solution method in water–isopropyl alcohol system at low temperature without the use of any surfactants which demonstrated a specific capacitance of 216 F/g at discharge current density of 0.15 A/g.[193] Im et al. achieved gravimetric capacitance of 473 F/g at 5 mV/s in CNT/MnO_2/GO ternary composite electrode prepared through redox reaction followed by co-filtration.[194] Holey graphene fiber, an outstanding candidate for the fiber-shaped supercapacitor, appears superior to other carbon-based fibers due to its large surface area, more active sites, and having edges for ion diffusion shortcuts. While incorporating MnO_2, it not only improves the conductivity and capacitance of the fiber-shaped supercapacitor but also retains the structural flexibility. Zhang et al. decorated δ-MnO_2 particles on the surface of HRGO. Due to the utilization of full active area of δ-MnO_2, a high specific capacitance of 245 F/g at 1 A/g in 1M Na_2SO_4 aqueous solution was obtained from as-prepared fiber electrode.[195]

Vanadium pentoxide (V_2O_5), an another transition metal oxide, has attracted interest of the researchers due to its intrinsic pseudocapacitive characteristics of high capacitance, multiple oxidation states (V^{2+}, V^{3+}, V^{4+}, and V^{5+}) in the central metal-layered structure, low cost, and low toxicity.[196, 197] Thangappan et al. used electrospinning technique to develop graphene oxide/vanadium pentoxide nanofibers electrode which showed an ultrahigh gravimetric capacitance of 453.824 F/g at 10 mV/s in aqueous electrolyte (2M KOH).[196] Li et al. developed a supercapacitor based on V_2O_5/rGO nanocomposite electrode by solvothermal technique that offered a greater energy density of 74.58 Wh/kg at 500 W/kg, higher gravimetric capacitance of 537 F/g at 1 A/g in aqueous electrolyte.[197] Again, Zhang et al. synthesized a bifunctional RGO/V_2O_5 composite hydrogel, yielding ultrahigh gravimetric capacitance of about 320 F/g at 1.0 A/g.[198] Saravanakumar et al. used co-precipitation method to synthesize 2D reduced graphene oxide (rGO) supported vanadium pentoxide (V_2O_5) networks which showed a maximum specific capacitance of 484 F/g at 0.5 A/g and 251 F/g at 10 A/g, resulting good rate capability and excellent cycling stability by retaining 83% of maximum capacitance after 1000 cycles. This electrode in an asymmetric cell with AC delivered a specific capacitance of 53 F/g at 0.5 A/g, energy density of 7.4 Wh/kg at 127 W/kg.[199] In another way, Li et al. solvothermally prepared V_2O_5/rGO nanocomposite using rod-like V_2O_5 nanocrystals and rGO resulting a capacitance value of 537 F/g at 1 A/g in neutral aqueous electrolyte.[200] Again, Ye et al. used hydrothermal process to develop V_2O_5 nanoribbon on the backbone of 3D rGO, showing a high specific capacitance of 437

F/g at 1 A/g and 234 F/g at 20 A/g resulting a good rate capability as well as excellent stability.[201]

Due to better conductivity, nanoparticle size, porosity, crystallinity, excellent electroactive nature, and ultrahigh theoretical specific capacitance (3560 F/g), cobalt oxide (Co_3O_4) composite with graphene shows remarkable electrochemical energy storage performance. This hybridization not only increases the conductivity of Co_3O_4 nanoparticles but also restricts its volumetric expansion/contraction during electrochemical reactions. Again, decoration of Co_3O_4 with graphene eliminates agglomeration of later part and hence rising the larger surface accessibility of electrolyte ions.[202] Subramani et al. showed a chemical route for direct decomposition of Co_3O_4 on RGO nanosheets without high temperature or high pressure treatment resulting in Co_3O_4/RGO nanocomposites which possess per unit mass capacitance of 784 F/g at 1 A/g in alkaline medium with 10% reduction after 5000 cycles.[203] According to Dong et al., Co_3O_4 nanowires in situ synthesized on 3D graphene foam exhibited a specific capacitance of ~ 1100 F/g at 10 A/g with good cycling stability.[204] Ma et al. used a facile refluxing method followed by a thermal annealing process to incorporate Co_3O_4 porous nanosheets on RGO to form Co_3O_4/RGO composite which showed a gravimetric capacitance of 518 F/g at 0.5 A/g with 7.6% decay of capacitance after 1500 GCD cycles when the current density was increased to 10 A/g.[205] Xie et al. prepared an asymmetric supercapacitor cell using Co_3O_4–rGO nanocomposites as anode, synthesized by co-precipitation followed by thermal treatment and activated carbon (AC) as cathode. The cell with potential window of 1.5 V delivered a gravimetric capacitance of 114.1 F/g, energy density of 35.7 Wh/kg at 225 W/kg, and a small decay of capacitance (5% after 1000 GCD cycles).[206]

11.3.3.2.2 Conducting Polymers Composite with Graphene Oxide/Reduced Graphene Oxide and Its Macrostructural Complexities

Instead of poor long-term stability, high conductive polymer (CP) nanomaterials play an important role to enhance specific capacitance and thereby energy density of the high power graphene derivatives through their composite structure. Among graphene family, GO/RGO has been demonstrated to be good filler for polymer nanocomposites. PANI is one of the most extensively used candidates among all CPs due to its low cost, easy synthesis, multiple redox state, high doping-de doping rate during charge–discharge, and controllable electrical conductivity. Hence, its composite structure with GO/RGO nanomaterials shows excellent response as an electrode in energy storage cells.[207–211]

Dinari et al. used in situ electrochemical polymerization technique to form PANI–GQD composite film on titanium electrode that showed excellent gravimetric capacitance than pure PANI.[212] Though graphene–CP nanocomposites show high capacitance, but reduced graphene oxide–polymer composites exceed in excellence as RGO possesses a good electrical conductivity and high surface area suited for high-performance energy storage electrode.[213, 214] Kazemi et al. proposed a ternary nanocomposite material based on manganese dioxide (MnO_2)-reduced graphene oxide quantum dot (rGOQD)–polyaniline (PANI) synthesized chemically in which rGOQD improves the conductivity and specific surface area, MnO_2 surpluses pseudocapacitance through Faradaic redox reaction, and PANI prevents the aggregation of rGOQD and MnO_2 nanoparticles. Therefore, the MGP ternary material exhibited an excellent electrochemical behavior as an electrode having high gravimetric capacitance of 423 F/g at 5.71 A/g.[215] Zhang et al. proposed CP–RGO nanocomposite through in situ polymerization process using ethanol. Their ass-prepared RGO–PANI exhibited specific capacitance of 361 F/g at 0.3 A/g. Moreover, RGO–PPy and RGO–PEDOT showed specific capacitance of 248 F/g and 108 F/g, respectively, at same current density.[216] Again, excellent properties of graphene and advantages of PANI are incorporated by Wang et al. to prepare graphene/PANI hybrid nanocomposite material using three-step synthesis route of in situ polymerization–reduction/dedoping–redoping process and thus yielding a specific capacitance of 1126 F/g.[217]

3D graphene macrostructure such as graphene hydrogel or aerogel prepared chemically is a layered structure overlapped in a 3D space which not only eliminates the agglomeration of graphene sheet but also allows electrolyte ions to freely diffuse inside and through the network. Wang et al. developed free-standing and flexible 3D reduced graphene (rGN)/PANI composite film by the means of dilute polymerization, yielding a specific capacitance of 740 F/g at 0.5 A/g with 87% of the initial value was retained after 1000 charge/discharge cycles at 10 A/g. Additionally, a maximum energy density of 65.94 Wh/kg was achieved at the power density of 0.2 kW/kg.[218] In another way, Liu et al. described a novel route to synthesize polyaniline-grafted reduced graphene oxide hybrid materials. As-prepared PANI/rGO 3D composite exhibited a maximum specific capacitance of 1045.51 F/g at 0.2 A/g., power density of 60 kW/kg and energy density of 8.3 Wh/kg.[219] Specific capacitance of bare PANI material was enhanced from 190 F/g to 303 F/g when it incorporated with rGO and gold nanoparticles by Shayeh et al. through electrochemical process. The gold nanoparticles modify the conductivity of the structure than PANI/rGO.[220]

Combine polypyrrole (PPy) with graphene-based material to overcome the limitation of PPy such as poor stability and volumetric degradation of electrodes, leading to the enhancement of electrochemical performance.[221] In situ polymerization was used by Konwer et al. to incorporate GO into PPy to form PPy/GO electrodes which offered a high gravimetric capacitance of 421.4 F/g as compared with pure PPy.[222] Again, Fan et al. showed a maximum gravimetric capacitance of 332.6 F/g in GO/PPy composite prepared by in situ oxidation polymerization of pyrrole in the presence of GO. This electrode when used in asymmetric cell with AC offered potential range of 0–16 V with energy density of 21.4 Wh/kg at a power density of 453.9 W/kg.[223] Another approach was demonstrated by Chang et al. to synthesize reduced graphene oxide/polypyrrole (rGO/PPy) film on gold plate by electropolymerization technique, yielding a capacitance of 424 F/g at 1 A/g.[224] Zhang et al. prepared PPy-pillared GO sheets which exhibited excellent specific capacitance of 500 F/g with good rate performance and cycling stability.[225]

PEDOT [poly (3,4-ethlenedioxythiophene)] and its derivatives displayed remarkable characteristics like high electrical conductivity, chemical and thermal stability, and electrochemical potential in a wide range of applications.[226] Lehtimaki et al. proposed that PEDOT and graphene oxide could be assimilated by a simple electropolymerization process on flexible substrates followed by the reduction of graphene oxide, yielding capacitance value of 14 F/cm^2.[227]

11.4 CONCLUSION

An extensive review on recent progress of synthesis, structural design, characterization, and electrochemical performances of graphene and its macrostructural complexities (such as 0D, 1D, 2D, and 3D graphene materials) as an energy storage electrodes are summarized. Various carbon-based materials are comprehensively used in energy storage devices. But among them, graphene has proved to be efficient because of its excellent physicochemical properties. Pure graphene nanomaterials and their derivatives without any functional groups/pseudocoating have been analyzed here for EDLC-type supercapacitor electrodes. Their charge storage mechanism which involves quick storing of charge physically at the electrode–electrolyte interface through non-Faradaic electrostatic process delivers high power density (hundred to thousand fold than battery) and excellent cyclic stability (95%~100% capacitance retention after 1000~10,000 cycles). Also, their quick charging–discharging time is incorporated in the energy recovery systems. But, due to lower specific capacitance in the range of 10–550 F/g, graphene-based EDLC unable to store equal amount of charge (3–30 times lower than battery) as compared with the batteries in the same volume. Hence, to increase the specific capacitance and as well as energy density of graphene-based EDLC-type supercapacitor, transition metal oxides or conducting polymer-based nanocomposite electrodes employed for supercapacitor are also reviewed. But, due to swelling of electrode caused by charge transfer reactions and reduction of pores by the pseudocapacitive nanomaterials during incorporation, rate capability and cyclic stability are greatly decayed. Hence, numerous research efforts have been devoted toward the different macrostructural complexities of graphene-based materials and their hybrids as supercapacitor electrodes for improving their electrochemical performances. The 0D graphene having controlled pore size less than 10 nm improves electrolyte ions accessibility, but conductivity remains low due to the presence of insulating polymer binder. 1D and 2D graphene and graphene nanocomposites have great attention for flexible, stretchable, and wearable electronics. On the other hand, 3D graphene alone as well as incorporation with pseudomaterials improves the gap between a supercapacitor and a battery. The various electrochemical properties along with specific surface area may be improved by modifying graphene-based electrodes using different synthesis techniques, thus making it suitable for supercapacitor electrodes. Mostly, researches must be carried out that enhance the electrochemical performances, especially energy density. These include elimination of re-stacking of graphene sheets, precision controlling of fabrication process for uniform graphene sheets, and proper interaction and intercalation of guest element into graphene for graphene nanocomposites. After critically reviewing, we can conclude that graphene and their nanocomposite-based electrodes for supercapacitors may only serve as standalone power sources in systems that require power delivery for a short duration or take a part of a hybrid system together with other high energy storage devices, such as batteries and fuel cells. But, their long-term standalone energy supply still remains challenging.

KEYWORDS

- **graphene**
- **supercapacitor**
- **specific energy**
- **power density**
- **storage mechanism**

REFERENCES

1. Burke, A. Ultracapacitors: Why, How, and Where is the Technology. *J. Power Sources* **2000**, *91* (1), 37–50.
2. Winter, M.; Brodd, R. J. What Are Batteries, Fuel Cells, and Supercapacitors? 2004.
3. Miller, J. R.; Simon, P. Electrochemical Capacitors for Energy Management. *Sci. Mag.* **2008**, *321* (5889), 651–652.
4. Burke, A. F. Batteries and Ultracapacitors for Electric, Hybrid, and Fuel Cell Vehicles. *Proc. IEEE* **2007**, *95* (4), 806–820.
5. Gidwani, M.; Bhagwani, A.; Rohra, N. Supercapacitors: The Near Future of Batteries. *Int. J. Eng. Invent.* **2014**, *4* (5), 22–23.
6. Miller, J. R.; Burke, A. F. Electrochemical Capacitors: Challenges and Opportunities for Real-world Applications. *Electrochem. Soc. Interface* **2008**, *17* (1), 53.
7. Burke, A.; Liu, Z.; Zhao, H. Review of the Present and Future Applications of Supercapacitors in Electric and Hybrid Vehicles. Research Report–UCD–ITS–RR–14-23. ITS–Institute of Transportation Studies. University of California, Davis (EEUU), 2014. Disponibleen: www. its. ucdavis. edu.
8. Zhang, L. L.; Zhao, 11. S. Carbon-based Materials as Supercapacitor Electrodes. *Chem. Soc. Rev.*, **2009**, *38* (9), 2520–2531.
9. An, K. H.; Kim, W. S.; Park, Y. S.; Moon, J. M.; Bae, D. J.; Lim, S. C.; ... & Lee, Y. H. Electrochemical Properties of High-power Supercapacitors Using Single-walled Carbon Nanotube Electrodes. *Adv. Funct. Mater.* **2001**, *11* (5), 387–392.
10. Dong, Z.; Jiang, C.; Cheng, H.; Zhao, Y.; Shi, G.; Jiang, L.; Qu, L. Facile Fabrication of Light, Flexible and Multifunctional Graphene Fibers. *Adv. Mater.* **2012**, *24* (14), 1856–1861.
11. Lee, C.; Wei, 11.; Kysar, J. W.; Hone, J. Measurement of the Elastic Properties and Intrinsic Strength of Monolayer Graphene. *Science* **2008**, *321*, 385–388. CAS PubMed Article.

12. Burke, A. R&D Considerations for the Performance and Application of Electrochemical Capacitors. *Electrochimica Acta* **2007**, *53* (3), 1083–1091.
13. German, J. Hybrid Vehicles: Technology Development and Cost Reduction. *Technical Brief* **2015**, *1*.
14. Zhang, J.; Zhao, X. S. On the Configuration of Supercapacitors for Maximizing Electrochemical Performance. *ChemSusChem* **2012**, *5* (5), 818–841.
15. Pandolfo, A. G.; Hollenkamp, A. F. Carbon Properties and Their Role in Supercapacitors. *J. Power Sources* **2006**, *157* (1), 11–27.
16. Dai, L.; Chang, D. W.; Baek, J. B.; Lu, W. Carbon Nanomaterials for Advanced Energy Conversion and Storage. *Small* **2012**, *8* (8), 1130–1166.
17. Chen, T.; Dai, L. Flexible Supercapacitors Based on Carbon Nanomaterials. *J. Mater. Chem. A* **2014**, *2* (28), 10756–10775.
18. Zhang, Y.; Feng, H.; Wu, X.; Wang, L.; Zhang, A.; Xia, T.; ... Zhang, L. Progress of Electrochemical Capacitor Electrode Materials: A Review. *Int. J. Hydrogen Energy* **2009**, *34* (11), 4889–4899.
19. Kuzmenko, A. B.; Van Heumen, E.; Carbone, F.; Van Der Marel, D. Universal Optical Conductance of Graphite. *Phys. Rev. Lett.* **2008**, *100* (11), 117401.
20. Xia, J.; Chen, F.; Li, J.; Tao, N. Measurement of the Quantum Capacitance of Graphene. *Nat. Nanotechnol.* **2009**, *4* (8), 505–509.
21. Booth, T. J.; Blake, P.; Nair, R. R.; Jiang, D.; Hill, E. W.; Bangert, U.; ... Geim, A. K. Macroscopic Graphene Membranes and Their Extraordinary Stiffness. *Nano Lett.* **2008**, *8* (8), 2442–2446.
22. Lee, C.; Wei, X.; Kysar, J. W.; Hone, J. Measurement of the Elastic Properties and Intrinsic Strength of Monolayer Graphene. *Science* **2008**, *321* (5887), 385–388.
23. Geim, A. K.; Novoselov, K. S. The Rise of Graphene. *Naturematerials* March **2007**, *6*, 183–191.
24. Nguyen, D. D.; Cao, T. T.; Le, P. H.; Phan, N. M. Recent Trends in Preparation and Application of Carbon Nanotube–graphene Hybrid Thin Films. *Adv. Nat. Sci.: Nanosci. Nanotechnol.* **2016**, *7* (3), 033002.
25. Wang, Y.; Shi, Z.; Huang, Y.; Ma, Y.; Wang, C.; Chen, M.; Chen, Y. Supercapacitor Devices Based on Graphene Materials. *J. Phys. Chem. C* **2009**, *113* (30), 13103–13107.
26. Bhuyan, M. S. A.; Uddin, M. N.; Islam, M. M.; Bipasha, F. A.; Hossain, S. S. Synthesis of Graphene. *Int. Nano Lett.* **2016**, *6* (2), 65–83.
27. Moon, J. S.; Gaskill, D. K. Graphene: Its Fundamentals to Future Applications. *IEEE Trans. Microwave Theory Techniques* **2011**, *59* (10), 2702–2708.
28. Nilsson, J.; Neto, A. C.; Guinea, F.; Peres, N. M. R. Electronic Properties of Bilayer and Multilayer Graphene. *Phys. Rev. B* **2008**, *78* (4), 045405.
29. Nakano, H.; Tetsuka, H.; Spencer, M. J.; Morishita, T. Chemical Modification of Group IV Graphene Analogs. *Sci. Technol. Adv. Mater.* **2018**, *19* (1), 76–100.
30. Yin, Z.; Sun, S.; Salim, T.; Wu, S.; Huang, X.; He, Q.; ... Zhang, H. Organic Photovoltaic Devices Using Highly Flexible Reduced Graphene Oxide Films as Transparent Electrodes. *ACS Nano* **2010**, *4* (9), 5263–5268.
31. Schedin, F.; Geim, A. K.; Morozov, S. V.; Hill, E. W.; Blake, P.; Katsnelson, M. I.; Novoselov, K. S. Detection of Individual Gas Molecules Adsorbed on Graphene. *Nat. Mater.* **2007**, *6* (9), 652–655.

32. Wang, X.; Zhi, L.; Müllen, K. Transparent, Conductive Graphene Electrodes for Dye-sensitized Solar Cells. Nano Lett. **2008**, *8* (1), 323–327.
33. He, Q.; Sudibya, H. G.; Yin, Z.; Wu, S.; Li, H.; Boey, F.; ... Zhang, H. Centimeter-long and Large-scale Micropatterns of Reduced Graphene Oxide Films: Fabrication and Sensing Applications. *ACS Nano* **2010**, *4* (6), 3201–3208.
34. Grande, M.; Bianco, G. V.; Vincenti, M. A.; De Ceglia, D.; Capezzuto, P.; Petruzzelli, V.; ... D'Orazio, A. Graphene-based Devices: A Platform for High Frequency Applications?. In *2016 18th International Conference on Transparent Optical Networks (ICTON)*, July 2016; pp. 1–4. IEEE.
35. Blake, P.; Brimicombe, P. D.; Nair, R. R.; Booth, T. J.; Jiang, D.; Schedin, F.; ... Geim, A. K. Graphene-based Liquid Crystal Device. *Nano Lett.* **2008**, *8* (6), 1704–1708.
36. Mitra, S.; Sampath, S. Electrochemical Capacitors Based on Exfoliated Graphite Electrodes. *Electrochem. Solid State Lett.* **2004**, *7* (9), A264.
37. Yang, W.; Ni, M.; Ren, X.; Tian, Y.; Li, N.; Su, Y.; Zhang, X. Graphene in Supercapacitor Applications. Curr. *Opin. Colloid Interface Sci.* **2015**, *20* (5–6), 416–428.
38. Roldán, S.; Barreda, D.; Granda, M.; Menéndez, R.; Santamaría, R.; Blanco, C. An Approach to Classification and Capacitance Expressions in Electrochemical Capacitors Technology. *Phys. Chem. Chem. Phys.* **2015,** *17* (2), 1084–1092.
39. Vangari, M.; Pryor, T.; Jiang, L. Supercapacitors: Review of Materials and Fabrication Methods. *J. Energy Eng.* **2013**, *139* (2), 72–79.
40. Libich, J.; Máca, J.; Vondrák, J.; Čech, O.; Sedlaříková, M. Supercapacitors: Properties and Applications. *J. Energy Storage* **2018**, *17*, 224–227.
41. Brousse, T.; Bélanger, D.; Long, J. W. To Be or Not to Be Pseudocapacitive. *J. Electrochem. Soc.* **2015**, *162* (5), A5185–A5189.
42. Qu, D. Studies of the Activated Carbons Used in Double-layer Supercapacitors. *J. Power Sources* **2002**, *109* (2), 403–411.
43. Shi, H. Activated Carbons and Double Layer Capacitance. *Electrochimica Acta* **1996**, *41* (10), 1633–1639.
44. Iro, Z. S.; Subramani, C.; Dash, S. S. A Brief Review on Electrode Materials for Supercapacitor. *Int. J. Electrochem. Sci.* **2016**, *11* (12), 10628–10643.
45. Naoi, K.; Simon, P. New Materials and New Configurations for Advanced Electrochemical Capacitors. *J. Electrochem. Soc. (JES)* **2008**, *17* (1), 34–37.
46. Stoller, M. D.; Park, S.; Zhu, Y.; An, J.; Ruoff, R. S. Graphene-based Ultracapacitors. *Nano Lett.* **2008**, *8* (10), 3498–3502.
47. Simon, P.; Gogotsi, Y. Materials for Electrochemical Capacitors. *Nanoscience and Technology: A Collection of Reviews from Nature Journals*, 2010; 320–329.
48. Zhang, L. L.; Zhou, R.; Zhao, X. S. Graphene-based Materials as Supercapacitor Electrodes. *J. Mater. Chem.* **2010**, *20* (29), 5983–5992.
49. Kötz, R.; Carlen, M. J. E. A. Principles and Applications of Electrochemical Capacitors. *Electrochimica Acta* **2000**, *45* (15–16), 2483–2498.
50. Vivekchand, S. R. C.; Rout, C. S.; Subrahmanyam, K. S.; Govindaraj, A.; Rao, C. N. R. Graphene-Based Electrochemical Supercapacitors. *J. Chem. Sci.* **2008**, *120* (1), 9–13.

51. Gogotsi, Y.; Penner, R. M. Energy Storage in Nanomaterials–capacitive, Pseudocapacitive, or Battery-Like?, 2018.
52. Laforgue, A.; Simon, P.; Fauvarque, J. F.; Mastragostino, M.; Soavi, F.; Sarrau, J. F.; ... Saguatti, S. Activated Carbon/conducting Polymer Hybrid Supercapacitors. *J. Electrochem. Soc.* **2003**, *150* (5), A645.
53. Arbizzani, C.; Mastragostino, M.; Soavi, F. New Trends in Electrochemical Supercapacitors. *J. Power Sources* **2001**, *100* (1–2), 164–170.
54. Cao, J.; Wang, Y.; Zhou, Y.; Ouyang, J. H.; Jia, D.; Guo, L. High Voltage Asymmetric Supercapacitor Based on MnO_2 and Graphene Electrodes. *J. Electroanal. Chem.* **2013**, *689*, 201–206.
55. Su, F.; Miao, M. Asymmetric Carbon Nanotube–MnO_2 Two-ply Yarn Supercapacitors for Wearable Electronics. *Nanotechnology* **2014**, *25* (13), 135401.
56. Maiti, S.; Pramanik, A.; Mahanty, S. Influence of Imidazolium-based Ionic Liquid Electrolytes on the Performance of Nano-structured MnO_2 Hollow Spheres as Electrochemical Supercapacitor. *RSC Adv.* **2015**, *5* (52), 41617–41626.
57. Nam, K. W.; Ma, S. B.; Yoon, W. S.; Yang, X. Q.; Kim, K. B. Novel Concept of Pseudocapacitor Using Stabilized Lithium Metal Powder and Non-lithiated Metal Oxide Electrodes in Organic Electrolyte. *Electrochem. Commun.* **2009,** *11* (6), 1166–1169.
58. Wang, P.; Liu, H.; Xu, Y.; Chen, Y.; Yang, J.; Tan, Q. Supported Ultrafine Ruthenium Oxides with Specific Capacitance up to 1099 F g−1 for a Supercapacitor. *Electrochimica Acta* **2016**, *194*, 211–218.
59. Maiti, S.; Pramanik, A.; Mahanty, S. Influence of Imidazolium-based Ionic Liquid Electrolytes on the Performance of Nano-structured MnO_2 Hollow Spheres as Electrochemical Supercapacitor. *RSC Adv.* **2015**, *5* (52), 41617–41626.
60. Ma, S. B.; Nam, K. W.; Yoon, W. S.; Yang, X. Q.; Ahn, K. Y.; Oh, K. H.; Kim, K. B. A Novel Concept of Hybrid Capacitor Based on Manganese Oxide Materials. *Electrochem. Commun.* **2007**, *9* (12), 2807–2811.
61. Perret, P.; Zohreh, K.; Thierry, B.; Daniel, B.; Daniel, G. Carbon/PbO_2 Asymmetric Electrochemical Capacitor Based on Methanesulfonic Acid Electrolyte. *Electrochimica Acta* **2011**, *56* (24), 8122–8128.
62. Liu, Q.; Yang, B.; Liu, J.; Yuan, Y.; Zhang, H.; Liu, L.; ... Li, R. Application of Chemical Doping and Architectural Design Principles to Fabricate Nanowire $Co_2Ni_3ZnO_8$ Arrays for Aqueous Asymmetric Supercapacitors. *ACS Appl. Mater. Interfaces* **2016**, *8* (31), 20157–20167.
63. Wang, P.; Liu, H.; Xu, Y.; Chen, Y.; Yang, J.; Tan, Q. Supported Ultrafine Ruthenium Oxides with Specific Capacitance up to 1099 F g−1 for a Supercapacitor. *Electrochimica Acta* **2016**, *194*, 211–218.
64. Peng, L.; Peng, X.; Liu, B.; Wu, C.; Xie, Y.; Yu, G. Ultrathin two-dimensional MnO_2/graphene Hybrid Nanostructures for High-performance, Flexible Planar Supercapacitors. *Nano Lett.* **2013**, *13* (5), 2151–2157.
65. Nam, K. W.; Ma, S. B.; Yoon, W. S.; Yang, X. Q.; Kim, K. B. Novel Concept of Pseudocapacitor Using Stabilized Lithium Metal Powder and Non-lithiated Metal Oxide Electrodes in Organic Electrolyte. *Electrochem. Commun.* **2009**, *11*, 1166e1169.
66. Chen, Z.; Augustyn, V.; Jia, X.; Xiao, Q.; Dunn, B.; Lu, Y. High-performance Sodium-ion Pseudocapacitors Based on Hierarchically Porous Nanowire Composites. *ACS Nano* **2012**, *6* (5), 4319–4327.

67. Ma, S. B.; Nam, K. W.; Yoon, W. S.; Yang, X. Q.; Ahn, K. Y.; Oh, K. H.; Kim, K. B. A Novel Concept of Hybrid Capacitor Based on Manganese Oxide Materials. *Electrochem. Commun.* **2007**, *9* (12), 2807–2811.
68. Perret, P.; Khani, Z.; Brousse, T.; Bélanger, D.; Guay, D. Carbon/PbO_2 Asymmetric Electrochemical Capacitor Based on Methanesulfonic Acid Electrolyte. *Electrochimica Acta* **2011**, *56* (24), 8122–8128.
69. Liu, Q.; Yang, B.; Liu, J.; Yuan, Y.; Zhang, H.; Liu, L.; ... Li, R. Application of Chemical Doping and Architectural Design Principles to Fabricate Nanowire $Co_2Ni_3ZnO_8$ Arrays for Aqueous Asymmetric Supercapacitors. *ACS Appl. Mater. Interfaces* **2016**, *8* (31), 20157–20167.
70. Lv, W.; Li, Z.; Deng, Y.; Yang, Q. H.; Kang, F. Graphene-based Materials for Electrochemical Energy Storage Devices: Opportunities and Challenges. *Energy Storage Mater.* **2016**, *2*, 107–138.
71. Wu, Z. S.; Sun, Y.; Tan, Y. Z.; Yang, S.; Feng, X.; Müllen, K. Three-dimensional Graphene-based Macro-and Mesoporous Frameworks for High-performance Electrochemical Capacitive Energy Storage. *J. Am. Chem. Soc.* **2012**, *134* (48), 19532–19535.
72. Pumera, M. Graphene-based Nanomaterials and Their Electrochemistry. *Chem. Soc. Rev.* **2010**, *39* (11), 4146–4157.
73. Xiong, G.; Meng, C.; Reifenberger, R. G.; Irazoqui, P. P.; Fisher, T. S. A Review of Graphene-based Electrochemical Microsupercapacitors. *Electroanalysis* **2014**, *26* (1), 30–51.
74. Horn, M.; Gupta, B.; MacLeod, J.; Liu, J.; Motta, N. Graphene-based Supercapacitor Electrodes: Addressing Challenges in Mechanisms and Materials. *Curr. Opin. Green Sustain. Chem.* **2019**, *17*, 42–48.
75. Zhou, C.; Zhang, Y.; Li, Y.; Liu, J. Construction of High-capacitance 3D CoO@ Polypyrrole Nanowire Array Electrode for Aqueous Asymmetric Supercapacitor. *Nano Lett.* **2013**, *13* (5), 2078–2085.
76. Tang, W.; Liu, L.; Tian, S.; Li, L.; Yue, Y.; Wu, Y.; Zhu, K. Aqueous Supercapacitors of High Energy Density Based on MoO_3 Nanoplates as Anode Material. *Chem. Commun.* **2011**, *47* (36), 10058–10060.
77. Hou, L.; Yuan, C.; Li, D.; Yang, L.; Shen, L.; Zhang, F.; Zhang, X. Electrochemically Induced Transformation of NiS Nanoparticles into Ni $(OH)_2$ in KOH Aqueous Solution Toward Electrochemical Capacitors. *Electrochimica Acta* **2011**, *56* (22), 7454–7459.
78. Pu, J.; Cui, F.; Chu, S.; Wang, T.; Sheng, E.; Wang, Z. Preparation and Electrochemical Characterization of Hollow Hexagonal $NiCo_2S_4$ Nanoplates as Pseudocapacitor Materials. *ACS Sustain. Chem. Eng.* **2014**, *2* (4), 809–815.
79. Ramya, R.; Sivasubramanian, R.; Sangaranarayanan, M. V. Conducting Polymers-based Electrochemical Supercapacitors—Progress and Prospects. *Electrochimica Acta* **2013**, *101*, 109–129.
80. Dai, C. S.; Chien, P. Y.; Lin, J. Y.; Chou, S. W.; Wu, W. K.; Li, P. H.; ... Lin, T. W. Hierarchically Structured Ni_3S_2/carbon Nanotube Composites as High Performance Cathode Materials for Asymmetric Supercapacitors. *ACS Appl. Mater. Interfaces* **2013**, *5* (22), 12168–12174.

81. Zheng, J. P.; Cygan, P. J.; Jow, T. R. Hydrous Ruthenium Oxide as an Electrode Material for Electrochemical Capacitors. *J. Electrochem. Soc.* **1995**, *142* (8), 2699.
82. Hu, C. C.; Wang, C. C. Nanostructures and Capacitive Characteristics of Hydrous Manganese Oxide Prepared by Electrochemical Deposition. *J. Electroch. Soc.* **2003**, *150* (8), A1079.
83. Zhang, Y. X. et al. One-pot Synthesis of Hierarchical MnO_2-modified Diatomites for Electrochemical Capacitor Electrodes. *J. Power Sources* **2014**, *246*, 449–456.
84. Li, J. M.; Chang, K. H.; Hu, C. C. A Novel Vanadium Oxide Deposit for the Cathode of Asymmetric Lithium-ion Supercapacitors. *Electrochem. Commun.* **2010**, *12* (12), 1800–1803.
85. Qu, Q. T.; Liu, L. L.; Wu, Y. P.; Holze, R. Electrochemical Behavior of $V_2O_5 \cdot 0.6\ H_2O$ Nanoribbons in Neutral Aqueous Electrolyte Solution. *Electrochimica Acta* **2013**, *96*, 8–12.
86. Stoller, M. D.; Park, S.; Zhu, Y.; An, J.; Ruoff, R. S. Graphene-based Ultracapacitors. *Nano Lett.* **2008**, *8* (10), 3498–3502.
87. Zhang, S.; Sui, L.; Dong, H.; He, W.; Dong, L.; Yu, L. High-performance Supercapacitor of Graphene Quantum Dots with Uniform Sizes. *ACS Appl. Mater. Interfaces* **2018**, *10* (15), 12983–12991.
88. Liu, W. W.; Feng, Y. Q.; Yan, X. B.; Chen, J. T.; Xue, Q. J. Superior Micro-supercapacitors Based on Graphene Quantum Dots. *Adv. Funct. Mater.* **2013**, *23* (33), 4111–4122.
89. Chen, L.; Liu, Y.; Zhao, Y.; Chen, N.; Qu, L. Graphene-based Fibers for Supercapacitor Applications. *Nanotechnology* **2015**, *27* (3), 032001.
90. Yang, Q.; Xu, Z.; Gao, C. Graphene Fiber Based Supercapacitors: Strategies and Perspective Toward High Performances. *J. Energy Chem.* **2018**, *27* (1), 6–11.
91. Lim, L.; Liu, Y.; Liu, W.; Tjandra, R.; Rasenthiram, L.; Chen, Z.; Yu, A. All-in-one Graphene Based Composite Fiber: Toward Wearable Supercapacitor. *ACS Appl. Mater. Interfaces* **2017**, *9* (45), 39576–39583.
92. Li, Z.; Huang, T.; Gao, W.; Xu, Z.; Chang, D.; Zhang, C.; Gao, C. Hydrothermally Activated Graphene Fiber Fabrics for Textile Electrodes of Supercapacitors. *ACS Nano* **2017**, *11* (11), 11056–11065.
93. Wang, Y.; Shi, Z.; Huang, Y.; Ma, Y.; Wang, C.; Chen, M.; Chen, Y. Supercapacitor Devices Based on Graphene Materials. *J. Phys. Chem. C* **2009**, *113* (30), 13103–13107.
94. Kim, S. J.; Choi, K.; Lee, B.; Kim, Y.; Hong, B. H. Materials for Flexible, Stretchable Electronics: Graphene and 2D Materials. *Ann. Rev. Mater. Res.* **2015**, *45*, 63–84.
95. Fang, Y.; Luo, B.; Jia, Y.; Li, X.; Wang, B.; Song, Q.; ... Zhi, L. Renewing Functionalized Graphene as Electrodes for High-performance Supercapacitors. *Adv. Mater.* **2012**, *24* (47), 6348–6355.
96. Du, Q.; Zheng, M.; Zhang, L.; Wang, Y.; Chen, J.; Xue, L.; ... Cao, J. Preparation of Functionalized Graphene Sheets by a Low-temperature Thermal Exfoliation Approach and Their Electrochemical Supercapacitive Behaviors. *Electrochimica Acta* **2010**, *55* (12), 3897–3903.
97. Lv, W.; Tang, D. M.; He, Y. B.; You, C. H.; Shi, Z. Q.; Chen, X. C.; ... Yang, Q. H. Low-temperature Exfoliated Graphenes: Cacuum-promoted Exfoliation and Electrochemical Energy Storage. *ACS Nano* **2009**, *3* (11), 3730–3736.

98. Yu, A.; Roes, I.; Davies, A.; Chen, Z. Ultrathin, Transparent, and Flexible Graphene Films for Supercapacitor Application. *Appl. Phys. Lett.* **2010**, *96* (25), 253105.
99. Lee, J. S.; Kim, S. I.; Yoon, J. C.; Jang, J. H. Chemical Vapor Deposition of Mesoporous Graphene Nanoballs for Supercapacitor. *ACS Nano* **2013**, *7* (7), 6047–6055.
100. Xu, Y.; Lin, Z.; Huang, X.; Liu, Y.; Huang, Y.; Duan, X. Flexible Solid-state Supercapacitors Based on Three-dimensional Graphene Hydrogel Films. *ACS Nano* **2013**, *7* (5), 4042–4049.
101. Xu, Y.; Lin, Z.; Huang, X.; Wang, Y.; Huang, Y.; Duan, X. Functionalized Graphene Hydrogel-Based High-performance Supercapacitors. Adv. Mater. **2013**, *25* (40), 5779–5784.
102. Zhang, W.; Lei, Y.; Jiang, Q.; Ming, F.; Costa, P. M.; Alshareef, H. N. 3D Laser Scribed Graphene Derived from Carbon Nanospheres: An Ultrahigh-Power Electrode for Supercapacitors. *Small Methods* **2019**, *3* (5), 1900005.
103. Ye, S.; Feng, J.; Wu, P. Deposition of Three-dimensional Graphene Aerogel on Nickel Foam as a Binder-free Supercapacitor Electrode. *ACS Appl. Mater. Interfaces* **2013**, *5* (15), 7122–7129.
104. Zhu, Y.; Murali, S.; Cai, W.; Li, X.; Suk, J. W.; Potts, J. R.; Ruoff, R. S. Graphene and Graphene Oxide: Synthesis, Properties, and Applications. *Adv. Mater.* **2010**, *22* (35), 3906–3924.
105. Rajagopalan, B.; Chung, J. S. Reduced Chemically Modified Graphene Oxide for Supercapacitor Electrode. *Nanoscale Res. Lett.* **2014**, *9* (1), 1–10.
106. Aboutalebi, S. H.; Jalili, R.; Esrafilzadeh, D.; Salari, M.; Gholamvand, Z.; Aminorroaya Yamini, S.; ... Innis, P. C. High-performance Multifunctional Graphene Yarns: Toward Wearable All-carbon Energy Storage Textiles. *ACS Nano* **2014**, *8* (3), 2456–2466.
107. Fan, T.; Zeng, W.; Niu, Q.; Tong, S.; Cai, K.; Liu, Y.; ... Epstein, A. J. Fabrication of High-quality Graphene Oxide Nanoscrolls and Application in Supercapacitor. *Nanoscale Res. Lett.* **2015**, *10* (1), 192.
108. Down, M. P.; Rowley-Neale, S. J.; Smith, G. C.; Banks, C. E. Fabrication of Graphene Oxide Supercapacitor Devices. *ACS Appl. Energy Mater.* **2018**, *1* (2), 707–714.
109. Cao, X.; Shi, Y.; Shi, W.; Lu, G.; Huang, X.; Yan, Q.; ... Zhang, H. Preparation of Novel 3D Graphene Networks for Supercapacitor Applications. *Small* **2011**, *7* (22), 3163–3168.
110. Xu, Y.; Chen, C. Y.; Zhao, Z.; Lin, Z.; Lee, C.; Xu, X.; ... Duan, X. Solution Processable Holey Graphene Oxide and Its Derived Macrostructures for High-performance Supercapacitors. *Nano Lett.* **2015**, *15* (7), 4605–4610.
111. Karthika, P.; Rajalakshmi, N.; Dhathathreyan, K. S. Functionalized Exfoliated Graphene Oxide as Supercapacitor Electrodes, 2012.
112. Skákalová, V.; Vretenár, V.; Kopera, Ľ.; Kotrusz, P.; Mangler, C.; Meško, M.; ... Hulman, M. Electronic Transport in Composites of Graphite Oxide with Carbon Nanotubes. *Carbon* **2014**, *72*, 224–232.
113. Zhu, Y.; Murali, S.; Cai, W.; Li, X.; Suk, J. W.; Potts, J. R.; Ruoff, R. S. Graphene and Graphene Oxide: Synthesis, Properties, and Applications. *Adv. Mater.* **2010**, *22* (35), 3906–3924.
114. Vivekchand, S. R. C.; Rout, C. S.; Subrahmanyam, K. S.; Govindaraj, A.; Rao, C. N. R. Graphene-based Electrochemical Supercapacitors. *J. Chem. Sci.* **2008**, *120* (1), 9–13.

115. Zhang, L. L.; Zhao, X.; Stoller, M. D.; Zhu, Y.; Ji, H.; Murali, S.; ... Ruoff, R. S. Highly Conductive and Porous Activated Reduced Graphene Oxide Films for High-power Supercapacitors. *Nano Lett.* **2012**, *12* (4), 1806–1812.
116. Yang, J.; Gunasekaran, S. Electrochemically Reduced Graphene Oxide Sheets for Use in High Performance Supercapacitors. *Carbon* **2013**, *51*, 36–44.
117. Ye, S.; Feng, J.; Wu, P. Deposition of Three-dimensional Graphene Aerogel on Nickel Foam as a Binder-free Supercapacitor Electrode. *ACS Appl. Mater. Interfaces* **2013**, *5* (15), 7122–7129.
118. Yu, H.; He, J.; Sun, L.; Tanaka, S.; Fugetsu, B. Influence of the Electrochemical Reduction Process on the Performance of Graphene-based Capacitors. *Carbon* **2013**, *51*, 94–101.
119. Chen, Y.; Zhang, X.; Zhang, D.; Yu, P.; Ma, Y. High Performance Supercapacitors Based on Reduced Graphene Oxide in Aqueous and Ionic Liquid Electrolytes. *Carbon* **2011**, *49* (2), 573–580.
120. Choi, W.; Lahiri, I.; Seelaboyina, R.; Kang, Y. S. Synthesis of Graphene and Its Applications: A Review. *Crit. Rev. Solid State Mater. Sci.* **2010**, *35* (1), 52–71.
121. Chabot, V.; Higgins, D.; Yu, A.; Xiao, X.; Chen, Z.; Zhang, J. A Review of Graphene and Graphene Oxide Sponge: Material Synthesis and Applications to Energy and the Environment. *Energy Environ. Sci.* **2014**, *7* (5), 1564–1596.
122. Diba, M.; Fam, D. W.; Boccaccini, A. R.; Shaffer, M. S. Electrophoretic Deposition of Graphene-related Materials: A Review of the Fundamentals. *Prog. Mater. Sci.* **2016**, *82*, 83–117.
123. Zhang, S.; Li, Y.; Pan, N. Graphene Based Supercapacitor Fabricated by Vacuum Filtration Deposition. *J. Power Sources* **2012**, *206*, 476–482.
124. Nazarova, M. Chemical Vapour Deposition (CVD) of Graphene on Transition Metals and Its Protective Properties. The University of Manchester (United Kingdom), 2019.
125. Zhamu, A.; Yu, Z.; Liu, C. G.; Jang, B. Z. U.S. Patent No. 8,315,039. Washington, DC: U.S. Patent and Trademark Office, 2012.
126. Wang, G.; Sun, X.; Lu, F.; Sun, H.; Yu, M.; Jiang, W.; ... Lian, J. Flexible Pillared Graphene-paper Electrodes for High-performance Electrochemical Supercapacitors. *Small* **2012**, *8* (3), 452–459.
127. Yan, J.; Wei, T.; Shao, B.; Ma, F.; Fan, Z.; Zhang, M.; ... Wei, F. Electrochemical Properties of Graphene Nanosheet/Carbon Black Composites as Electrodes for Supercapacitors. *Carbon* **2010**, *48* (6), 1731–1737.
128. Zheng, C.; Zhou, X.; Cao, H.; Wang, G.; Liu, Z. Synthesis of Porous Graphene/activated Carbon Composite with High Packing Density and Large Specific Surface Area for Supercapacitor Electrode Material. *J. Power Sources* **2014**, *258*, 290–296.
129. Zhu, Q.; Ma, L.; Wang, H.; Jia, M.; Guan, Y.; Xu, B. Activated Carbon/graphene Hybrid Aerogels as Electrode Materials for High Performance Supercapacitors. *ChemistrySelect* **2017**, *2* (16), 4456–4461.
130. Jeong, H. M.; Lee, J. W.; Shin, W. H.; Choi, Y. J.; Shin, H. J.; Kang, J. K.; Choi, J. W. Nitrogen-doped Graphene for High-performance Ultracapacitors and the Importance of Nitrogen-doped Sites at Basal Planes. *Nano Lett.* **2011**, *11* (6), 2472–2477.

131. Li, J.; Hao, H.; Wang, J.; Zhang, Y.; Wei, D. 3D Nitrogen-enriched Graphene Porous Graphene Hydrogel for High-performance Supercapacitor. In *IOP Conference Series: Materials Science and Engineering* (Vol. 490, No. 2, p. 022069). IOP Publishing, April 2019.
132. Si, Y.; Samulski, E. T. Exfoliated Graphene Separated by Platinum Nanoparticles. *Chem. Mater.* **2008**, *20* (21), 6792–6797.
133. Chen, J.; Bi, H.; Sun, S.; Tang, Y.; Zhao, W.; Lin, T.; ... Jiang, M. Highly Conductive and Flexible Paper of 1D Silver-nanowire-doped Graphene. *ACS Appl. Mater. Interfaces* **2013**, *5* (4), 1408–1413.
134. Zhi, J.; Zhao, W.; Liu, X.; Chen, A.; Liu, Z.; Huang, F. Highly Conductive Ordered Mesoporous Carbon Based Electrodes Decorated by 3D Graphene and 1D Silver Nanowire for Flexible Supercapacitor. *Adv. Funct. Mater.* **2014**, *24* (14), 2013–2019.
135. Cherusseri, J.; Kowsik, S. K.; Deepak, P.; Elizabeth, B.; Jayan, T. Vertically Aligned Graphene–Carbon Fiber Hybrid Electrodes with Superlong Cycling Stability for Flexible Supercapacitors. *Small* **2019**, *15* (44), 1902606.
136. Kou, L.; Huang, T.; Zheng, B.; Han, Y.; Zhao, X.; Gopalsamy, K.; ... Gao, C. Coaxial Wet-spun Yarn Supercapacitors for High-energy Density and Safe Wearable Electronics. *Nat. Commun.* **2014**, *5* (1), 1–10.
137. Cheng, H.; Dong, Z.; Hu, C.; Zhao, Y.; Hu, Y.; Qu, L.; ... Dai, L. Textile Electrodes Woven by Carbon Nanotube–graphene Hybrid Fibers for Flexible Electrochemical Capacitors. *Nanoscale* **2013**, *5* (8), 3428–3434.
138. Sun, H.; You, X.; Deng, J.; Chen, X.; Yang, Z.; Ren, J.; Peng, H. Novel graphene/carbon Nanotube Composite Fibers for Efficient Wire-shaped Miniature Energy Devices. *Adv. Mater.* **2014**, *26* (18), 2868–2873.
139. Qiu, L.; Yang, X.; Gou, X.; Yang, W.; Ma, Z. F.; Wallace, G. G.; Li, D. Dispersing Carbon Nanotubes with Graphene Oxide in Water and Synergistic Effects between Graphene Derivatives. *Chem.– Eur. J.* **2010**, *16* (35), 10653–10658.
140. Huang, Z. D.; Zhang, B.; Liang, R.; Zheng, Q. B.; Oh, S. W.; Lin, X. Y.; ... Kim, J. K. Effects of Reduction Process and Carbon Nanotube Content on the Supercapacitive Performance of Flexible Graphene Oxide Papers. *Carbon* **2012**, *50* (11), 4239–4251.
141. Xu, Y.; Li, X.; Hu, G.; Wu, T.; Luo, Y.; Sun, L.; ... Li, M. Graphene Oxide Quantum Dot-derived Nitrogen-enriched Hybrid Graphene Nanosheets by Simple Photochemical Doping for High-Performance Supercapacitors. *Appl. Surface Sci.* **2017**, *422*, 847–855.
142. Kumar, A.; Kumar, N.; Sharma, Y.; Leu, J.; Tseng, T. Y. Synthesis of Free-standing Flexible rGO/MWCNT Films for Symmetric Supercapacitor Application. *Nanoscale Res. Lett.* **2019**, *14* (1), 1–17.
143. Yu, D.; Goh, K.; Wang, H.; Wei, L.; Jiang, W.; Zhang, Q.; ... Chen, Y. Scalable Synthesis of Hierarchically Structured Carbon Nanotube–graphene Fibres for Capacitive Energy Storage. *Nat. Nanotechnol.* **2014**, *9* (7), 555.
144. Zhang, K.; Mao, L.; Zhang, L. L.; Chan, H. S. O.; Zhao, X. S.; Wu, J. Surfactant-intercalated, Chemically Reduced Graphene Oxide for High Performance Supercapacitor Electrodes. *J. Mater. Chem.* **2011**, *21* (20), 7302–7307.
145. Yoon, Y. S.; Cho, W. I.; Lim, J. H.; Choi, D. J. Solid-state Thin-film Supercapacitor with Ruthenium Oxide and Solid Electrolyte Thin Films. *J. Power Sources* **2001**, *101* (1), 126–129.

146. Lin, N.; Tian, J.; Shan, Z.; Chen, K.; Liao, W. Hydrothermal Synthesis of Hydrous Ruthenium Oxide/graphene Sheets for High-performance Supercapacitors. *Electrochimica Acta* **2013**, *99*, 219–224.
147. Bianco, A.; Chen, Y.; Chen, Y.; Ghoshal, D.; Hurt, R. H.; Kim, Y. A.; ... Terrones, M. A Carbon Science Perspective in 2018: Current Achievements and Future Challenges. *Carbon* **2018**, *132*, 785–801.
148. Kou, L.; Huang, T.; Zheng, B.; Han, Y.; Zhao, X.; Gopalsamy, K.; ... Gao, C. Coaxial Wet-spun Yarn Supercapacitors for High-energy Density and Safe Wearable Electronics. *Nat. Commun.* **2014**, *5* (1), 1–10.
149. Li, X.; Liu, D.; Yin, X.; Zhang, C.; Cheng, P.; Guo, H.; ... Wang, J. Hydrated Ruthenium Dioxides@ Graphene Based Fiber Supercapacitor for Wearable Electronics. *J. Power Sources* **2019**, 440, 227143.
150. Lin, N.; Tian, J.; Shan, Z.; Chen, K.; Liao, W. Hydrothermal Synthesis of Hydrous Ruthenium Oxide/graphene Sheets for High-performance Supercapacitors. *Electrochimica Acta* **2013**, *99*, 219–224.
151. Thangappan, R.; Arivanandhan, M.; Kumar, R. D.; Jayavel, R. Facile Synthesis of RuO_2 Nanoparticles Anchored on Graphene Nanosheets for High Performance Composite Electrode for Supercapacitor Applications. *J. Phys. Chem. Solids* **2018**, *121*, 339–349.
152. Wang, W.; Guo, S.; Lee, I.; Ahmed, K.; Zhong, J.; Favors, Z.; ... Ozkan, C. S. Hydrous Ruthenium Oxide Nanoparticles Anchored to Graphene and Carbon Nanotube Hybrid Foam for Supercapacitors. *Sci. Rep.* **2014**, *4*, 4452.
153. Yu, G.; Hu, L.; Liu, N.; Wang, H.; Vosgueritchian, M.; Yang, Y.; ... Bao, Z. Enhancing the Supercapacitor Performance of Graphene/MnO_2 Nanostructured Electrodes by Conductive Wrapping. *Nano Lett.* **2011**, *11* (10), 4438–4442.
154. Ma, W.; Chen, S.; Yang, S.; Chen, W.; Cheng, Y.; Guo, Y.; ... Zhu, M. Hierarchical MnO_2 Nanowire/graphene Hybrid Fibers with Excellent Electrochemical Performance for Flexible Solid-state Supercapacitors. *J. Power Sources* **2016**, *306*, 481–488.
155. Chen, Q.; Meng, Y.; Hu, C.; Zhao, Y.; Shao, H.; Chen, N.; Qu, L. MnO_2-modified Hierarchical Graphene Fiber Electrochemical Supercapacitor. *J. Power Sources* **2014**, *247*, 32–39.
156. Yan, J.; Fan, Z.; Wei, T.; Qian, W.; Zhang, M.; Wei, F. Fast and Reversible Surface Redox Reaction of Graphene–MnO_2 Composites as Supercapacitor Electrodes. *Carbon* **2010**, *48* (13), 3825–3833.
157. Peng, L.; Peng, X.; Liu, B.; Wu, C.; Xie, Y.; Yu, G. Ultrathin Two-dimensional MnO_2/graphene Hybrid Nanostructures for High-performance, Flexible Planar Supercapacitors. *Nano Lett.* **2013**, *13* (5), 2151–2157.
158. Sadak, O.; Wang, W.; Guan, J.; Sundramoorthy, A. K.; Gunasekaran, S. MnO_2 Nanoflowers Deposited on Graphene Paper as Electrode Materials for Supercapacitors. *ACS Appl. Nano Mater.* **2019**, *2* (7), 4386–4394.
159. He, Y.; Chen, W.; Li, X.; Zhang, Z.; Fu, J.; Zhao, C.; Xie, E. Freestanding Three-dimensional Graphene/MnO_2 Composite Networks as Ultralight and Flexible Supercapacitor Electrodes. *ACS Nano* **2013**, *7* (1), 174–182.
160. Zhang, J. L.; Xing, L. B.; Liu, T. Z.; Qin, K.; Zhou, J.; Cui, H.; ... Si, W. Three-Dimensional Reduced Graphene Hydrogels Using Various Carbohydrates for High Performance Supercapacitors. *J. Nanosci. Nanotechnol.* **2017**, *17* (2), 1099–1107.

161. Meng, X.; Lu, L.; Sun, C. Green Synthesis of Three-dimensional MnO_2/graphene Hydrogel Composites as a High-performance Electrode Material for Supercapacitors. *ACS Appl. Mater. Interfaces* **2018**, *10* (19), 16474–16481.
162. Choudhury, A.; Bonso, J. S.; Wunch, M.; Yang, K. S.; Ferraris, J. P.; Yang, D. J. In-situ Synthesis of Vanadium Pentoxide Nanofibre/exfoliated Graphene Nanohybrid and Its Supercapacitor Applications. *J. Power Sources* **2015**, *287*, 283–290.
163. Ndiaye, N. M.; Ngom, B. D.; Sylla, N. F.; Masikhwa, T. M.; Madito, M. J.; Momodu, D.; ... Manyala, N. Three Dimensional Vanadium Pentoxide/graphene Foam Composite as Positive Electrode for High Performance Asymmetric Electrochemical Supercapacitor. *J. Colloid Interface Sci.* **2018**, *532*, 395–406.
164. He, G.; Li, J.; Chen, H.; Shi, J.; Sun, X.; Chen, S.; Wang, X. Hydrothermal Preparation of Co_3O_4@ Graphene Nanocomposite for Supercapacitor with Enhanced Capacitive Performance. *Mater. Lett.* **2012**, *82*, 61–63.
165. Liao, Q.; Li, N.; Jin, S.; Yang, G.; Wang, C. All-solid-state Symmetric Supercapacitor Based on Co_3O_4 Nanoparticles on Vertically Aligned Graphene. *ACS Nano* **2015**, *9* (5), 5310–5317.
166. Dong, X. C.; Xu, H.; Wang, X. W.; Huang, Y. X.; Chan-Park, M. B.; Zhang, H.; ... Chen, P. 3D Graphene–cobalt Oxide Electrode for High-performance Supercapacitor and Enzymeless Glucose Detection. *ACS Nano* **2012**, *6* (4), 3206–3213.
167. Bao, L.; Li, T.; Chen, S.; Peng, C.; Li, L.; Xu, Q.; ... Xu, W. 3D Graphene Frameworks/ Co_3O_4 Composites Electrode for High-Performance Supercapacitor and Enzymeless Glucose Detection. *Small* **2017**, *13* (5), 1602077.
168. Snook, G. A.; Kao, P.; Best, A. S. Conducting-polymer-based Supercapacitor Devices and Electrodes. *J. Power Sources* **2011**, *196* (1), 1–12.
169. Eftekhari, A.; Li, L.; Yang, Y. Polyaniline Supercapacitors. *J. Power Sources* **2017**, *347*, 86–107.
170. Mondal, S.; Rana, U.; Malik, S. Graphene Quantum Dot-doped Polyaniline Nanofiber as High Performance Supercapacitor Electrode Materials. *Chem. Commun.* **2015**, *51* (62), 12365–12368.
171. Hu, Y.; Cheng, H.; Zhao, F.; Chen, N.; Jiang, L.; Feng, Z.; Qu, L. All-in-one Graphene Fiber Supercapacitor. *Nanoscale* **2014**, *6* (12), 6448–6451.
172. Zhao, X.; Zheng, B.; Huang, T.; Gao, C. Graphene-based Single Fiber Supercapacitor with a Coaxial Structure. *Nanoscale* **2015**, *7* (21), 9399–9404.
173. Shao, Y.; El-Kady, M. F.; Wang, L. J.; Zhang, Q.; Li, Y.; Wang, H.; ... Kaner, R. B. Graphene-based Materials for Flexible Supercapacitors. *Chem. Soc. Rev.* **2015**, *44* (11), 3639–3665.
174. Chen, L.; Liu, Y.; Zhao, Y.; Chen, N.; Qu, L. Graphene-based Fibers for Supercapacitor Applications. *Nanotechnology* **2015**, *27* (3), 032001.
175. Zhang, K.; Zhang, L. L.; Zhao, X. S.; Wu, J. Graphene/polyaniline Nanofiber Composites as Supercapacitor Electrodes. *Chem. Mater.* **2010**, *22* (4), 1392–1401.
176. Mensing, J. P.; Wisitsoraat, A.; Phokharatkul, D.; Lomas, T.; Tuantranont, A. Novel Surfactant-stabilized Graphene-polyaniline Composite Nanofiber for Supercapacitor Applications, *Compos. B. Eng.* **2015,** *77,* 93–99.
177. Meng, C.; Liu, C.; Chen, L.; Hu, C.; Fan, S. Highly Flexible and All-solid-state Paperlike Polymer Supercapacitors. *Nano Lett.* **2010**, *10* (10), 4025–4031.

178. Yan, J.; Wei, T.; Shao, B.; Fan, Z.; Qian, W.; Zhang, M.; Wei, F. Preparation of a Graphene Nanosheet/polyaniline Composite with High Specific Capacitance. *Carbon* **2010**, *48* (2), 487–493.
179. Kulkarni, S. B.; Patil, U. M.; Shackery, I.; Sohn, J. S.; Lee, S.; Park, B.; Jun, S. High-performance Supercapacitor Electrode Based on a Polyaniline Nanofibers/3D Graphene Framework as an Efficient Charge Transporter. *J. Mater. Chem. A* **2014**, *2* (14), 4989–4998.
180. Wu, Q.; Xu, Y.; Yao, Z.; Liu, A.; Shi, G. Supercapacitors Based on Flexible Graphene/polyaniline Nanofiber Composite Films. *ACS Nano* **2010**, *4* (4), 1963–1970.
181. Sawangphruk, M.; Suksomboon, M.; Kongsupornsak, K.; Khuntilo, J.; Srimuk, P.; Sanguansak, Y.; ... Chiochan, P. High-performance Supercapacitors Based on Silver Nanoparticle–polyaniline–graphene Nanocomposites Coated on Flexible Carbon Fiber Paper. *J. Mater. Chem. A* **2013**, *1* (34), 9630–9636.
182. Bose, S.; Kim, N. H.; Kuila, T.; Lau, K. T.; Lee, J. H. Electrochemical Performance of a Graphene–Polypyrrole Nanocomposite as a Supercapacitor Electrode. *Nanotechnology* **2011**, *22* (29), 295202.
183. Zhang, D.; Zhang, X.; Chen, Y.; Yu, P.; Wang, C.; Ma, Y. Enhanced Capacitance and Rate Capability of Graphene/polypyrrole Composite as Electrode Material for Supercapacitors. *J. Power Sources* **2011**, *196* (14), 5990–5996.
184. Ding, X.; Zhao, Y.; Hu, C.; Hu, Y.; Dong, Z.; Chen, N.; ... Qu, L. Spinning Fabrication of Graphene/polypyrrole Composite Fibers for All-solid-state, Flexible Fibriform Supercapacitors. *J. Mater. Chem. A* **2014**, *2* (31), 12355–12360.
185. Zhang, J.; Yu, Y.; Liu, L.; Wu, Y. Graphene–hollow PPy Sphere 3D-nanoarchitecture with Enhanced Electrochemical Performance. *Nanoscale* **2013**, *5* (7), 3052–3057.
186. Zheng, J. P.; Cygan, P. J.; Jow, T. R. Hydrous Ruthenium Oxide as an Electrode Material for Electrochemical Capacitors. *J. Electrochem. Soc.* **1995**, *142* (8), 2699.
187. Zhai, S.; Wang, C.; Karahan, H. E.; Wang, Y.; Chen, X.; Sui, X.; ... Chen, Y. Nano-RuO_2-Decorated Holey Graphene Composite Fibers for Micro-Supercapacitors with Ultrahigh Energy Density. *Small* **2018**, *14* (29), 1800582.
188. Chen, Y.; Zhang, X.; Zhang, D.; Ma, Y. One-pot Hydrothermal Synthesis of Ruthenium Oxide Nanodots on Reduced Graphene Oxide Sheets for Supercapacitors. *J. Alloys Comp.* **2012**, *511* (1), 251–256.
189. Amir, F. Z.; Pham, V. H.; Dickerson, J. H. Facile Synthesis of Ultra-small Ruthenium Oxide Nanoparticles Anchored on Reduced Graphene Oxide Nanosheets for High-performance Supercapacitors. *RSC Adv.* **2015**, *5* (83), 67638–67645.
190. Yang, Y.; Liang, Y.; Zhang, Y.; Zhang, Z.; Li, Z.; Hu, Z. Three-dimensional Graphene Hydrogel Supported Ultrafine RuO_2 Nanoparticles for Supercapacitor Electrodes. *New J. Chem.* **2015**, *39* (5), 4035–4040.
191. Yang, Y.; Liang, Y.; Zhang, Y.; Zhang, Z.; Li, Z.; Hu, Z. Three-dimensional Graphene Hydrogel Supported Ultrafine RuO_2 Nanoparticles for Supercapacitor Electrodes. *New J. Chem.* **2015**, *39* (5), 4035–4040.
192. Chen, H.; Zhou, S.; Chen, M.; Wu, L. Reduced Graphene oxide–MnO_2 Hollow Sphere Hybrid Nanostructures as High-performance Electrochemical Capacitors. *J. Mater. Chem.* **2012**, *22* (48), 25207–25216.
193. Chen, S.; Zhu, J.; Wu, X.; Han, Q.; Wang, X. Graphene Oxide− MnO_2 Nanocomposites for Supercapacitors. *ACS Nano* **2010**, *4* (5), 2822–2830.

194. Im, C.; Yun, Y. S.; Kim, B.; Park, H. H.; Jin, H. J. Amorphous Carbon Nanotube/MnO_2/graphene Oxide Ternary Composite Electrodes for Electrochemical Capacitors. *J. Nanosci. Nanotechnol.* **2013**, *13* (3), 1765–1768.
195. Zhang, J.; Yang, X.; He, Y.; Bai, Y.; Kang, L.; Xu, H.; ... Liu, Z. H. δ-MnO_2/holey Graphene Hybrid Fiber for All-solid-state Supercapacitor. *J. Mater. Chem. A* **2016**, *4* (23), 9088–9096.
196. Thangappan, R.; Kalaiselvam, S.; Elayaperumal, A.; Jayavel, R. Synthesis of Graphene Oxide/vanadium Pentoxide Composite Nanofibers by Electrospinning for Supercapacitor Applications. *Solid State Ionics* **2014**, *268*, 321–325.
197. Li, M.; Sun, G.; Yin, P.; Ruan, C.; Ai, K. Controlling the Formation of Rodlike V_2O_5 Nanocrystals on Reduced Graphene Oxide for High-performance Supercapacitors. *ACS Appl. Mater. Interfaces* **2013**, *5* (21), 11462–11470.
198. Zhang, H.; Xie, A.; Wang, C.; Wang, H.; Shen, Y.; Tian, X. Bifunctional Reduced Graphene Oxide/V_2O_5 Composite Hydrogel: Fabrication, High Performance as Electromagnetic Wave Absorbent and Supercapacitor. *ChemPhysChem* **2014**, *15* (2), 366–373.
199. Saravanakumar, B.; Purushothaman, K. K.; Muralidharan, G. Fabrication of Two-dimensional Reduced Graphene Oxide Supported V_2O_5 Networks and Their Application in Supercapacitors. *Mater. Chem. Phys.* **2016**, 170, 266–275.
200. Li, M.; Sun, G.; Yin, P.; Ruan, C.; Ai, K. Controlling the Formation of Rodlike V_2O_5 Nanocrystals on Reduced Graphene Oxide for High-performance Supercapacitors. *ACS Appl. Mater. Interfaces* **2013**, *5* (21), 11462–11470.
201. Ye, G.; Gong, Y.; Keyshar, K.; Husain, E. A.; Brunetto, G.; Yang, S.; ... Ajayan, P. M. 3D Reduced Graphene Oxide Coated V_2O_5 Nanoribbon Scaffolds for High-Capacity Supercapacitor Electrodes. *Particle Particle Syst. Characterization* **2015**, *32* (8), 817–821.
202. Naveen, A. N.; Manimaran, P.; Selladurai, S. Cobalt Oxide (Co_3O_4)/graphene Nanosheets (GNS) Composite Prepared by Novel Route for Supercapacitor Application. *J. Mater. Sci.: Mater. Electron.* **2015**, *26* (11), 8988–9000.
203. Subramani, K.; Kowsik, S.; Sathish, M. Facile and Scalable Ultra–fine Cobalt Oxide/reduced Graphene Oxide Nanocomposites for High Energy Asymmetric Supercapacitors. *ChemistrySelect* **2016**, *1* (13), 3455–3467.
204. Dong, X. C.; Xu, H.; Wang, X. W.; Huang, Y. X.; Chan-Park, M. B.; Zhang, H.; ... Chen, P. 3D Graphene–cobalt Oxide Electrode for High-performance Supercapacitor and Enzymeless Glucose Detection. *ACS Nano* **2012**, *6* (4), 3206–3213.
205. Ma, L.; Zhou, H.; Shen, X.; Chen, Q.; Zhu, G.; Ji, Z. Facile Synthesis of Co_3O_4 Porous Nanosheets/reduced Graphene Oxide Composites and Their Excellent Supercapacitor Performance. *RSC Adv.* **2014**, *4* (95), 53180–53187.
206. Xie, L. J.; Wu, J. F.; Chen, C. M.; Zhang, C. M.; Wan, L.; Wang, J. L.; ... Sun, G. H. A Novel Asymmetric Supercapacitor with an Activated Carbon Cathode and a Reduced Graphene Oxide–cobalt Oxide Nanocomposite Anode. *J. Power Sources* **2013**, *242*, 148–156.
207. Shown, I.; Ganguly, A.; Chen, L. C.; Chen, K. H. Conducting Polymer-based Flexible Supercapacitor. *Energy Sci. Eng.* **2015**, *3* (1), 2–26.
208. Gómez, H.; Ram, M. K.; Alvi, F.; Villalba, P.; Stefanakos, E. L.; Kumar, A. Graphene-conducting Polymer Nanocomposite as Novel Electrode for Supercapacitors. *J. Power Sources* **2011**, *196* (8), 4102–4108.

209. Moussa, M.; El-Kady, M. F.; Abdel-Azeim, S.; Kaner, R. B.; Majewski, P.; Ma, J. Compact, Flexible Conducting Polymer/graphene Nanocomposites for Supercapacitors of High Volumetric Energy Density. *Compos. Sci. Technol.* **2018**, *160*, 50–59.
210. Meng, Q.; Cai, K.; Chen, Y.; Chen, L. Research Progress on Conducting Polymer Based Supercapacitor Electrode Materials. *Nano Energy* **2017**, *36*, 268–285.
211. Abdah, M. A. A. M.; Azman, N. H. N.; Kulandaivalu, S.; Sulaiman, Y. Review of the Use of Transition-metal-oxide and Conducting Polymer-based Fibres for High-performance Supercapacitors. *Mater. Design* **2020**, *186*, 108199.
212. Dinari, M.; Momeni, M. M.; Goudarzirad, M. Nanocomposite Films of Polyaniline/graphene Quantum Dots and Its Supercapacitor Properties. *Surface Eng.* **2016**, *32* (7), 535–540.
213. Zhang, J.; Zhao, X. S. Conducting Polymers Directly Coated on Reduced Graphene Oxide Sheets as High-performance Supercapacitor Electrodes. *J. Phys. Chem. C* **2012**, *116* (9), 5420–5426.
214. Wang, Z.; Zhang, Q. E.; Long, S.; Luo, Y.; Yu, P.; Tan, Z.; ... Zhou, H. Three-dimensional Printing of Polyaniline/reduced Graphene Oxide Composite for High-performance Planar Supercapacitor. *ACS Appl. Mater. Interfaces* **2018**, *10* (12), 10437–10444.
215. Kazemi, S. H.; Aghdam, S. A. High-Performance Asymmetric Supercapacitor Based on Ternary MnO 2-Polyaniline-Reduced Graphene Oxide Quantum Dots Nanocomposite Electrode. *J. Electron. Mater.* **2019**, *48* (8), 5088–5098.
216. Zhang, J.; Zhao, X. S. Conducting Polymers Directly Coated on Reduced Graphene Oxide Sheets as High-performance Supercapacitor Electrodes. *J. Phys. Chem. C* **2012**, *116* (9), 5420–5426.
217. Wang, H.; Hao, Q.; Yang, X.; Lu, L.: Wang, X. A Nanostructured Graphene/polyaniline Hybrid Material for Supercapacitors. *Nanoscale* **2010**, *2* (10), 2164–2170.
218. Wang, S.; Ma, L.; Gan, M.; Fu, S.; Dai, W.; Zhou, T.; ... Wang, H. Free-standing 3D Graphene/polyaniline Composite Film Electrodes for High-performance Supercapacitors. *J. Power Sources* **2015**, *299*, 347–355.
219. Liu, X.; Shang, P.; Zhang, Y.; Wang, X.; Fan, Z.; Wang, B.; Zheng, Y. Three-dimensional and Stable Polyaniline-grafted Graphene Hybrid Materials for Supercapacitor Electrodes. *J. Mater. Chem. A* **2014**, *2* (37), 15273–15278.
220. Shayeh, J. S.; Ehsani, A.; Ganjali, M. R.; Norouzi, P.; Jaleh, B. Conductive Polymer/reduced Graphene oxide/Au Nano Particles as Efficient Composite Materials in Electrochemical Supercapacitors. *Appl. Surface Sci.* **2015**, *353*, 594–599.
221. Zhu, J.; Feng, T.; Du, X.; Wang, J.; Hu, J.; Wei, L. High Performance Asymmetric Supercapacitor Based on Polypyrrole/graphene Composite and Its Derived Nitrogen-doped Carbon Nano-sheets. *J. Power Sources* **2017**, *346*, 120–127.
222. Konwer, S.; Boruah, R.; Dolui, S. K. Studies on Conducting Polypyrrole/graphene Oxide Composites as Supercapacitor Electrode. *J. Electron. Mater.* **2011**, *40* (11), 2248.
223. Fan, L. Q.; Liu, G. J.; Wu, J. H.; Liu, L.; Lin, J. M.; Wei, Y. L. Asymmetric Supercapacitor Based on Graphene Oxide/polypyrrole Composite and Activated Carbon Electrodes. *Electrochimica Acta* **2014**, *137*, 26–33.
224. Chang, H. H.; Chang, C. K.; Tsai, Y. C.; Liao, C. S. Electrochemically Synthesized Graphene/polypyrrole Composites and Their Use in Supercapacitor. *Carbon* **2012**, *50* (6), 2331–2336.

225. Zhang, L. L.; Zhao, S.; Tian, X. N.; Zhao, X. S. Layered Graphene Oxide Nanostructures with Sandwiched Conducting Polymers as Supercapacitor Electrodes. *Langmuir* **2010**, *26* (22), 17624–17628.
226. Groenendaal, L.; Jonas, F.; Freitag, D.; Pielartzik, H.; Reynolds, J. R. Poly (3, 4-ethylenedioxythiophene) and Its Derivatives: Past, Present, and Future. *Adv. Mater.* **2000**, *12* (7), 481–494.
227. Lehtimäki, S.; Suominen, M.; Damlin, P.; Tuukkanen, S.; Kvarnström, C.; Lupo, D. Preparation of Supercapacitors on Flexible Substrates with Electrodeposited PEDOT/graphene Composites. *ACS Appl. Mater. Interfaces* **2015**, *7* (40), 22137–22147.

CHAPTER 12

Negative Capacitance Field Effect Transistors for Future Low Power Electronics

D. NIRMAL[1*] and J. AJAYAN[2]

[1]*Department of Electronics and Communication Engineering, Karunya Institute of Technology and Science, Coimbatore, India*

[2]*Department of Electronics and Communication Engineering, SNS College of Technology, Coimbatore, India*

[*]*Corresponding author. E-mail: dnirmalphd@gmail.com*

ABSTRACT

Traditional silicon-based CMOS field-effect transistors (FETs) cannot satisfy the requirements of future high-speed low-power integrated circuits due to their ultra–high-power consumption at deep-nanometer regime. Moreover, conventional FETs cannot overcome the subthreshold swing (SS) limit of 60mV/decade. Negative capacitance field-effect transistors (NC-FETs) are considered as one of the most promising transistor technologies for future high-speed and low–power-integrated circuits due to their unique features, such as high ON current (I_{ON}), low OFF current (I_{OFF}), SS limit of sub-60 mV/decade, low-threshold voltage (V_{th}), ultra–low-power consumption, and high-switching current ratio (I_{ON}/I_{OFF}). NC-FETs can be realized using ferroelectric materials. By varying the ferroelectric properties, the DC and RF characteristics of the transistor can be tuned to the desirable level. This chapter deals with the various architectures of NC-FETs, performance-enhancement techniques of NC-FETs, scalability of NC-FETs, ferroelectric materials for NC-FETs, FSOI-NC-FETs, 2-dimentional NC-FETs, analog/RF performance of NC-FETs, DC characteristics of NC-FETs, effect of traps on the performance of NC-FETs, work-function engineering in NC-FETs, negative capacitance carbon nanotube FETs, impact of gate-stack structure on NC-FET performance, NC-FinFET, switching-speed limitation of NC-FETs, gate-stress engineering in NC-FETs, impact of temperature on NC-FET performance, and modeling of drain current in NC-FETs.

12.1 INTRODUCTION

The global semiconductor business has reached $500 billion. Steep subthreshold slope or low-voltage logic transistors are required to continue the growth of semiconductor industry. Transistor scaling has been done in the past decades to reduce the power supply voltage and to improve the speed of the logic circuits. However, thermionic emission-dependent transport mechanism of the conventional MOSFETs limits its subthreshold swing to greater than or equal to 60 mV/dec at 300 K. Negative capacitance field effect transistors (NCFETs) have been considered as one of the promising devices to overcome the aforementioned limitations.[1–11] The switching performance of NCFETs depends on the modulation of capacitance in the device. The basic structure of a NCFET and the various types of capacitances present in NCFET are illustrated in Figure 12.1.

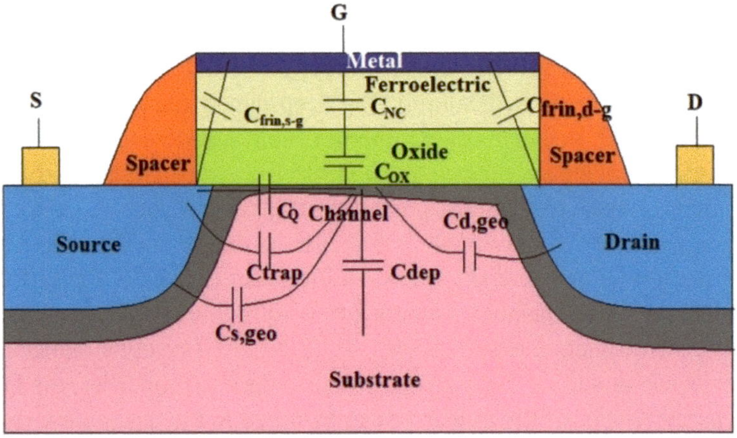

FIGURE 12.1 The basic structure of a NCFET and the various types of capacitances present in NCFET (C_Q: Quantum capacitance, C_{de}: Depletion capacitance, $C_{S,\ geo}$: Source geometrical capacitance, $C_{d,\ geo}$: Drain geometrical capacitance, C_{trap}: Trap induced capacitance, C_{NC}: Negative capacitance, C_{OX}: Oxide capacitance, $C_{frin,g-s}$: Source-gate fringing capacitance, $C_{frin,d-g}$: Drain-gate fringing capacitance).[1]

C_Q can be computed as[1]

$$C_Q = \frac{dQ_e}{d\phi_{ch}} \tag{12.1}$$

Q_e: Electron charge density
Φ_{ch}: Channel potential

NCFETs consist of a negative capacitance (NC) layer formed of ferroelectric (FE) materials. The NC layer has S-shaped polarization, which is the main reason for this negative capacitance. This negative capacitance in the NC layer is unstable and which can be stabilized by providing a traditional oxide layer below the NC layer.[1]

12.2 NEGATIVE CAPACITANCE-FULLY DEPLETED SILICON ON INSULATOR FETS (NC-FDSOI FET)

The device structure of NC-FDSOI FET with single and double FE layers is shown in Figure 12.2. The voltage across the ferroelectric capacitor (V_{FE}) can be calculated as[12]

$$V_{FE} = \left(2\alpha Q + 4\beta Q^3 + 6\gamma Q^5 + \rho\frac{\partial Q}{\partial t}\right)T_{FE} \tag{12.2}$$

α, β, g, and ρ are material parameters of ferroelectric layer. Q and T_{FE} represent the ferroelectric charge and ferroelectric layer thickness, respectively. The capacitance of the ferroelectric layer (C_{FE}) can be calculated as[12–14]

$$C_{FE} = \frac{dQ}{dV_{FE}} \approx \frac{2}{3\sqrt{3}} \times \frac{P_r}{E_C.T_{FE}} \tag{12.3}$$

$$\alpha = -\frac{3\sqrt{3}}{2}\frac{E_C T_{FE}}{P_r} \tag{12.4}$$

$$\beta = \frac{3\sqrt{3}}{2}\frac{E_C T_{FE}}{P_r^3} \tag{12.5}$$

E_C and P_r represents coercive field and remnant polarization, respectively. The drain of the NC-FDSOI FET can be computed as[13]

$$I_{DS} = \mu\frac{W}{L}\left[\frac{Q_{inv,s} + Q_{inv,d}}{2}\left(\psi_{1,s} - \psi_{1,d}\right) + \eta V_t\left(Q_{inv,s} - Q_{inv,d}\right)\right] \tag{12.6}$$

$\Psi_{1,s}$: Surface potential at source side
$\Psi_{1,d}$: Surface potential at drain side

$Q_{inv,s}$: Inversion charge density at source side
$Q_{inv,d}$: Inversion charge density at drain side
V_t: Thermal voltage

The impact of P_r on the drain current (I_{DS}) and ferroelectric charge (Q) are depicted in Figure 12.3a and b, respectively. A high value of P_r is preferable for achieving high drain current and low subthreshold leakage current. Higher P_r leads to the increase of ferroelectric charge and gate capacitance.[15] The impact of E_C on the I_{DS} and gate capacitance of the NC-FDSOI FET are depicted in Figure 12.3c and d, respectively. A large value of E_C leads to the enhancement of I_{DS} (Figure 12.3c) due to improved gate capacitance (Figure 12.3d). The impact of single- and double-ferroelectric layers on the I_{ON}/I_{OFF} ratio of NC-FDSOI FET is illustrated in Figure 12.4a and the variation of SS of NC-FDSOI FET with channel-length scaling is illustrated in Figure 12.4b. The influence of single- and double-ferroelectric layers on the I_{DS} and gate capacitance of NC FDSOI FET is demonstrated in Figure 12.5. NC-FDSOI FET with double-ferroelectric layers exhibits superior performance compared with single-ferroelectric layer-based NC-FDSOI FETs.

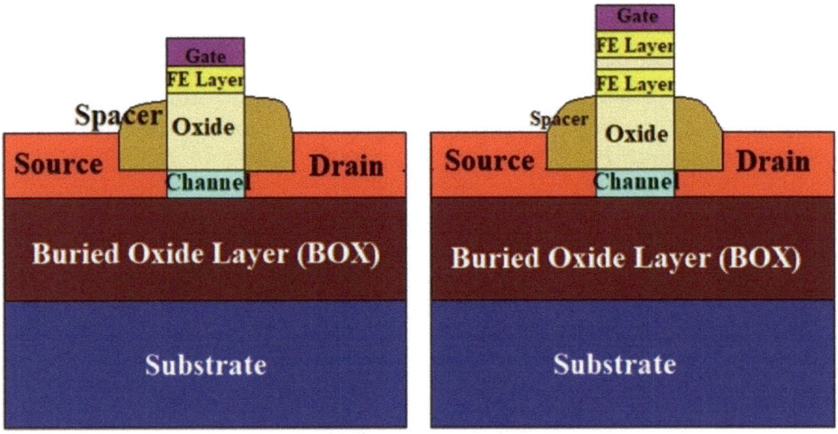

FIGURE 12.2 NC-FDSOI FETs with (a) single- and (b) double-FE layers.[12]

NC-FDSOI FETs with double-ferroelectric layer exhibits higher I_{ON}/I_{OFF} ratio compared with mono-ferroelectric layer-based NCFETs. In addition, SS of the NC-FDSOI FETs are found to be increasing along with downscaling of channel length.

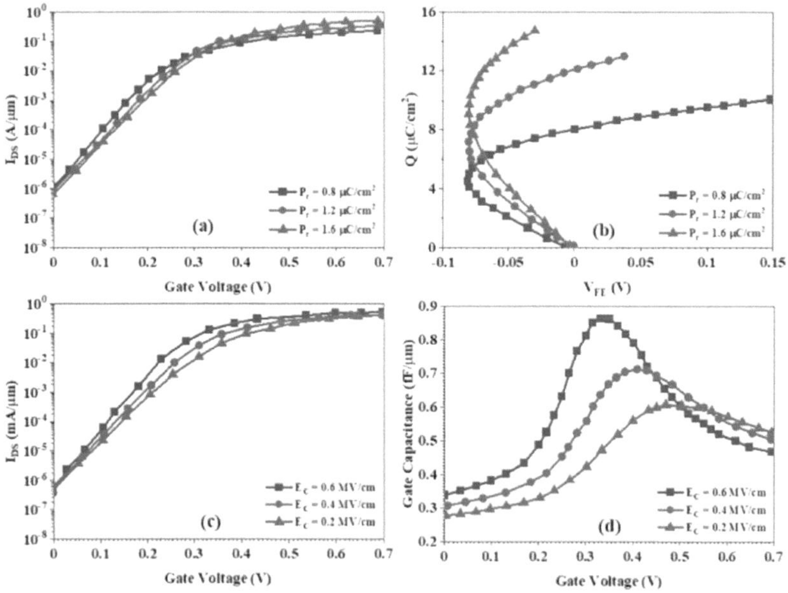

FIGURE 12.3 (a) The effect of P_r on I_{DS} with constant E_C. (b) The effect of P_r on the ferroelectric charge at constant E_C. (c) The effect of E_C on I_{DS} with constant P_r. (d) The effect of E_C on gate capacitance with constant P_r.[12]

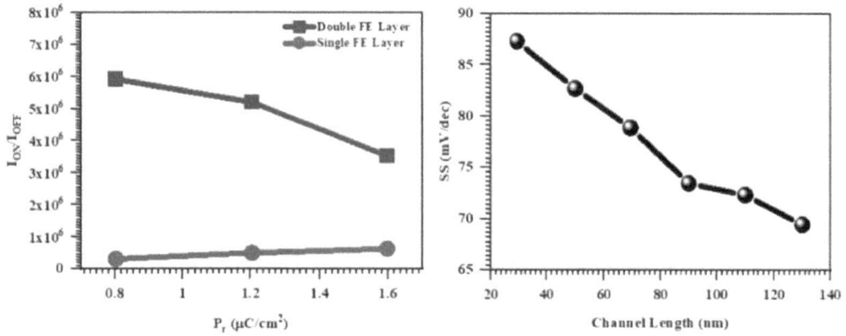

FIGURE 12.4 (a) The impact of single- and double-ferroelectric layers on the I_{ON}/I_{OFF} ratio of NC-FDSOI FET.[12] (b) the variation of SS of NC-FDSOI FET with channel-length scaling.[16]

12.3 SILICON ON INSULATOR JUNCTIONLESS NEGATIVE CAPACITANCE-FET (SOI JL NC-FET)

The device structure of a SOI JL NC-FET is shown in Figure 12.6. It consists of a p-type substrate, BOX layer and a channel made of Si, SiO_2

layer, a top and suspended metal gate,s and a ferroelectric layer. The impacts of temperature and ferroelectric layer thickness (T_{FE}) on the I_{DS} of SOI JL NC-FET are illustrated in Figure 12.7.

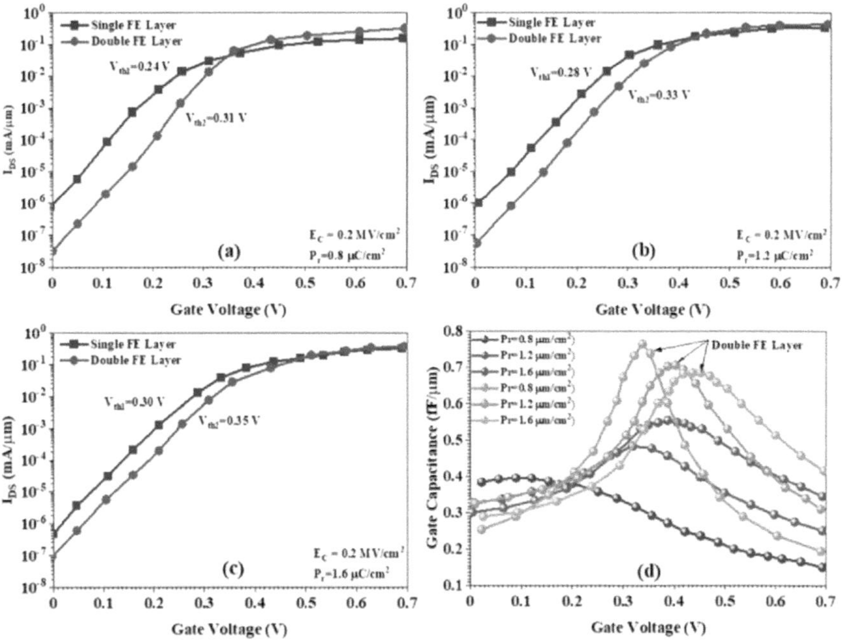

FIGURE 12.5 The influence of single- and double-ferroelectric layers on the I_{DS} and gate capacitance of NC FDSOI FET.[12]

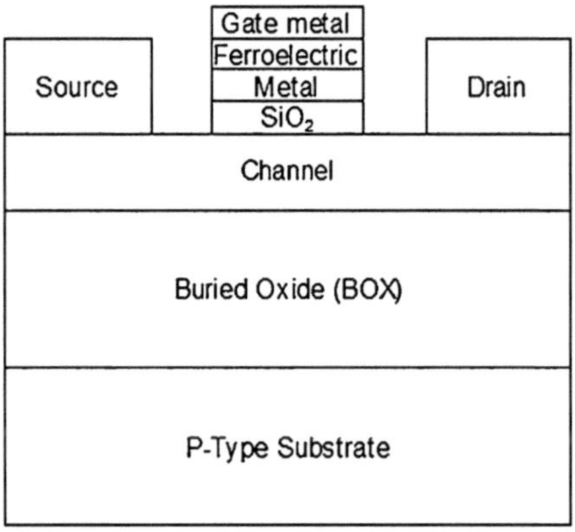

FIGURE 12.6 The device structure of a SOI JL NC-FET.[17]

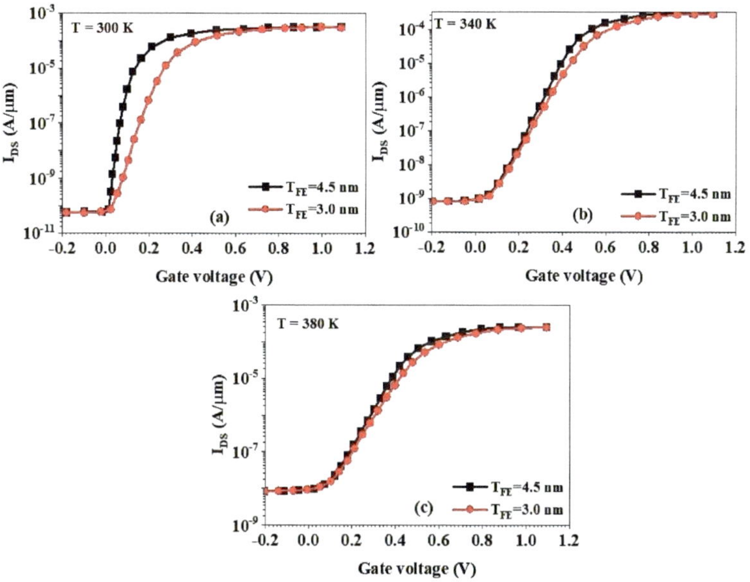

FIGURE 12.7 The impacts of temperature and ferroelectric layer thickness (T_{FE}) on the I_{DS} of SOI JL NC-FET.[17]

A large T_{FE} is preferable to achieve high I_{DS}. When the temperature is increased from 300 K to 380 K, the ON current is found to be decreasing and OFF current is found to be increasing. The impact of temperature and T_{FE} on the SS, I_{ON}/I_{OFF} ratio and cut off frequency (f_T) of SOI JL NC-FET are illustrated in Figure 12.8a–c, respectively. SS is found to be increasing with increase in temperature and decrease in T_{FE}. I_{ON}/I_{OFF} ratio and the cut-off frequency of the SOI JL NC-FETs increases with increase in T_{FE}. When temperature is increased from 300 K to 380 K, both I_{ON}/I_{OFF} ratio and cut off frequency of the SOI JL NC-FETs are found to be decreasing. HfO_2, $Hf_{1-x}Zr_xO_2$ (HZO), $SrTiO_3$ (STO), $BiFeO_3$, PVDF, and $Pb_{0.2}Zr_{0.8}TiO_3$ (PZT) are the widely used ferroelectric materials for the development of NCFETs.[19,20] Physical vapor deposition (PVD) can be used to grow these ferroelectric layers.

12.4 GATE ALL AROUND NANOWIRE NEGATIVE CAPACITANCE FET (GAA NW NCFET)

In 2020, Shen-Yang Lee et al.[21] reported the successful fabrication of a poly-Si GAA NW NCFET with metal ferroelectric insulator semiconductor

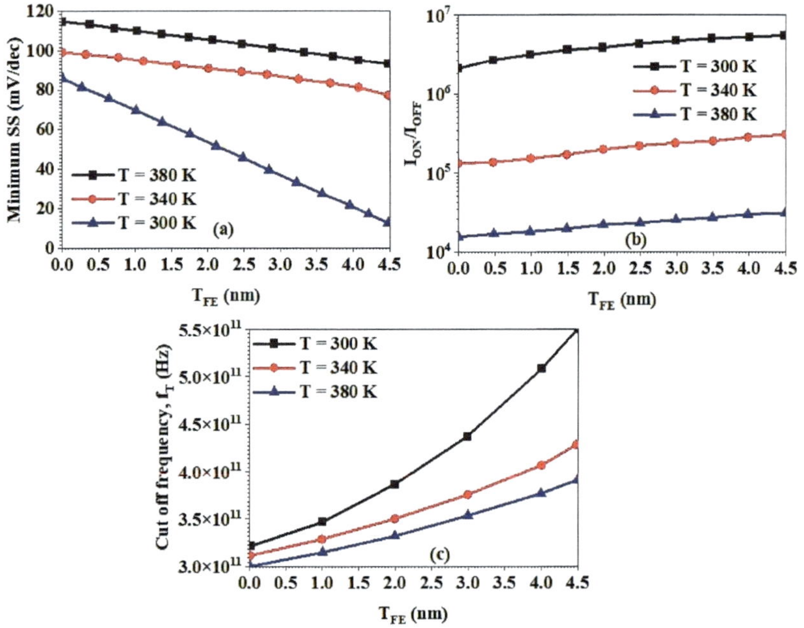

FIGURE 12.8 The impact of temperature and T_{FE} on the SS, I_{ON}/I_{OFF} ratio, and cut-off frequency (f_T) of SOI JL NC-FET.[17]

(MFIS) and metal ferroelectric metal insulator semiconductor (MFMIS) structures. The cross-sectional views of GAA NW NCFETs are shown in Figure 12.9. The impacts of Tri-gate(TG) and GAA structures on the NCFET output curves are demonstrated in Figure 12.10a. GAA NW NCFET shows superior drain current compared with TG NCFETs due to better gate control over the channel.[22, 23] The comparison of output characteristics of MFMIS and MFIS GAA NW NCFETs are depicted in Figure 12.10b. MFMIS GAA NW NCFETs are found to be effective in achieving higher drain current compared with MFIS GAA NW NCFETs due to reduction in parasitic, higher transconductance, and improved gate capacitance. The effect of ferroelectric thickness (T_{FE}) and gate length (L) scaling on the drain current of conventional MFMIS structure-based NCFETs are illustrated in Figure 12.10c and d, respectively. A large T_{FE} and low L are preferable for achieving higher drain current. The conventional MFMIS NCFET structure is shown in Figure 12.11. The GAA NW NCFETs are capable of providing better drain current compared with conventional NCFETs. T_{FE} significantly affects the threshold voltage of

NCFETs. The threshold voltage of the NCFETs decreases with increase in T_{FE} (Figure 12.12a). A large T_{FE} results in the reduction of SS which, in turn, reduces the threshold voltage. A large T_{FE} also results in the improvement in gate capacitance, which helps to improve the drain current. Figure 12.12b illustrates the impact of single-and double-layer channels on the output curves of GAA NW NCFETs.

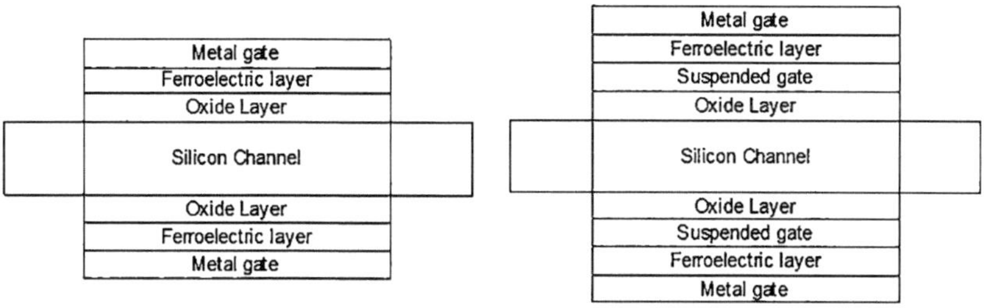

FIGURE 12.9 (a) GAA NW NCFET with MFIS structure. (b) GAA NW NCFET with MFMIS structure.

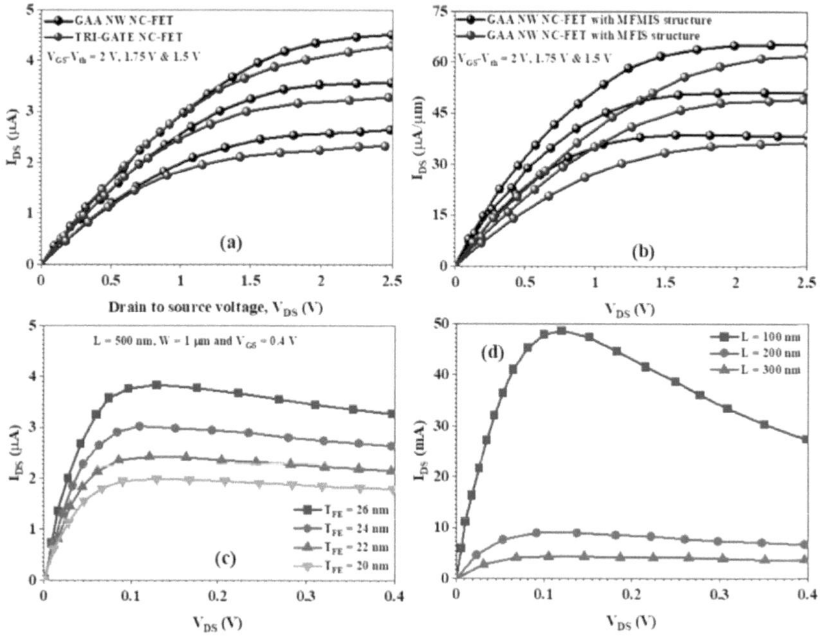

FIGURE 12.10 (a) The impact of Tri-gate (TG) and GAA structures on the NCFET output curves.[21] (b) The comparison of output characteristics of MFMIS and MFIS GAA NW NCFETs.[21] (c) The impact of T_{FE} scaling on I_{DS} of conventional MFMIS NCFETs.[10] (d) The impact of scaling of L on I_{DS} of conventional MFMIS NCFETs.[10]

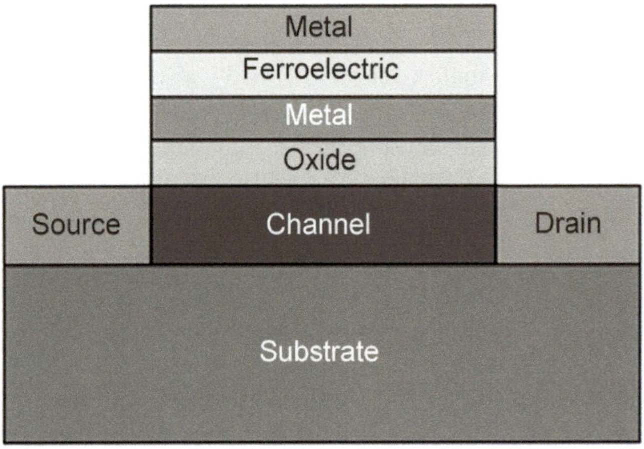

FIGURE 12.11 Conventional MFMIS NCFET.[10]

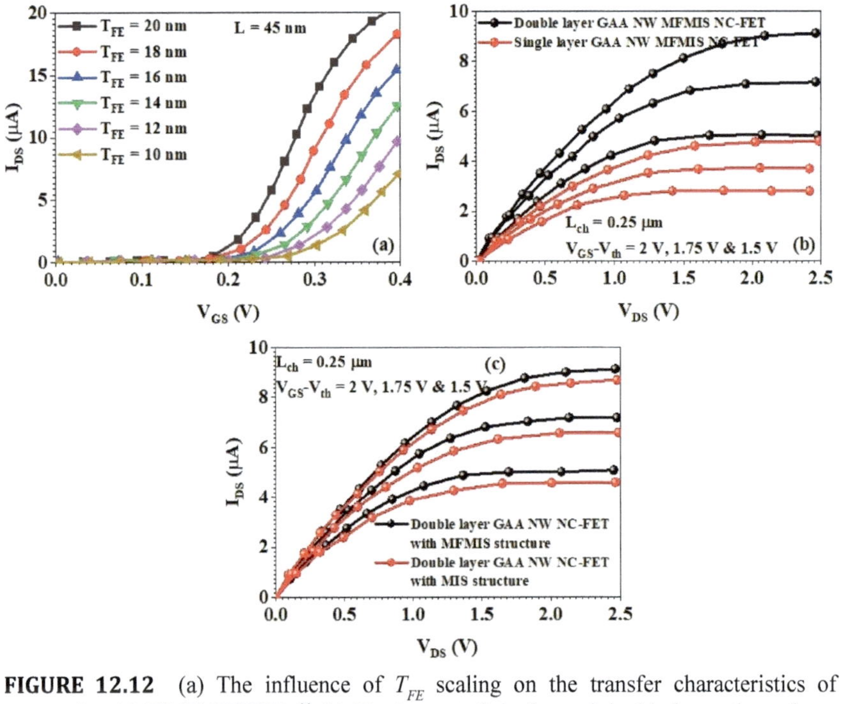

FIGURE 12.12 (a) The influence of T_{FE} scaling on the transfer characteristics of conventional MFMIS NCFETs.[10] (b) The impact of single- and double-layer channels on the output curves of GAA NW NCFETs.[24] (c) The output curves of MFMIS and MFIS double layer GAA NCFETs.[24]

Stacked channel structures in GAA NW NCFETs help to significantly improve the ON current. The output curves of MFMIS and MFIS double-layer GAA NCFETs are depicted in Figure 12.12c. Raised drain and source structures are found to be effective in reducing parasites and thereby improving device performance. Ion-implantation technique can be used to grow raised drain and source structures in GAA NW NCFETs. The drain current increases with an increase in the diameter of GAA NW NCFETs.[22] However, a large diameter of the GAA NW NCFETs leads to the drastic increase in subthreshold leakage current. Therefore, a smaller diameter is preferable for achieving a large I_{ON}/I_{OFF} ratio in GAA NW NCFETs. In 2017, Kyungmin Jang et al.[22] investigated the scalability and I_{ON}/I_{OFF} ratio improvement of GAA NW NCFETs.

12.5 RECONFIGURABLE NEGATIVE CAPACITANCE FETS (R-NCFETS)

The structure of a reconfigurable double gate NCFET is shown in Figure 12.13. This device has NiSi drain/source contacts, intrinsic silicon channel, an interfacial layer, a ferroelectric layer, and two metal gates namely control gate (CG) and polarity gate (PG). The polarity gate controls the schottky barrier height at drain/source-channel interfaces and the control gate takes the electrostatic control of the channel. The I_{ON}/I_{OFF} ratio of R-NCFETs can be improved by using spacer engineering techniques. The net channel charge density (Q_G) of R-NCFETs can be computed as[25]

$$Q_G = Q_{MOS} + C_{PD}\left(V_{GS} - V_{FE} - V_{DS}\right) + C_{PS}\left(V_{GS} - V_{FE} - V_S\right) \qquad (12.7)$$

Q_{MOS} : Base line transistor charge density
C_{PS} : Source parasitic capacitance
C_{PD} : Drain parasitic capacitance
V_{FE} : Voltage across the ferroelectric layer

The output curves (I_{DS} Vs V_{DS}) of both n-type and p-type R-NCFETs are demonstrated in Figure 12.14a and b, respectively. For the same-device dimensions, p-R-NCFETs offer high drain current compared with n-R-NCFETs. The drain current of R-NCFETs increases with increase in over-drive voltage (V_{OV}).The temperature dependence of maximum gain and minimum SS of both n-type and p-type R-NCFETs are illustrated in Figure 12.14c and d, respectively.

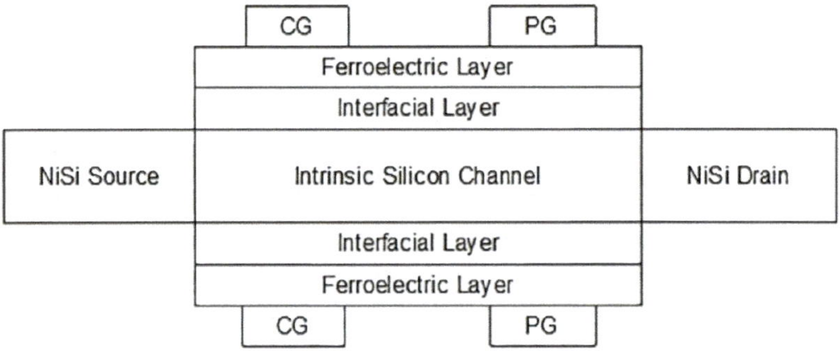

FIGURE 12.13 The structure of a reconfigurable double gate NCFET.[26]

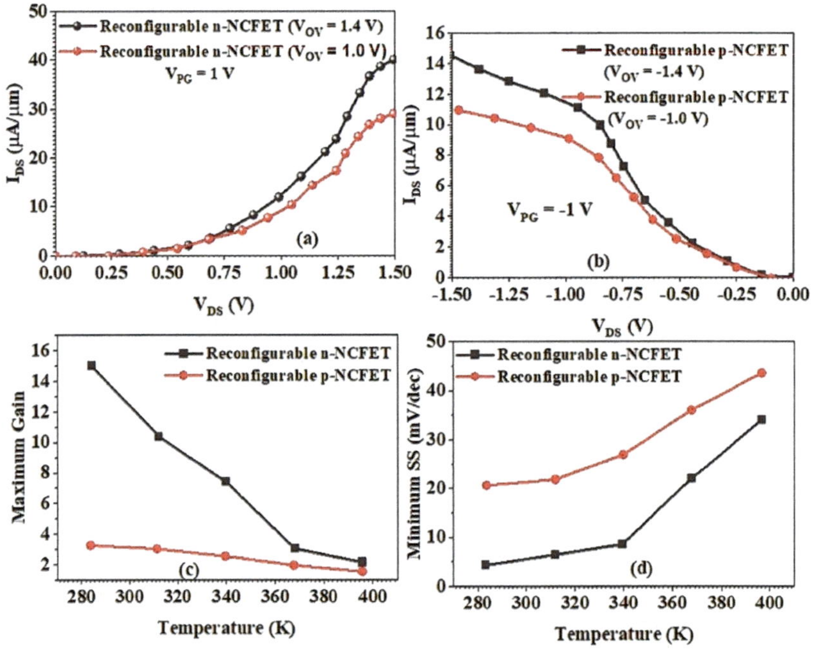

FIGURE 12.14 I_{DS} *Vs* V_{DS} plot for n-R-NCFET (b) I_{DS} *Vs* V_{DS} plot for p-R-NCFET. (c) Temperature dependence of maximum gain for both n-type and p-type R-NCFETs. (d) Temperature dependence of minimum SS of both n-type and p-type R-NCFETs.[26]

n-R-NCFETs exhibits maximum gain and lower minimum SS compared with p-R-NCFETs. Also, the maximum gain of both n and p type R-NCFETs decreases with increase in temperature. On the other hand, the minimum SS of both n and p type R-NCFETs increases drastically with increase in temperature due to the increase of OFF current. The temperature dependence of threshold voltages of both n-type and p-type R-NCFETs are illustrated in Figure 12.15a and b, respectively. The threshold voltages of both n and p type R-NCFETs are found to be decreasing with respect to increase in temperature. The variation of DIBL and I_{ON}/I_{OFF} ratio of R-NCFETs with temperature are illustrated in Figure 12.15c and d, respectively. Both DIBL and I_{ON}/I_{OFF} ratio of n and p type R-NCFETs degrades severely with increase in temperature.

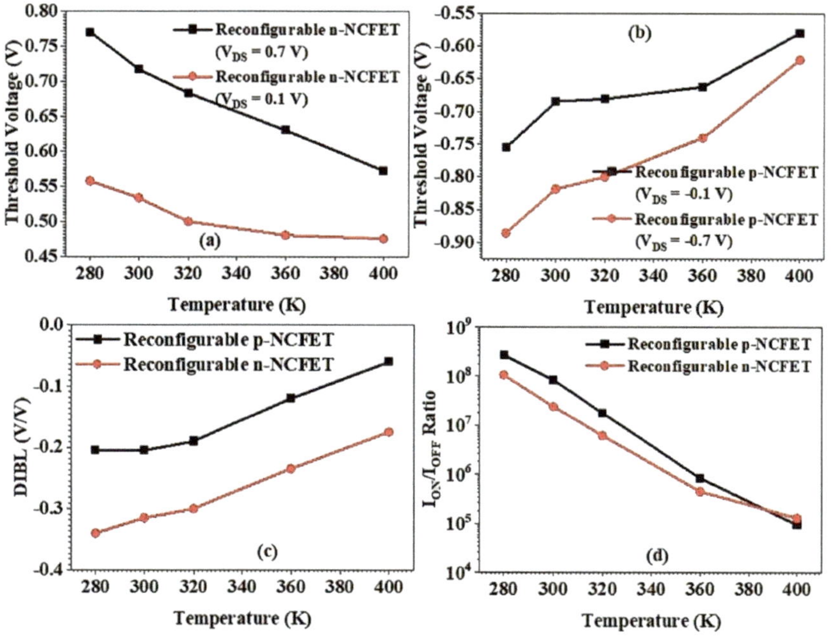

FIGURE 12.15 (a) Threshold voltage versus temperature plot for n-R-NCFET. (b) Threshold voltage versus temperature plot for P-R-NCFET. (c) DIBL versus temperature plot for R-NCFETs. (d) I_{ON}/I_{OFF} ratio versus temperature plot for R-NCFETs.[26]

The variation of peak g_m with temperature for both n and p type R-NCFETs are demonstrated in Figure 12.16a. It is observed that for a temperature range from 280 K to 350 K, n-R-NCFET exhibits higher peak g_m compared with n-R-NCFET for the same-device dimensions. However, p-R-NCFET exhibits superior peak g_m for a temperature range from 360 K to 400 K. The variation of peak g_d with temperature for both *n* and *p* type

R-NCFETs are demonstrated in Figure 12.16b. It is found that, the peak g_d of both *n* and *p*-type R-NCFETs decreases with increase in temperature from 280 K to 400 K. Also, n-R-NCFET exhibits higher peak g_d compared with p-R-NCFETs. The temperature dependence of cut-off frequency (f_T) of both n and p-type R-NCFETs are illustrated in Figure 12.16c and d, respectively. The f_T of both n and p-type R-NCFETs are found to be increasing with increase in temperature.

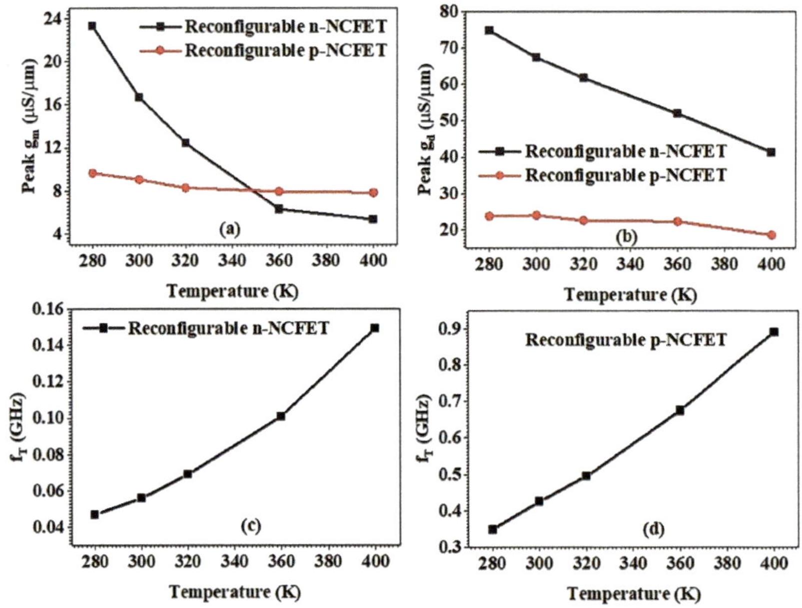

FIGURE 12.16 (a) Peak g_m versus temperature plot for R-NCFETs. (b) Peak g_d versus temperature plot for R-NCFETs. (c) f_T versus temperature plot for n-R-NCFETs. (d) f_T versus temperature plot for p-R-NCFETs.[26]

12.6 NEGATIVE CAPACITANCE TUNNEL FET (NC-TFETS)

The structures of negative capacitance vertical TFETs (NC V-TFETs) with and without N^{++} doping concentration are shown in Figures 12.17 and 12.18, respectively. The MFIS and MFMIS structures of GAA NC-TFETs are shown in Figure 12.19a and b, respectively. Saving of power is one of the most critical issues in future ultra-low power internet-of-things (IoT) applications, logic integrated circuit (IC) applications, and wearable devices.[27–45] For saving power in the aforementioned applications, transistors with steep SS are essential.[46] Compound semiconductors can provide superior ON and OFF state performances compared with traditional Silicon technology-based transistors. In Figures 12.17 and 12.18, $GaAs_{0.51}Sb_{0.49}$ is used as source material and $In_{0.53}Ga_{0.47}As$ is used as drain and channel materials.

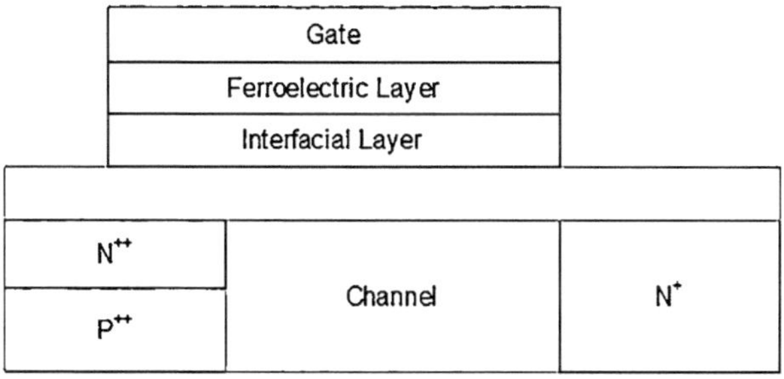

FIGURE 12.17 The structure of NC V-TFET with N^{++} doping concentration.[47]

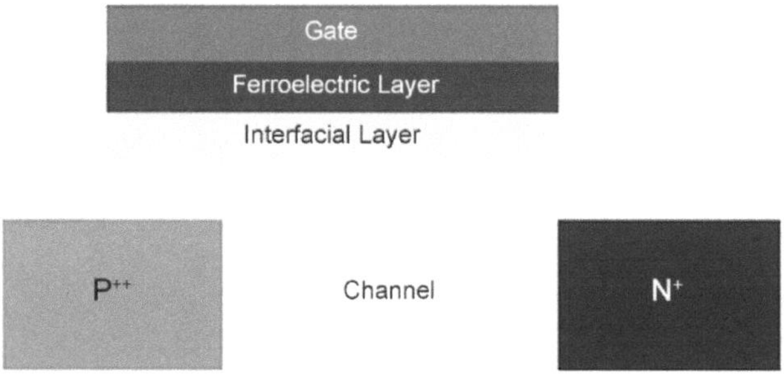

FIGURE 12.18 The structure of NC V-TFET without N^{++} doping concentration.[47]

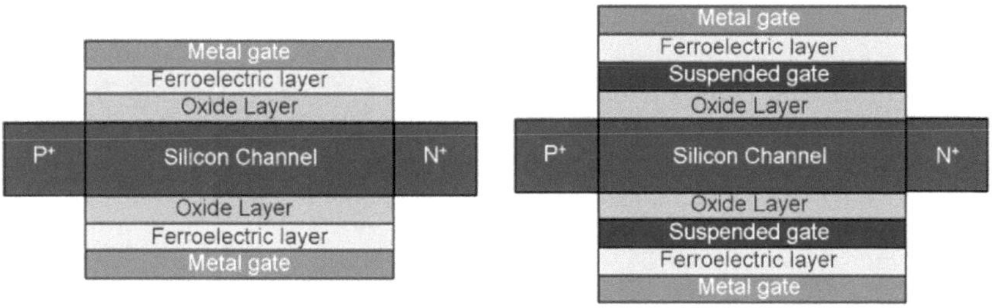

FIGURE 12.19 (a) MFIS structure-based GAA NC TFET.[23] (b) MFMIS structure-based GAA NC TFET.[23]

The impact of source–gate overlap length (L_{SOV}) on the I_{ON} and SS of NC V-TFET is illustrated in Figure 12.20a and b, respectively. The I_{ON} is found to be increasing with increase in L_{SOV}. In addition, the NC V-TFET provides higher I_{ON} compared with traditional TFETs. Moreover, the SS of NC V-TFETs decreases with increase in L_{SOV}. That is, NC V-TFETs with large L_{SOV} exhibits better immunity against short-channel effects (SCEs). NC V-TFETs are attractive compared with conventional TFET technology due to their steep SS. The influence of N^{++} doping concentration on the transfer curves of NC V-TFETs are illustrated in Figure 12.20c and the influence of T_{FE} scaling on the transfer curves of NC V-TFET without N^{++} doping is illustrated in Figure 12.20d.

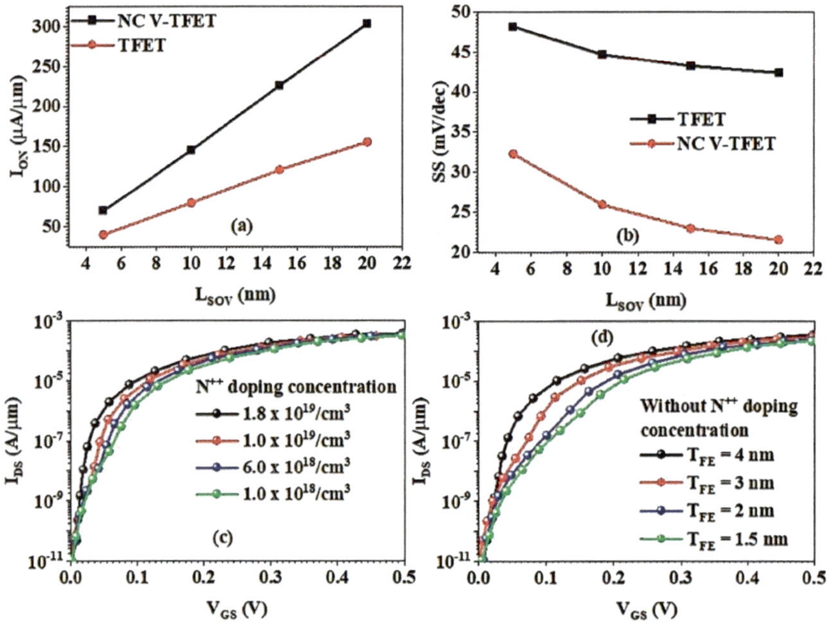

FIGURE 12.20 (a) Influence of L_{SOV} scaling on the I_{ON} of NC-V-TFET. (b) Influence of L_{SOV} scaling on the SS of NC-V-TFET. (c) The influence of N^{++} doping concentration on the transfer curves of NC V-TFETs. (d) The influence of T_{FE} scaling on the transfer curves of NC V-TFET without N^{++} doping.[47]

NC V-TFET with higher N^{++} doping concentration exhibits higher drain current and transcondcutance along with lower threshold voltage. The T_{FE} also significantly influences the transfer characteristics of the

NC V-TFETs and it is found that the NC V-TFET with large T_{FE} exhibits superior drive current. This is due to the large gate capacitance created by the large T_{FE}. The influence of coercive field (E_C) and remnant polarization (P_r) on the transfer curves of NC V-TFET without N^{++} doping is depicted in Figure 12.21a and b, respectively. NC V-TFET with large E_C and small P_r is preferable for achieving high drive current and lower operating voltage. In addition, the SS of the NC V-TFET is directly proportional to P_r. Therefore, a lower P_r is essential for improving I_{ON}, transconductance and to reduce SS and threshold voltage.

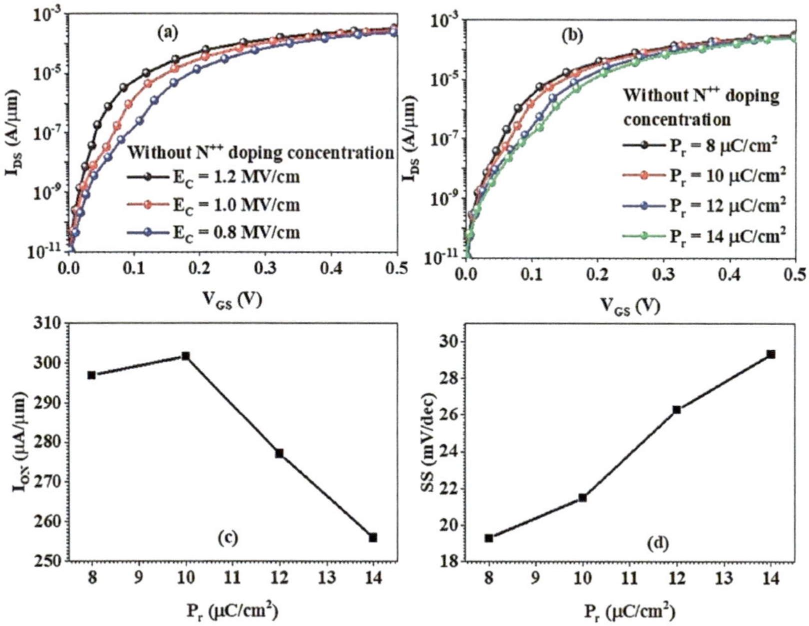

FIGURE 12.21 (a) The influence of coercive field (E_C) on the transfer curves of NC V-TFET without N^{++} doping. (b) The influence of remnant polarization (P_r) on the transfer curves of NC V-TFET without N^{++} doping. (c) The influence of remnant polarization (P_r) on the I_{ON} of NC V-TFET. (d) The influence of remnant polarization (P_r) on the SS of NC V-TFET.[47]

NCFETs are found to be effective in boosting the performance of analog circuits due to the steep switching characteristics. NCFETS are also attractive for memory design and ultra-low power digital logic integrated circuits.[48] Negative capacitance of NCFETs enables the scaling of power supply by improving the drain current at a specific supply voltage. NCFETs also show excellent immunity against SCEs.[49, 50] GAA structure also improves the scalability of NC-TFETs by providing superior gate control of the channel.[23]

12.7 GRADED CHANNEL NEGATIVE CAPACITANCE SILICON ON INSULATOR FET (GCNCSOI FET)

A combination of graded channel and SOI technology offers several advantages such as low hot-carrier effects, better gate control, and excellent scalability. The structure of a GCNC SOI FET is shown in Figure 12.22. The capacitance equivalent network (Refer[51]) of GCNCSOI FET consist of ferroelectric capacitance, source parasitic capacitance, drain parasitic capacitance, interfacial layer capacitance, and Si channel capacitance.[51] The output characteristic of GCNCSOI FET as a function of temperature (T) is depicted in Figure 12.23a. The I_{DS} is found to be degrading severely with increase in T. The influence of parasitic capacitance (C_p) and T on the I_{OFF} of GCNCSOI FET is demonstrated in Figure 12.23b. I_{OFF} of GCNCSOI FET is found to be increasing with increase in T and decreasing with increase in C_p.

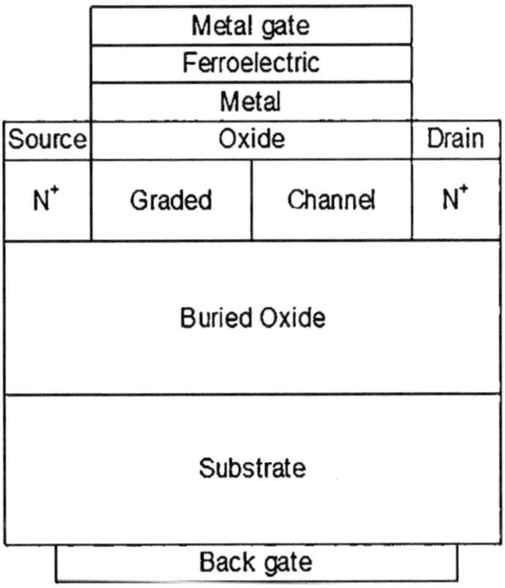

FIGURE 12.22 The structure of a GCNC SOI FET.[51]

The influence of parasitic capacitance (C_p) and T on the I_{ON} of GCNCSOI FET is demonstrated in Figure 12.23c. I_{ON} of GCNCSOI FET is found to be decreasing with increase in T and C_p. The influence of parasitic capacitance (C_p) and T on the I_{ON}/I_{OFF} ratio of GCNCSOI FET is demonstrated in Figure 12.23d. A low temperature and high C_p are essential for achieving a large I_{ON}/I_{OFF} ratio and low SS in GCNCSOI FETs.[51] Moreover, at low temperature, GCNCSOI FETs exhibits high-threshold voltage.

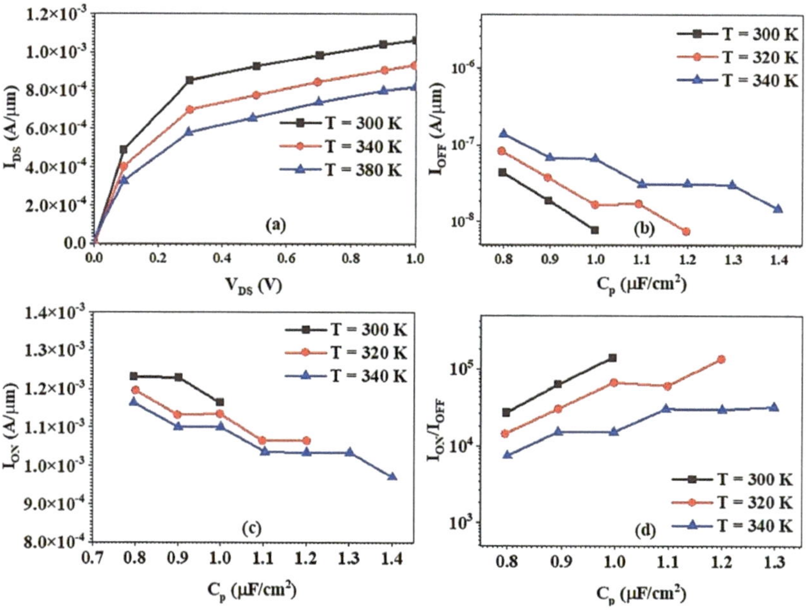

FIGURE 12.23 (a) The output curves of GCNC SOI FET. (b) The variation of IOFF of GCNC SOI FET with C_p and T. (c) The variation of I_{ON} of GCNC SOI FET with C_p and T. (d) The variation of I_{ON}/I_{OFF} ratio of GCNC SOI FET with C_p and T.[51]

12.8 2D MATERIALS-BASED NCFETS

The structure of a long-channel 2D NCFET is shown in Figure 12.24. The heavily doped substrate also acts as back gate. The 2D semiconductor material acts as channel. Gate is not shown in Figure 12.24. 2D materials are considered as attractive materials for developing sub-10 nm transistors for ultra–low-power applications. Influence of interface trap density (D_{it}) on the transfer characteristics of 2D-NCFET with T_{FE} = 20 nm and

T_{FE} = 50 nm are demonstrated in Figure 12.25a and b, respectively. The increase in D_{it} severely degrades the I_{DS} of 2D-NCFETs. A large T_{FE} helps to improve the I_{ON} as well as reduction of I_{OFF}. The influence of D_{it} and T_{FE} on the I_{ON}/I_{OFF} ratio of 2D-NCFET is illustrated in Figure 12.25c. I_{ON}/I_{OFF} ratio of 2D-NCFET degrades severely with increase in D_{it}. A lower D_{it} and higher T_{FE} is required for achieving high I_{ON}/I_{OFF} ratio in 2D-NCFETs. The influence of D_{it} and T_{FE} on the SS of 2D-NCFET is illustrated in Figure 12.25d.

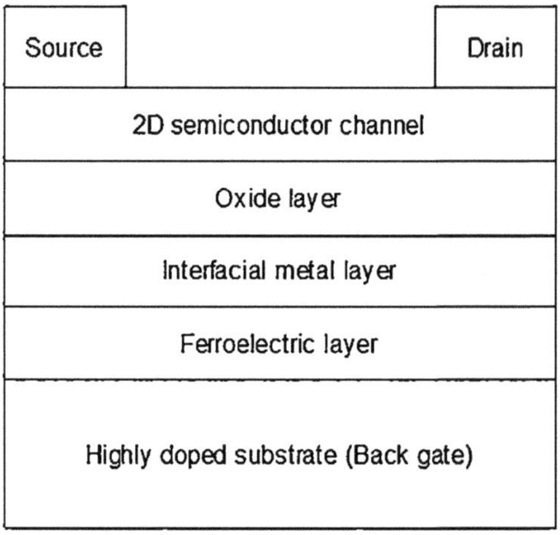

FIGURE 12.24 The structure of a long channel 2D NCFET.[52, 53]

SS of 2D-NCFET can be computed as

$$SS = \frac{2.3k_BT}{q} . \left(1 + \frac{C_{it} + C_{2D}}{C_{OX}}\right) . \left(1 - \frac{C_{device}}{|C_{FE}|}\right) \tag{12.8}$$

$$C_{it} = q^2 D_{it} \tag{12.9}$$

$$C_{device} = 2C_p + \frac{(C_{2D} + C_{it}).C_{OX}}{C_{2D} + C_{it} + C_{OX}} \tag{12.10}$$

MoS_2 is considered as the most attractive 2D channel material for developing 2D-NCFETs.[54, 55] In NCFETs, in saturation region, the drain current decreases with increase in V_{DS}. This results in the creation of a negative differential resistance (NDR). NDR is a useful property for the design of oscillator circuits.[56] The matching of capacitance between oxide layer and ferroelectric layer is essential for achieving good DC and RF performance.[57] NCFET with internal gate (MFMIS) exhibits superior performance compared with NCFET without internal gate (MFIS) due to proper capacitance matching between oxide layer and ferroelectric layer.[58] The polarization (*P*) versus electric field (*E*) characteristics of ferroelectric material in NCFET is demonstarted in Figure 12.26.

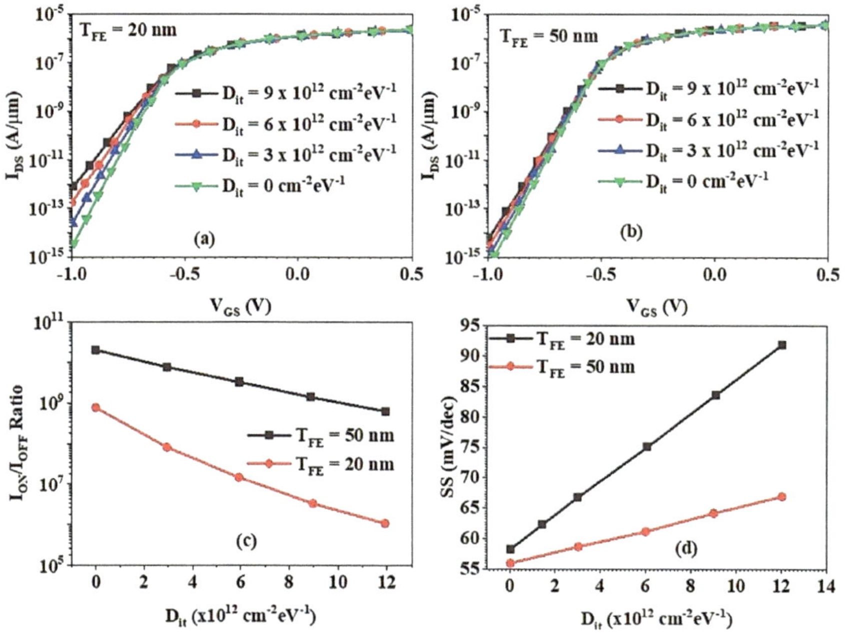

FIGURE 12.25 (a) Influence of D_{it} on the transfer characteristics of 2D-NCFET with T_{FE} = 20 nm. (b) Influence of D_{it} on the transfer characteristics of 2D-NCFET with T_{FE} = 50 nm. (c) Influence of D_{it} and T_{FE} on the I_{ON}/I_{OFF} ratio of 2D-NCFET (d) Influence of D_{it} and T_{FE} on the SS of 2D-NCFET.[52]

The S-shaped curve helps to achieve negative capacitance in the ferroelectric layer. Gate stress engineering techniques are found to be effective in NCFETs to improve negative capacitance in ferroelectric layer. In 2018, Chun-Hu Cheng et al.[60] reported gate stress engineering in NCFETs with $HfAlO_x$ ferroelectric material by adding nitrogen (N_2) into TaN gate. Stressed-gate NCFET with high N_2 incorporated TaN gate exhibits superior negative capacitance. The structure of a 2D-SOI NC transition metal dichalcogenide (TMD) FET is shown in Figure 12.27. MoS_2 is used as channel material in 2D-SOI NC TMD FET. The bandgap, electron affinity, dielectric constant, effective electron mass, and effective hole mass of MoS_2 are 1.67 eV, 4.28 eV, 4.8, O.467, and 0.544 respectively.

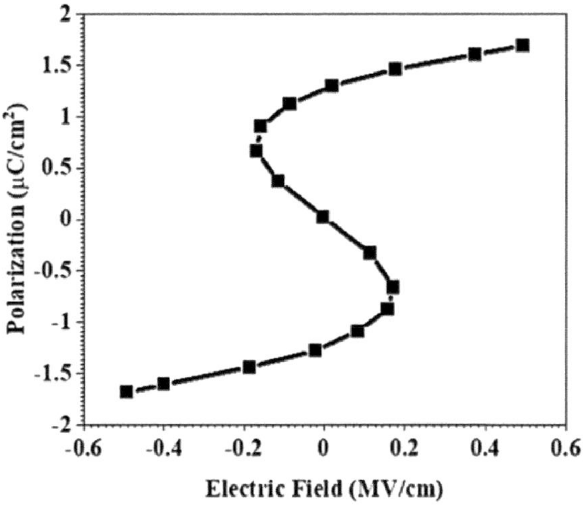

FIGURE 12.26 The polarization versus electric field characteristics of ferroelectric material in NCFET.[59]

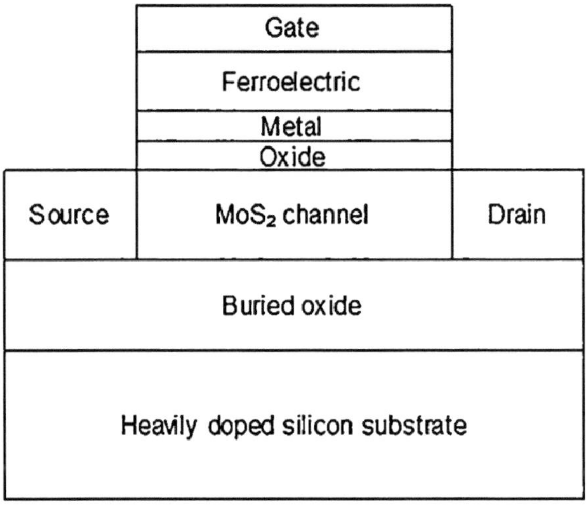

FIGURE 12.27 Structure of a 2D-SOI NC TMD FET.[61, 62]

The I_{DS} of 2D-SOI NC TMD FET can be computed as[61]

$$I_{DS} = \frac{W \mu_n kT \left(n_i^2 / N_{ch}\right)\left[1-\exp\left(-qV_{DS}/kT\right)\right]}{\int_0^{L_g} \exp\left[-q\varphi_s(y)/kT\right]dy} \tag{12.11}$$

μ_n : Electron mobility
n_i : Intrinsic carrier concentration
N_{ch} : Channel charge density
$\phi_s(y)$: Channel surface potential

The impact of gate length (L_g) scaling on the transfer curves of 2D-SOI NC TMD FET is depicted in Figure 12.28a. In 2D-SOI NC TMD FETs, down scaling of L_g improves I_{ON} and suppress I_{OFF} due to the reduction of parasitic components. The impact of L_g and T_{FE} scaling on the SS of 2D-SOI NC TMD FET is depicted in Figure 12.28b. A combination of small L_g and large T_{FE} can effectively reduce the SS of 2D-SOI NC TMD FETs. The impact of L_g and T_{ox} scaling on the SS of 2D-SOI NC TMD FET is illustrated in Figure 12.28c. A low-oxide thickness (T_{OX}) is highly preferable for achieving low SS. The impact of high-K dielectrics on the transfer curves of 2D-SOI NC TMD FET is demonstrated in Figure 12.28d. Gate oxides with high dielectric constants significantly reduces the OFF current. L_g, T_{FE}, T_{OX}, and permittivity of gate oxides are the major factors that affects the ON and OFF currents and short-channel behavior of 2D-SOI NC TMD FETs. The structure of a double-gate (DG) 2D MFIS NCFET is shown in Figure 12.29.

The impact of L_g and T_{FE} scaling on the threshold voltage (V_{th}) of 2D-SOI NC TMD FET is depicted in Figure 12.30a. T_{FE} and L_g significantly affect the V_{th} of 2D-SOI NC TMD FETs. V_{th} is found to be increasing with decrease in L_g. A large T_{FE} is required to achieve smaller V_{th}. The impact of L_g and T_{OX} scaling on the V_{th} of 2D-SOI NC TMD FET is depicted in Figure 12.30b. A lower T_{OX} is found to be effective for achieving low V_{th}. The impact of L_g scaling and permittivity of gate oxides on the SS of 2D-SOI NC TMD FET is illustrated in Figure 12.30c. The polarization versus V_{FE} plot for both 2D-SOI NC TMD FET and 2D-DG MFIS NCFETs are demonstarted in Figure 12.30d. 2D-SOI NCFETs are found to be effective in achieving high I_{ON}/I_{OFF} ratio with low SS compared with 2D-DG NCFETs due to higher negaive capacitance. Poly (vinylidene diflouride-triflouroethylene) can also be used as a ferroelectric material for the development of 2D-NCFETs.[64] Cr/Au combination can be used for making drain/source contacts for 2D-NCFETs.

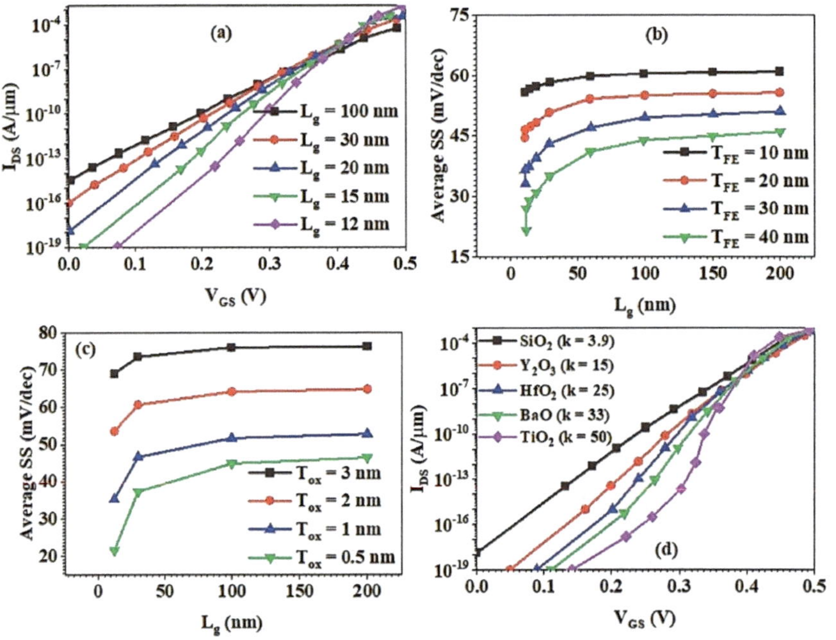

FIGURE 12.28 (a) The impact of L_g scaling on the transfer curves of 2D-SOI NC TMD FET. (b) The impact of L_g and T_{FE} scaling on the SS of 2D-SOI NC TMD FET. (c) The impact of L_g and T_{ox} scaling on the SS of 2D-SOI NC TMD FET. (d) The impact of high-K dielectrics on the transfer curves of 2D-SOI NC TMD FET.[61]

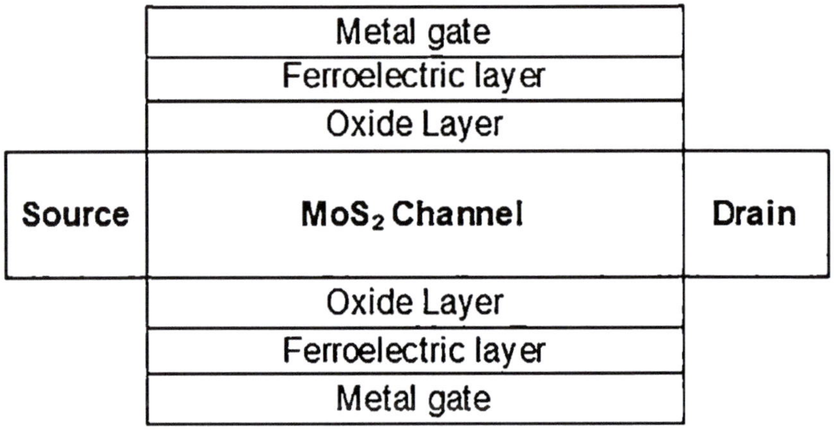

FIGURE 12.29 The structure of a double-gate (DG) 2D MFIS NCFET.[63]

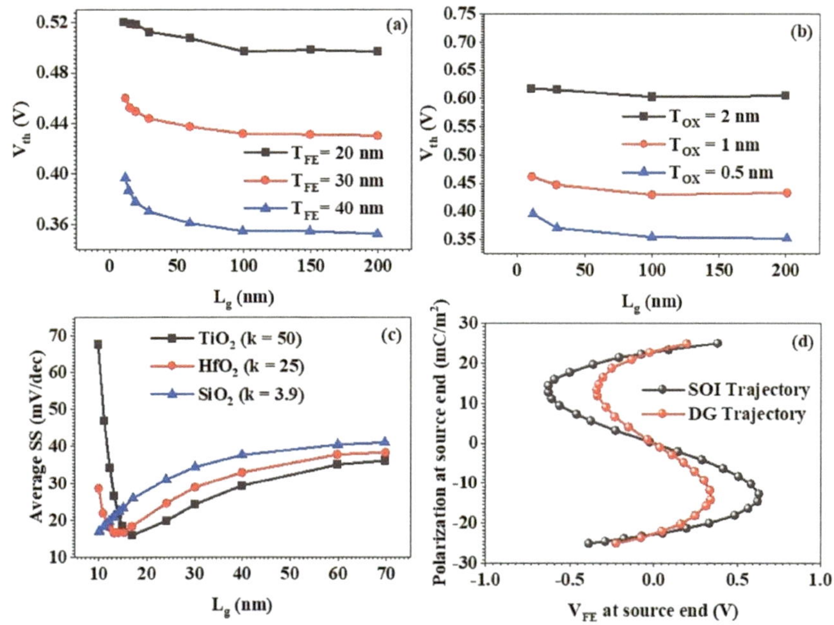

FIGURE 12.30 (a) The impact of L_g and T_{FE} scaling on the V_{th} of 2D-SOI NC TMD FET.[61] (b) The impact of L_g and T_{OX} scaling on the threshold voltage (V_{th}) of 2D-SOI NC TMD FET.[61] (c) The impact of L_g scaling and permittivity of gate oxides on the SS of 2D-SOI NC TMD FET.[61] (d) The polarization versus V_{FE} plot for 2D-SOI NC TMD FET and 2D-DG MFIS NCFET.[63]

12.9 NEGATIVE CAPACITANCE GE FETS (NC GE-FETS)

Si, Ge, GeSn, SiGe, InGaAs, and InAs are considered as attractive channel materials for for the development of NCFETs.[65–67] The structure of a p-type NC Ge-FET with internal metal gate is shown in Figure 12.31. The structures of MFIS and MFMIS type NC Ge-p-FETs are shown in Figure 12.32. I_{DS} of the p-type NC Ge-FET can be computed as[65]

$$I_{DS} = \mu_{eff} \frac{W}{L} \int_{0}^{V_{DS}} [Q_i(V)]dV \tag{12.12}$$

μ_{eff} : Effective hole mobility
V : Quasi-Fermi potential at a point in the channel
W: Channel width
L: Gate length

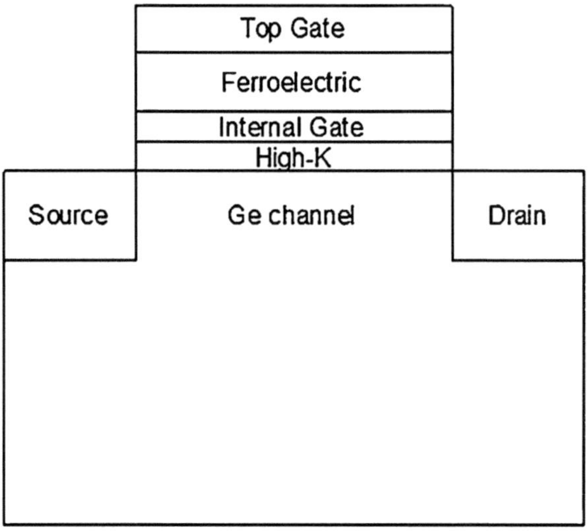

FIGURE 12.31 The structure of a p-type NC Ge-FET with internal metal gate.[65]

Channel charge $Q_i\,(V)$ can be calculated as[67]

$$Q_i(V) = \pm\sqrt{2\xi_s kTN_d}\left[\left(e^{-\frac{q\varphi_s}{kT}} + \frac{q\varphi_s}{kT} - 1\right) + \frac{n_i^2}{N_d^2}\left(e^{-\frac{qV}{kT}}\left(e^{\frac{q\varphi_s}{kT}} - 1\right) - \frac{q\varphi_s}{kT}\right)\right]^{1/2} \quad (12.13)$$

ϕ_s : Channel surface potential
ξ_s : Semiconductor permittivity
n_i : Intrinsic carrier concentration
N_d : Substrate doping density

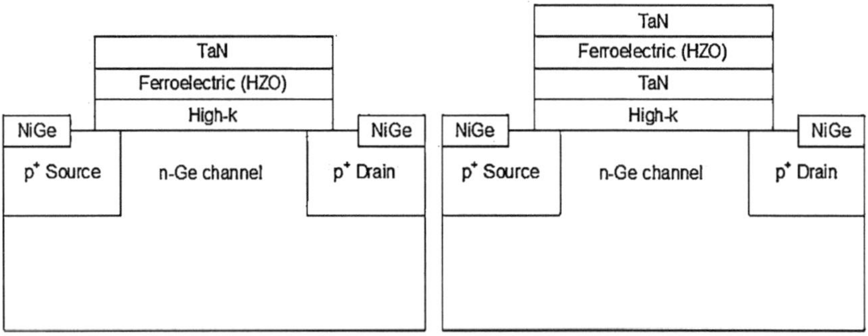

FIGURE 12.32 (a) MFIS NC-Ge-p-FET (b) MFMIS NC-Ge-p-FET.[68–70]

I_{DS}–V_{DS} curves of MFIS NC p-Ge-FET is illustrated in Figure 12.33a. I_{DS}–V_{DS} curves of MFIS NC p-Ge-FET exhibits NDR due to the reduction of I_{DS} after a particular V_{DS}. The I_{DS} of NC p-Ge-FETs depends on T_{FE}, E_C, and P_r. A large value of T_{FE} is required to improve I_{DS} in NC p-Ge-FET (Figure 12.33b). The I_{DS} of NC p-Ge-FET also increases with increase in E_C (Figure 12.33c). The variation of I_{DS} with P_r change is illustrated in Figure 12.33d. TaN can be used as gate material and NiGe can be used to form drain and source electrodes. HfZrOx (HZO) is considered as a choice for ferroelectric material in NC-p-Ge-FETs. The NC p-Ge-FETs exhibits hysteresis in its transfer curves.[70–73] Hysteresis is nothing but the separation between curves for the reverse and forward sweeping of biasing voltages. In 2017, Jing Li et al.[73] reported the fabrication of a MFMIS NC p-Ge FET with Si_2H_6 passivation. Drain current and peak g_m of the NC p-Ge FETs are found to increase with increase in gate capacitance. The use of a large T_{FE} results in the increase of C_G, which, in turn, leads to the improvement of I_{DS}.[74–77] Ultra–thin body (UTB) designs are suitable for the development of short channel hysteresis free NCFETs.[78–83] DIBL of NCFETs increases with reducing the channel length of the device.[81] In contrast, V_{th} of NCFETs decrease with channel length down scaling. ZrO_2 orthorhombic nanocrystals embedded in amorphous Al_2O_3 insulator can also be used as ferroelectric materials in NCFETs.[82]

For a non-hysteric operation,

$$T_{FE} \le \frac{2V_t C_{OX} + Q_G}{C_{OX}\left(2|\alpha|Q_G - 12\beta Q_G^3\right)} \tag{12.14}$$

In weak inversion and depletion regions, Q_G can be computed as[14]

$$Q_G \approx \gamma C_{OX}\sqrt{\psi_S + V_t\left(e^{-\frac{\psi_S}{V_t}} - 1\right)} \tag{12.15}$$

In strong inversion region, Q_G can be computed as[14]

$$Q_G = \gamma C_{OX}\sqrt{\psi_S + V_t e^{\frac{\psi_S - 2\varphi_F - V_C}{V_t}}} \tag{12.16}$$

$V_t = \frac{kT}{q}$ is the thermal voltage and Q_G represents the gate charge density.

$\gamma = \sqrt{2q\varepsilon_s N_A}$ is the body coefficient and ψ_S, N_A, V_C and ϕ_F represents surface potential, body doping density, difference between the hole and electron quasi-Fermi levels, and Fermi potential. respectively.

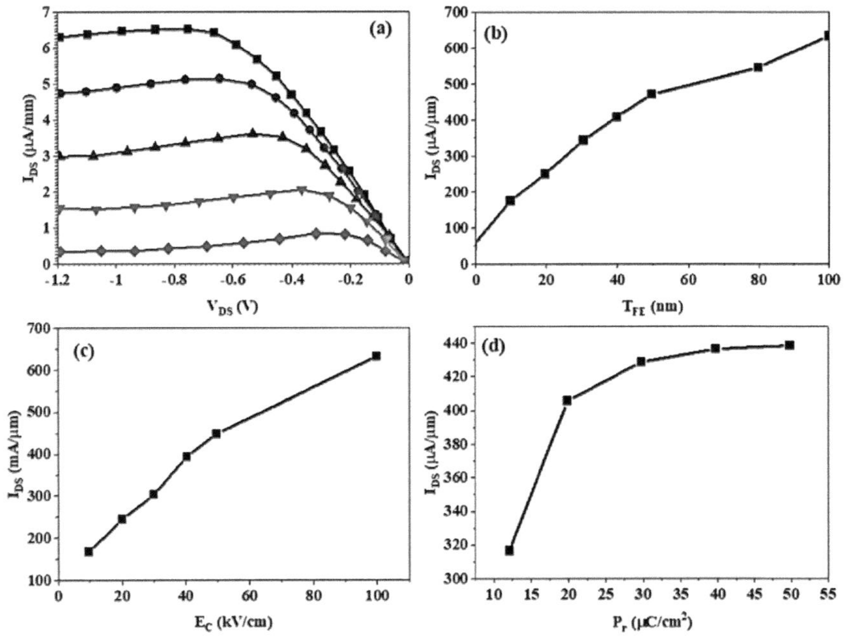

FIGURE 12.33 (a) I_{DS} - V_{DS} curves of MFIS NC p-Ge-FET.[68] (b) I_{DS} for different T_{FE} for MFMIS NC-p-Ge-FET.[65] (c) I_{DS} for different E_C for MFMIS NC-p-Ge-FET.[65] (d) I_{DS} for different P_r for MFMIS NC-p-Ge-FET.[65]

12.10 SUMMARY

NCFETs have gained tremendous attention in the field of non-volatile energy efficient counting. The performance of NCFETs can be further improved by using work-function engineering. Carbon nanotube (CNT) can also be used as a channel material for the development of NCFETs to enhance ON currents. A higher T_{FE} and lower T_{OX} are required for achieving low SS, high I_{ON}, low I_{OFF}, and high I_{ON}/I_{OFF} ratio. The use of 2D channel materials can effectively improve the scalability of NCFETs. NCFETs are considered to be one of the most promising technologies for future ultra–low-power applications.

KEYWORDS

- **2D-materials**
- **carbon nanotube (CNT)**
- **FDSOI**
- **gate-all-around (GAA)**
- **nanowire**
- **subthreshold swing (SS)**

REFERENCES

1. Cao, W.; Banerjee, K. Is Negative Capacitance FET a Steep-slope Logic Switch? *Nat. Commun.* **2020**, *11*, 196.
2. Hong, Y.; Choi, Y.; Shin, C. NCFET-Based 6-T SRAM: Yield Estimation Based on Variation-Aware Sensitivity. *IEEE J. Electr. Devices Soc.* **2020**, *8*, 182–188.
3. Liang, Y.; Li, X.; Gupta, S. K.; Datta, S.; Narayanan, V. Analysis of DIBL Effect and Negative Resistance Performance for NCFET Based on a Compact SPICE Model. *IEEE Trans. Electr. Devices* December **2018**, *65* (12), 5525–5529.
4. Jo, J.; Shin, C. Impact of Temperature on Negative Capacitance Field-effect Transistor. *Electron. Lett.* **2015**, *51* (1), 106–108.
5. Jain, A.; Alam, M. A. Stability Constraints Define the Minimum Subthreshold Swing of a Negative Capacitance Field-Effect Transistor. *IEEE Trans. Electr. Devices* July **2014**, *61* (7), 2235–2242.
6. Yuan, Z. C.; Rizwan, S.; Wong, M.; Holland, K.; Anderson, S.; Hook, T. B.; Kienle, D.; Gadelrab, S.; Gudem, P. S.; Vaidyanathan, M. Switching-Speed Limitations of Ferroelectric Negative-Capacitance FETs. *IEEE Trans. Electr. Devices* October **2016**, *63* (10), 4046–4052.
7. Khandelwal, S.; Duarte, J. P.; Khan, A. I.; Salahuddin, S.; Hu,C. Impact of Parasitic Capacitance and Ferroelectric Parameters on Negative Capacitance FinFET Characteristics. *IEEE Electr. Device Lett.* January **2017**, *38* (1), 142–144.
8. Sharma, A.; Roy, K. Design Space Exploration of Hysteresis-Free HfZrOx-Based Negative Capacitance FETs. *IEEE Electr. Device Lett.* August **2017**, *38* (8), 1165–1167.
9. Pahwa, G.; Dutta, T.; Agarwal, A.; Khandelwal, S.; Salahuddin, S.; Hu, C.; Chauhan, Y. S. Analysis and Compact Modeling of Negative Capacitance Transistor with High ON-Current and Negative Output Differential Resistance—Part I: Model Description. *IEEE Trans. Electr. Devices* December **2016**, *63* (12), 4981–4985.
10. Liang, Y.; Zhu, Z.; Li, X.; Gupta, S. K.; Datta, S.; Narayanan, V. Mismatch of Ferroelectric Film on Negative Capacitance FETs Performance. *IEEE Trans. Electr. Devices* March **2020**, *67* (3), 1297–1304.

11. Pahwa, G.; Dutta, T.; Agarwal, A.; Khandelwal, S.; Salahuddin, S.; Hu, C.; Chauhan, Y. S. Analysis and Compact Modeling of Negative Capacitance Transistor with High ON-Current and Negative Output Differential Resistance—Part II: Model Validation. *IEEE Trans. Electr. Devices* December **2016**, *63* (12), 4986–4992.
12. Yu, T.; Lü, W.; Zhao, Z.; Si, P.; Zhang, K. Effect of Different Capacitance Matching on Negative Capacitance FDSOI Transistors. *Microelectronics J.* **2020**, *98*, 104730.
13. Dabhi, C. K.; Parihar, S. S.; Dasgupta, A.; Chauhan, Y. S. Compact Modeling of Negative-Capacitance FDSOI FETs for Circuit Simulations. *IEEE Trans. Electr. Devices* July **2020**, *67* (7), 2710–2716.
14. Pahwa, G.; Dutta, T.; Agarwal, A.; Chauhan, Y. S. Compact Model for Ferroelectric Negative Capacitance Transistor with MFIS Structure. *IEEE Trans. Electr. Devices* March **2017**, *64* (3), 1366–1374.
15. Lin, C.; Khan, A. I.; Salahuddin, S.; Hu, C. Effects of the Variation of Ferroelectric Properties on Negative Capacitance FET Characteristics. *IEEE Trans. Electr. Devices* May **2016**, *63* (5), 2197–2199.
16. Kwon, D.; Chatterjee, K.; Tan, A. J.; Yadav, A. K.; Zhou, H.; Sachid, A. B.; dos Reis, R.; Hu, C.; Salahuddin, S. Improved Subthreshold Swing and Short Channel Effect in FDSOI n-Channel Negative Capacitance Field Effect Transistors. *IEEE Electr. Device Lett.* February **2018**, *39* (2), 300–303.
17. Moparthi, S.; Adarsh, K.P.; Tiwari, P. K.; Saramekala, G. K. Analog and RF Performance Evaluation of Negative Capacitance SOI Junctionless Transistor. *Int. J. Electron. Commun. (AEÜ)* **2020**, *122*, 153243.
18. Kwon, D.; Cheema, S.; Lin, Y-K.; Liao, Y-H.; Chatterjee, K.; Tan, A. J.; Hu, C.; Salahuddin, S. Near Threshold Capacitance Matching in a Negative Capacitance FET With 1 nm Effective Oxide Thickness Gate Stack. *IEEE Electr. Device Lett.* January **2020**, *41* (1), 179–182.
19. Zhou, J.; Han, G.; Peng, Y.; Liu, Y.; Zhang, J.; Sun, Q-Q.; Zhang, D. W.; Hao, Y. Ferroelectric Negative Capacitance GeSn PFETs With Sub-20 mV/decade Subthreshold Swing. *IEEE Electr. Device Lett.* August **2017**, *38* (8), 1157–1160.
20. Dasgupta, S.; Dasgupta, S.; Rajashekhar, A.; Majumdar, K.; Agrawal, N.; Razavieh, A.; Trolier-Mckinstry, S.; Datta, S. Sub-kT/q Switching in Strong Inversion in PbZr0.52Ti0.48O3 Gated Negative Capacitance FETs. *IEEE J. Exploratory Solid-State Comput. Devices Circuits* December **2015**, *1*, 43–48.
21. Lee, S.; Chen, H-W.; Shen, C-H.; Kuo, Po-Yi; Chung, C-C.; Huang, Yu-En; Chen, Hsin-Yu; Chao, T-S. Effect of Seed Layer on Gate-All-Around Poly-Si Nanowire Negative-Capacitance FETs With MFMIS and MFIS Structures: Planar Capacitors to 3-D FETs. *IEEE Trans. Electr. Devices* February **2020**, *67* (2), 711–716.
22. Jang, K.; Saraya, T.; Kobayashi, M.; Hiramoto, T. Ion/Ioff Ratio Enhancement and Scalability of Gate-All-Around Nanowire Negative-Capacitance FET with Ferroelectric HfO_2. *Solid-State Electron.* **2017**, *136*, 60–67.
23. Jiang, C.; Liang, R.; Xu, J. Investigation of Negative Capacitance Gate-all-Around Tunnel FETs Combining Numerical Simulation and Analytical Modeling. *IEEE Trans. Nanotechnol.* January **2017**, *16* (1), 58–67.
24. Lee, S.; Chen, H-W; Shen, C-H; Kuo, Po-Yi; Chung, C-C; Huang, Yu-En; Chen, H-Yu; Chao, T-S. Experimental Demonstration of Stacked Gate- All-Around Poly-Si Nanowires Negative Capacitance FETs With Internal Gate Featuring Seed Layer and

Free of Post-Metal Annealing Process. *IEEE Electr. Device Lett.* November **2019**, *40* (11), 1708–1711.
25. Dong, Z.; Guo, J. A Simple Model of Negative Capacitance FET with Electrostatic Short Channel Effects. *IEEE Trans. Electr. Devices* July **2017**, *64* (7), 2927–2934.
26. Pandey, P.; Kaur, H. Improved Temperature Resilience and Device Performance of Negative Capacitance Reconfigurable Field Effect Transistors. *IEEE Trans. Electr. Devices* February **2020**, *67* (2), 738–744.
27. Ajayan, J.; Ravichandran, T.; Mohankumar, P.; Prajoon, P.; Pravin, J. C.; Nirmal, D. Investigation of DC and RF Performance of Novel MOSHEMT on Silicon Substrate for Future Submillimetre Wave Applications. *Semiconductors* **2018**, *52* (16), 1991–1997.
28. Ajayan, J.; Nirmal, D.; Dheena, K.; Mohankumar, P.; Arivazhagan, L.; Augustine Fletcher, A. S.; Subash, T. D.; Saravanan, M. Investigation of Impact of Gate Underlap/overlap on the Analog/RF Performance of Composite Channel Double Gate MOSFETs. *J. Vacuum Sci. Technol. B* **2019**, *37* (6), 06221.
29. Ajayan, J.; Nirmal, D.; Mohankumar, P.; Arivazhagan, L.; Saravanan, M.; Saravanan, S. LG = 20 nm High Performance GaAs Substrate Based Metamorphic Metal Oxide Semiconductor High Electron Mobility Transistor for Next Generation High Speed Low Power Applications. *J. Nanoelectron. Optoelectron.* **2019**, *14* (8), 1133–1142.
30. Ajayan, J.; Nirmal, D. 20-nm T-gate Composite Channel Enhancement-mode Metamorphic HEMT on GaAs Substrates for Future THz Applications. *J. Computat. Electron.* **2016**, *15*, 1291–1296.
31. Ajayan, J.; Nirmal, D. 20 nm High Performance Enhancement Mode InP HEMT with Heavily Doped S/D Regions for Future THz Application. *Superlattices Microstruct.* **2016**, *100*, 526–534.
32. Ajayan, J.; Ravichandran, T.; Mohankumar, P.; Prajoon, P.; Pravin, J. C.; Nirmal, D. Investigation of DC-RF and Breakdown Behaviour in Lg = 20 nm Novel Asymmetric GaAs MHEMTs for Future Submillimetre Wave Applications. *AEU- Int. J. Electron. Commun.* **2018**, *84*, 387–393.
33. Ajayan, J.; Nirmal, D. A Review of InP/InAlAs/InGaAs Based Transistors for High Frequency Applications. *Superlattices Microstruct.* **2015**, *86*, 1–19.
34. Ajayan, J.; Subash, T. D.; Kurian, D. 20 nm High Performance Novel MOSHEMT on InP Substrate for Future High Speed Low Power Applications. *Superlattices Microstruct.* **2017**, *109*, 183–193.
35. Ajayan, J.; Ravichandran, T.; Prajoon, P.; Pravin, J. C.; Nirmal, D. Investigation of Breakdown Performance in Lg = 20 nm Novel Asymmetric InP HEMTs for Future High-speed High-power Applications. *J. Computat. Electron.* **2018**, *17* (1), 265–272.
36. Ajayan, J.; Nirmal, D.; Prajoon, P.; Pravin, J. C. Analysis of Nanometer-Scale InGaAs/InAs/InGaAs composite Channel MOSFETs Using High-K Dielectrics For High Speed Applications. *AEU Int. J. Electron. Commun.* **2017**, *79*, 151–157.
37. Ajayan, J.; Nirmal, D. 22 nm In0:75Ga0: 25As Channel-based HEMTs on InP/GaAs Substrates for Future THz Applications. *J. Semiconductors* **2017**, *38*, 27–32.
38. Ajayan, J.; Nirmal, D.; Ravichandran, T.; Mohankumar, P.; Prajoon, P.; Arivazhagan, L.; Chandan, K. S. InP High Electron Mobility Transistors for Submillimetre Wave and Terahertz Frequency Applications: A Review. *Int. J. Electron. Commun.* **2018**, *94*, 199–214.

39. Ajayan, J.; Nirmal, D. 20-nm Enhancement-mode Metamorphic GaAs HEMT with Highly Doped InGaAs Source/drain Regions for High-frequency Applications. *Int. J. Electron.* **2017**, *104*, 504–512.
40. Ajayan, J.; Nirmal, D.; Mohankumar, P.; Dheena, K.; Augustine, F.; Arivazhagan, L.; Santhosh Kumar, B. GaAs Metamorphic High Electron Mobility Transistors for Future Deep Space-biomedical-millitary and Communication System Applications: A Review. *Microelectron. J.* **2019**, *92*, 104604.
41. Charles Pravin, J.; Nirmal, D.; Prajoon, P.; Ajayan, J. Implementation of Nanoscale Circuits Using Dual Metal Gate Engineered Nanowire MOSFET with High-k Dielectrics for Low Power Applications. *Phys. E: Low-dimensional Syst. Nanostruct.* **2016**, *83*, 95–100.
42. Charles Pravin, J.; Nirmal, D.; Prajoon, P.; Mohan Kumar, N.; Ajayan, J. Investigation of 6T SRAM Memory Circuit Using High-k Dielectrics Based Nano Scale Junctionless Transistor. *Superlattices Microstruct.* **2017**, *104*, 470–476.
43. Ajayan, J.; Nirmal, D.; Mohankumar, P.; Arivazhagan, L. Investigation of Impact of Passivation Materials on the DC/RF Performances of InP-HEMTs for Terahertz Sensing and Imaging. *Silicon* **2020**, *12*, 1225–1230.
44. Ajayan, J.; Ravichandran, T.; Mohankumar, P.; Prajoon, P.; Charles Pravin, J.; Nirmal, D. Investigation of RF and DC Performance of E-Mode $In_{0.80}Ga_{0.20}As/InAs/In_{0.80}Ga_{0.20}as$ Channel based DG-HEMTs for Future Submillimetre Wave and THz Applications. *IETE J. Res.* **2018**. doi: 10.1080/03772063.2018.1553641.
45. Alam, M. A.; Si, M.; Ye, P. D. A Critical Review of Recent Progress on Negative Capacitance Field-effect Transistors. *Appl. Phys. Lett.* **2019**, *114*, 090401. doi: 10.1063/1.5092684.
46. Zhou, J.; Han, G.; Xu, N.; Li, J.; Peng, Y.; Liu, Y.; Zhang, J.; Sun, Q-Q.; Wei Zhang, D.; Hao, Y. Experimental Validation of Depolarization Field Produced Voltage Gains in Negative Capacitance Field-Effect Transistors. *IEEE Trans. Electr. Devices* October **2019**, *66* (10), 4419–4424.
47. Hu, V. P.; Lin, H.; Lin, Y.; Hu, C. Optimization of Negative-Capacitance Vertical-Tunnel FET (NCVT-FET). *IEEE Trans. Electr. Devices* June **2020**, *67* (6), 2593–2599.
48. Liang, Y.; Zhu, Z.; Li, X.; Gupta, S. K.; Datta, S.; Narayanan, V. Utilization of Negative-Capacitance FETs to Boost Analog Circuit Performances. *IEEE Trans. Very Large Scale Integration (VLSI) Syst.* December **2019,** *27* (12), 2855–2860.
49. Saeidi, A.; Jazaeri, F.; Stolichnov. Negative Capacitance as Universal Digital and Analog Performance Booster for Complementary MOS Transistors. *Sci Rep.* **2019**, *9,* 9105. https://doi.org/10.1038/s41598-019-45628-8.
50. Hu, V. P.; Chiu, P.; Lu, Y. Impact of Work Function Variation, Line-Edge Roughness, and Ferroelectric Properties Variation on Negative Capacitance FETs. *IEEE J. Electr. Devices Soc.* **2019**, *7*, 295–302.
51. Mehta, H.; Kaur, H. Study on Impact of Parasitic Capacitance on Performance of Graded Channel Negative Capacitance SOI FET at High Temperature. *IEEE Trans. Electr. Devices* July **2019**, *66* (7), 2904–2909.
52. Jiang, C.; Zhong, L.; Xie, L. Impact of Interface Trap Charges on the Electrical Characteristics of Back-Gated 2D Negative Capacitance (NC) FET. *IEEE Trans. Nanotechnol.* **2019**, *18*, 1204–1207.

53. Jiang, C.; Si, M.; Liang, R.; Xu, J.; Ye, P. D.; Alam, M. A. A Closed Form Analytical Model of Back-Gated 2-D Semiconductor Negative Capacitance Field Effect Transistors. *IEEE J. Electr. Devices Soc.* **2018**, *6*, 189–194.
54. Lin, Y.; McGuire, F.; Noyce, S.; Williams, N.; Cheng, Z.; Andrews, J. Effects of Gate Stack Composition and Thickness in 2-D Negative Capacitance FETs. *IEEE J. Electr. Devices Soc.* **2019**, *7*, 645–649.
55. Wang, X.; Yu, P.; Lei, Z. Van der Waals Negative CapaciY. K.; Sachid, A. B.; Kao, M-Y.; Chang, H. L.; Salahuddin, S.; Hu, C. Engineering Negative Differential Resistance in NCFETs for Analog Applications. *IEEE Trans. Electr. Devices* May **2018**, *65* (5), 2033–2039.
56. Agarwal, H.; Kushwaha, P.; Lin, Y-K.; Kao, M-Yen; Liao, Yu-Hung; Duarte, J-P.; Salahuddin, S.; Hu, C. NCFET Design Considering Maximum Interface Electric Field. *IEEE Electr. Device Lett.* August **2018**, *39* (8), 1254–1257.
57. Kao, M.; Lin, Y-K; Agarwal, H.; Liao, Yu-Hung; Kushwaha, P.; Dasgupta, A.; Salahuddin, S.; Hu, C. Optimization of NCFET by Matching Dielectric and Ferroelectric Nonuniformly Along the Channel. *IEEE Electr. Device Lett.* May **2019**, *40* (5), 822–825.
58. Zhang, X.; Gong, X.; Liang, G. Analysis on Performance of Ferroelectric NC-FETs Based on Real-Space Gibbs-Free Energy with Atomic Channel Structure. *IEEE Trans. Electr. Devices* February **2019**, *66* (2), 1100–1106.
59. Cheng, C.; Fan, C-Chi; Liu, C.; Hsu, H-H.; Chen, H-H.; Hsu, C-C.; Wang, S-An; Chang, C-Yen. Investigation of Gate-Stress Engineering in Negative Capacitance FETs Using Ferroelectric Hafnium Aluminum Oxides. *IEEE Trans. Electr. Devices* February **2019**, *66* (2), 1082–1086.
60. You, W.; Tsai, C.; Su, P. Short-Channel Effects in 2D Negative-Capacitance Field-Effect Transistors. *IEEE Trans. Electr. Devices*, April **2018**, *65* (4), 1604–1610.
61. You, W.; Su, P. Design Space Exploration Considering Back-Gate Biasing Effects for 2D Negative-Capacitance Field-Effect Transistors. *IEEE Trans. Electr. Devices* August **2017**, *64* (8), 3476–3481.
62. You, W.; Su, P. Intrinsic Difference Between 2-D Negative-Capacitance FETs with Semiconductor-on-Insulator and Double-Gate Structures. *IEEE Trans. Electr. Devices* October **2018**, *65* (10), 4196–4201.
63. Wang, X.; Chen, Y.; Wu, G. Two-dimensional Negative Capacitance Transistor with Polyvinylidene Fluoride-based Ferroelectric Polymer Gating. *npj 2D Mater. Appl.* **2017**, *1*, 38. https://doi.org/10.1038/s41699-017-0040-4.
64. Peng, Y.; Han, G.; Chen, Z.; Li, Q.; Zhang, J.; Hao, Y. Analytical Calculation of Influence of Ferroelectric Properties on Electrical Characteristics Negative Capacitance Germanium FETs. *IEEE J. Electr. Devices Soc.* **2018**, *6*, 233–239.
65. Zhou, J.; Han, G.; Li, J.; Peng, Y.; Liu, Y.; Zhang, J.; Sun, Q-Q.; Zhang, D. W.; Hao, Y. Comparative Study of Negative Capacitance Ge pFETs With HfZrOx Partially and Fully Covering Gate Region. *IEEE Trans. Electr. Devices* December **2017**, *64* (12), 4838–4843.
66. Chen, H.-P.; Lee, V. C.; Ohoka, A.; Xiang, J.; Taur, Y. Modeling and Design of Ferroelectric MOSFETs. *IEEE Trans. Electr. Devices* August **2011**, *58* (8), 2401–2405.
67. Wu, J.; Kanyang, R.; Han, G.; Zhou, J.; Liu, Y.; Wang, Y.; Peng, Y.; Zhang, J.; Sun, Q-Q.; Wei Zhang, D.; Hao, Y. Nonideality of Negative Capacitance Ge Field-Effect

Transistors Without Internal Metal Gate. *IEEE Electr. Device Lett.* April **2018**, *39* (4), 614–617.

68. Zhou, J.; Han, G.; Li, J.; Liu, Y.; Peng, Y.; Zhang, J.; Sun, Q-Q; Zhang, D. W.; Hao, Y. Negative Differential Resistance in Negative Capacitance FETs. *IEEE Electr. Device Lett.* April **2018**, *39* (4), 622–625.
69. Li, J.; Zhou, J.; Han, G.; Liu, Y.; Peng, Y.; Zhang, J.; Sun, Q-Q.; Zhang, D. W.; Hao, Y. Negative Capacitance Ge PFETs for Performance Improvement: Impact of Thickness of HfZrOx. *IEEE Trans. Electr. Devices* March **2018**, *65* (3), 1217–1222.
70. Bidenko, P.; Lee, S.; Han, J.; Song, J. D.; Kim, S. Simulation Study on the Design of Sub- kT/q Non-Hysteretic Negative Capacitance FET Using Capacitance Matching. *IEEE J. Electr. Devices Soc.* **2018**, *6*, 910–921.
71. Rollo, T.; Esseni, D. New Design Perspective for Ferroelectric NC-FETs. *IEEE Electr. Device Lett.* April **2018**, *39* (4), 603–606.
72. Li, J.; Zhou, J.; Han, G.; Liu, Yan; Peng, Yue; Zhang, J.; Sun, Q-Q.; Zhang, D. W.; Hao, Yue. Correlation of Gate Capacitance with Drive Current and Transconductance in Negative Capacitance Ge PFETs. *IEEE Electr. Device Lett.* October **2017**, *38* (10), 1500–1503.
73. Li, X.; Sampson, J. Enabling Energy-Efficient Nonvolatile Computing with Negative Capacitance FET. *IEEE Trans. Electr. Devices* August **2017**, *64* (8), 3452–3458.
74. Khan, A. I.; Radhakrishna, U.; Salahuddin, S.; Antoniadis, D. Work Function Engineering for Performance Improvement in Leaky Negative Capacitance FETs. *IEEE Electr. Device Lett.* September **2017**, *38* (9), 1335–1338.
75. Srimani, T.; Hills, Gage. Negative Capacitance Carbon Nanotube FETs. *IEEE Electr. Device Lett.* February **2018**, *39* (2), 304–307.
76. Liu, F.; Zhou, Y.; Wang, Y. Negative Capacitance Transistors with Monolayer Black Phosphorus. *npj Quant Mater*. **2016**, *1*, 16004. https://doi.org/10.1038/npjquantmats.2016.4.
77. Seo, J.; Lee, J.; Shin, M. Analysis of Drain-Induced Barrier Rising in Short-Channel Negative-Capacitance FETs and Its Applications. *IEEE Trans. Electr. Devices* April **2017**, *64* (4), 1793–1798.
78. Saeidi, A.; Jazaeri, F.; Stolichnov, I.; Ionescu, A. M. Double-Gate Negative-Capacitance MOSFET With PZT Gate-Stack on Ultra Thin Body SOI: An Experimentally Calibrated Simulation Study of Device Performance. *IEEE Trans. Electr. Devices* December **2016**, *63* (12), 4678–4684.
79. Jimenez, D.; Miranda, E.; Godoy, A. Analytic Model for the Surface Potential and Drain Current in Negative Capacitance Field-Effect Transistors. *IEEE Trans. Electr. Devices* October **2010**, *57* (10), 2405–2409.
80. Kwon, D.; Cheema, S. Negative Capacitance FET With 1.8-nm-Thick Zr-Doped HfO2 Oxide. *IEEE Electr. Device Lett.* June **2019**, *40* (6), 993–996.
81. Peng, Y.; Xiao, W. Nanocrystal-Embedded-Insulator Ferroelectric Negative Capacitance FETs With Sub-kT/q Swing. *IEEE Electr. Device Lett.* January **2019**, *40* (1), 9–12.
82. Zhou, J.; Han, G. Effects of the Variation of V_{GS} Sweep Range on the Performance of Negative Capacitance FETs. *IEEE Electr. Device Lett.* April **2018**, *39* (4), 618–621.

Index

T